# Signaling Mechanisms Regulating T Cell Diversity and Function

# METHODS IN SIGNAL TRANSDUCTION SERIES

Joseph Eichberg, Jr. and Michael X. Zhu
Series Editors

## Published Titles

*Signaling Mechanisms Regulating T Cell Diversity and Function*,
Jonathan Soboloff and Dietmar J. Kappes

*Gap Junction Channels and Hemichannels*, Donglin Bai and Juan C. Sáez

*Cyclic Nucleotide Signaling*, Xiaodong Cheng

*TRP Channels*, Michael Xi Zhu

*Lipid-Mediated Signaling*, Eric J. Murphy and Thad A. Rosenberger

*Signaling by Toll-Like Receptors*, Gregory W. Konat

*Signal Transduction in the Retina*, Steven J. Fliesler and Oleg G. Kisselev

*Analysis of Growth Factor Signaling in Embryos*, Malcolm Whitman and
Amy K. Sater

*Calcium Signaling, Second Edition*, James W. Putney, Jr.

*G Protein–Coupled Receptors: Structure, Function, and Ligand Screening*,
Tatsuya Haga and Shigeki Takeda

*G Protein–Coupled Receptors*, Tatsuya Haga and Gabriel Berstein

*Signaling Through Cell Adhesion Molecules*, Jun-Lin Guan

*G Proteins: Techniques of Analysis*, David R. Manning

*Lipid Second Messengers*, Suzanne G. Laychock and Ronald P. Rubin

# Signaling Mechanisms Regulating T Cell Diversity and Function

Edited by
Jonathan Soboloff
Dietmar J. Kappes

CRC Press
Taylor & Francis Group
Boca Raton  London  New York

CRC Press is an imprint of the
Taylor & Francis Group, an **informa** business

CRC Press
Taylor & Francis Group
6000 Broken Sound Parkway NW, Suite 300
Boca Raton, FL 33487-2742

First issued in paperback 2020

© 2017 by Taylor & Francis Group, LLC
CRC Press is an imprint of Taylor & Francis Group, an Informa business

No claim to original U.S. Government works

ISBN-13: 978-1-4987-0508-0 (hbk)
ISBN-13: 978-0-367-65812-0 (pbk)

### Library of Congress Cataloging-in-Publication Data

Names: Soboloff, Jonathan, editor. | Kappes, Dietmar, editor.
Title: Signaling mechanisms regulating T cell diversity and function /
[edited by] Jonathan Soboloff  nd Dietmar Kappes.
Description: Boca Raton : Taylor & Francis, 2017. | Includes bibliographical references.
Identifiers: LCCN 2016040600| ISBN 9781498705080 (hardback :alk. paper) |
ISBN 9781498705097 (e-book)
Subjects: | MESH: T-Lymphocytes--immunology | Receptors, Antigen,
T-Cell--immunology | TCF Transcription Factors | Signal Transduction |
Thymus Gland--secretion
Classification: LCC QR185.8.T2 | NLM QW 568 | DDC 616.07/97--dc23
LC record available at https://lccn.loc.gov/2016040600

**Visit the Taylor & Francis Web site at**
**http://www.taylorandfrancis.com**

**and the CRC Press Web site at**
**http://www.crcpress.com**

# Contents

# Series Preface

The concept of signal transduction is now long established as a central tenet of biological sciences. Since the inception of the field close to fifty years ago, the number and variety of signal transduction pathways, cascades, and networks have steadily increased and now constitute what is often regarded as a bewildering array of mechanisms by which cells sense and respond to extracellular and intracellular environmental stimuli. It is not an exaggeration to state that virtually every cell function is dependent on the detection, amplification, and integration of these signals. Moreover, there is increasing appreciation that in many disease states, aspects of signal transduction are critically perturbed.

Our knowledge of how information is conveyed and processed through these cellular molecular circuits and biochemical switches has increased enormously in scope and complexity since this series was initiated fifteen years ago. Such advances would not have been possible without the supplementation of older technologies, drawn chiefly from cell and molecular biology, biochemistry, physiology, and pharmacology, with newer methods that make use of sophisticated genetic approaches as well as structural biology, imaging, bioinformatics, and systems biology analysis.

The overall theme of this series continues to be the presentation of the wealth of up-to-date research methods applied to the many facets of signal transduction. Each volume is assembled by one or more editors who are preeminent in their specialty. In turn, the guiding principle for editors is to recruit chapter authors who will describe procedures and protocols with which they are intimately familiar in a reader-friendly format. The intent is to ensure that each volume will be of maximum practical value to a broad audience, including students and researchers just entering an area, as well as seasoned investigators.

It is hoped that the information contained in the books of this series will constitute a useful resource to the life sciences research community well into the future.

**Joseph Eichberg**
**Michael Xi Zhu**
*Series Editors*

# Preface

Considering the major industry that the study of T cell biology has now become, it is interesting to realize that even the existence of T cells was not recognized until relatively recently. Thus, the first proof that the thymus had an important immunological function came in 1961 from a seminal study by Jacques Miller showing that neonatally thymectomized mice failed to reject allogeneic skin grafts.[1] Before that the thymus was generally viewed as an obscure evolutionary vestige like the appendix. The next major advance, also by Miller and colleagues, was the recognition that circulating lymphocytes were not a uniform population of cells but comprised distinct T (thymus-derived) and B (bone marrow-derived) cell subsets, responsible for cell-mediated and humoral immunity, respectively.[2] By cotransfer of thymus-derived and bone marrow-derived cells into adoptive hosts, it was shown that the former were necessary for development of an antibody response, giving rise to the concept of "helper" T cells. Thus began the splitting of lymphocytes into ever smaller functional subsets that continues to this day.

The next major breakthrough in T cell biology came in the 1970s, again using adoptive transfer and grafting approaches, with the discovery that T cells recognize specific major histocompatibility complex (MHC) alleles (MHC restriction) and that this restriction is imposed during development in the thymus.[3,4] Around the same time, antibody-mediated depletion studies demonstrated that cell-mediated helper and cytotoxic activities were mediated by distinct Ly1+ (CD4) and Ly2+ (CD8) subsets.[5,6] Thus, the phenotypic and functional separation of CD4 and CD8 T cells became established.

Although it was by then well-appreciated that T cells used a clonotypic antigen-specific receptor to recognize target cells, the nature of this T cell receptor (TCR) remained obscure until the early 1980s when protein biochemical approaches led to the identification of the disulfide-linked TCRab heterodimer on the surface of various T cell hybridomas and tumors.[7–9] This was followed shortly by the cloning of the corresponding genes.[10–12] The cloning of the TCR chains in turn allowed the generation of TCR transgenic mice expressing a single clonotypic TCR, which proved critical to the dissection of thymic development and establishing the affinity model of thymocyte selection. Thus, depending on the specificity and affinity of the transgenic TCR, thymocytes were shown to undergo selection into the CD4 or CD8 lineages by the process of positive selection[13–16] or undergo deletion via negative selection.[17,18] Subsequently, it has been established that differences in TCR affinity determine not only thymic development of conventional CD4 and CD8 T cell subsets, but also of other more specialized subsets including γδ T cells,[19,20] Tregs,[21] and iNKT cells.[22]

Around the mid-1980s the first evidence emerged that mature CD4 T cells continue their differentiation in the periphery and diverge into multiple functionally distinct subsets in response to differential stimulation by antigens and soluble factors. Mosmann et al. were the first to show in 1986 that CD4 T cells could be subdivided into IFNγ and IL-4 producing subsets,[23] what we now know as Th1 and Th2 cells. Since then, multiple additional T helper subsets have been reported with different cytokine profiles, and effector functions, including Th9,[24] Th17,[25] and Th22.[26] The processes by which naive CD4 T cells differentiate into these distinct functional subtypes requires TCR engagement as well as the positive feedback loop triggered by a major cytokine product of the differentiated cell itself to enforce a strong degree of polarization.

The elucidation of the key signaling pathways mediated by TCR engagement began with the realization that the TCR heterodimer lacks inherent signaling capacity and is instead connected to the intracellular signaling machinery through an associated complex of CD3 subunits.[27] The CD3 subunits were found to contain ITAM motifs that are targets for phosphorylation by the SRC family kinase p56-LCK, which is brought into the vicinity of the TCR complex upon interaction with peptide-MHC complexes via its association with the CD4 and CD8 coreceptors.[28,29] Many other key components of the TCR signaling pathway have been subsequently identified. Finally, in the mid-2000s several master transcriptional regulators of different T lineages have been identified, including Foxp3, ThPOK and Plzf for Treg, CD4, and NKT lineages, that are selectively induced in and critical for both development and function of their respective lineages.[30–32] Similarly, T-bet, Gata3, and RORγt were identified as the master regulators of Th1, Th2, and Th17 mature T cell subtypes.[33–35] Understanding the regulation of these factors during T cell development and differentiation, in particular how they are controlled by different TCR signals, remains a subject of intense interest.

The importance of signaling via the TCR and coreceptors for T cell development and function cannot be overemphasized, and has led to an intense research effort to understand this process that continues to the present day. In the present volume, we have attempted to give a broad overview of the field, spanning the gamut from earliest stages of thymic development to memory T cell differentiation. Hence, after implantation of lymphoid progenitors in the thymus, early T cell development is primarily driven by the interactions of a series of thymic epithelial cell-derived growth factors and morphogens (discussed in Chapter 1). As these early T cells progress through development, their maintenance switches to being predominantly dependent on TCR–MHC interactions with intermediate signals driving development (positive selection; discussed in Chapter 2) and the strongest signals leading to apoptosis (negative selection; discussed in Chapters 3 and 4). Chapter 5 discusses the role of ThPOK and TCR signaling in driving cells to the cd4 versus cd8 lineages, while Chapter 6 considers the control of αβ/δγ T lineage choice by TCR signals and other factors. Next, Chapter 7 discusses the predominant features of regulatory T cell development and function, while Chapter 8 focuses on features defining NKT cell development. Once these various T cell types have matured, activation is controlled via a complex signaling network. Hence, several concluding chapters focus on regulation and function of mature T cells. Chapter 9 discusses the role of the GRB2 family members in TCR signaling, and Chapter 10 considers the role of $Ca^{2+}$ signaling in T cell activation. Chapter 11 covers the role of epigenetic regulation in control of T cell memory, and Chapter 12 describes the contribution of regulatory T cell types in control of autoimmunity.

In conclusion, we wish to thank all of the authors for their contributions to this volume. We hope that it will be useful for anyone wishing to understand the current state of the field of T cell diversity and function.

# REFERENCES

1. Miller JFAP. Immunological function of the thymus. *Lancet* 1961; 2: 748–749.
2. Mitchell GF, Miller JFAP. Cell to cell interaction in the immune response. II. The source of hemolysin-forming cells in irradiated mice given bone marrow and thymus or thoracic duct lymphocytes. *J Exp Med* 1968; 128: 821–837.
3. Zinkernagel RM, Doherty PC. Restriction of in vitro T cell-mediated cytotoxicity in lymphocytic choriomeningitis within a syngeneic or semiallogeneic system. *Nature* 1974; 248: 701–702.
4. Zinkernagel RM et al. On the thymus in the differentiation of "H-2 self-recognition" by T cells: Evidence for dual recognition? *J Exp Med* 1978: 147: 882–896.
5. Cantor H et al. Functional subclasses of T-lymphocytes bearing different Ly antigens. I. The generation of functionally distinct T-cell subclasses is a differentiative process independent of antigen. *J Exp Med* 1975; 141: 1376–1389.
6. Kisielow P et al. Ly antigens as markers for functionally distinct subpopulations of thymus-derived lymphocytes of the mouse. *Nature* 1975; 253: 219–220.
7. Allison JP et al. Tumor-specific antigen of murine T-lymphoma defined with monoclonal antibody. *J Immunol* 1982; 129: 2293.
8. Haskins K et al. The major histocompatibility complex-restricted antigen receptor on T cells. I. Isolation with a monoclonal antibody. *J Exp Med* 1983; 157: 1149–1169.
9. Meuer SC et al. Clonotypic structures involved in antigen-specific human T cell function. Relationship to the T3 molecular complex. *J Exp Med* 1983; 157: 705–719.
10. Chien Y et al. A third type of murine T-cell receptor gene. *Nature* 1984; 312: 31–35.
11. Hedrick SM et al. Isolation of cDNA clones encoding T cell-specific membrane-associated proteins. *Nature* 1984; 308: 149–153.
12. Yanagi Y et al. A human T cell-specific cDNA clone encodes a protein having extensive homology to immunoglobulin chains. *Nature* 1984; 308: 145–149.
13. Berg LJ et al. Antigen/MHC-specific T cells are preferentially exported from the thymus in the presence of their MHC ligand. *Cell* 1989; 58: 1035–1046.
14. Kaye J et al. Selective development of CD4+ T cells in transgenic mice expressing a class II MHC-restricted antigen receptor. *Nature* 1989; 341: 746–749.
15. Kisielow P et al. Positive selection of antigen-specific T cells in thymus by restricting MHC molecules. *Nature* 1988a; 335: 730–733.
16. Teh HS et al. Thymic major histocompatibility complex antigens and the alpha beta T-cell receptor determine the CD4/CD8 phenotype of T cells. *Nature* 1988; 335: 229–233.
17. Kisielow P et al. Tolerance in T-cell-receptor transgenic mice involves deletion of nonmature CD4+8+ thymocytes. *Nature* 1988b; 333: 742–746.
18. Sha WC et al. Positive and negative selection of an antigen receptor on T cells in transgenic mice. *Nature* 1988; 336: 73–76.
19. Haks MC et al. Attenuation of gammadelta TCR signaling efficiently diverts thymocytes to the alphabeta lineage. *Immunity* 2005; 22: 595–606.
20. Pereira P et al. Blockade of transgenic gamma delta T cell development in beta 2-microglobulin deficient mice. *EMBO J* 1992; 11: 25–31.
21. Jordan MS et al. Thymic selection of CD4+CD25+ regulatory T cells induced by an agonist self-peptide. *Nat Immunol* 2001; 2: 301–306.
22. Moran AE et al. T cell receptor signal strength in Treg and iNKT cell development demonstrated by a novel fluorescent reporter mouse. *J Exp Med* 2011; 208: 1279–1289.
23. Mosmann TR et al. Two types of murine helper T cell clone. I. Definition according to profiles of lymphokine activities and secreted proteins. *J Immunol* 1986; 136: 2348–2357.
24. Schmitt E et al. IL-9 production of naive CD4+ T cells depends on IL-2, is synergistically enhanced by a combination of TGF-β and IL-4, and is inhibited by IFNγ. *J Immunol* 1994; 153: 3989–3996.

25. Harrington LE et al. Interleukin 17-producing CD4+ effector T cells develop via a lineage distinct from the T helper type 1 and 2 lineages. *Nat Immunol* 2005; 6: 1123–1132.

26. Eyerich S et al. Th22 cells represent a distinct human T cell subset involved in epidermal immunity and remodeling. *J Clin Invest* 2009; 119: 3573–3585.

27. Letourneur F, Klausner RD. Activation of T cells by a tyrosine kinase activation domain in the cytoplasmic tail of CD3 epsilon. *Science* 1992; 255: 79–82.

28. Barber EK et al. The CD4 and CD8 antigens are coupled to a protein-tyrosine kinase (p56lck) that phosphorylates the CD3 Complex. *PNAS* 1989; 86: 3277–3281.

29. Rudd CE et al. The CD4 receptor is complex in detergent lysates to a protein-tyrosine kinase (pp58) from human T lymphocytes. *PNAS* 1988; 85: 5190–5194.

30. He X et al. The zinc finger transcription factor TH POK regulates CD4 versus CD8 T lineage commitment. *Nature* 2005; 433: 826–833.

31. Hori S et al. Control of regulatory T cell development by the transcription factor Foxp3. *Science* 2003; 299: 1057–1061.

32. Kovalovsky D et al. The BTB-zinc finger transcriptional regulator PLZF controls the development of invariant natural killer T cell effector functions. *Nat Immunol* 2008; 9: 1055–1064.

33. Ivanov II et al. The orphan nuclear receptor RORgammat directs the differentiation program of proinflammatory IL-17+ T helper cells. *Cell* 2006; 126: 1121–1133.

34. Szabo SJ et al. A novel transcription factor, T-bet, directs Th1 lineage commitment. *Cell* 2000; 100: 655–669.

35. Zheng W, Flavell RA. The transcription factor GATA-3 is necessary and sufficient for Th2 cytokine gene expression in CD4 T cells. *Cell* 1997; 89: 587–596.

# Contributors

**Dorina Avram**
College of Medicine
University of Florida
Gainesville, Florida

**John B. Barnett**
Department of Microbiology,
    Immunology and Cell Biology
West Virginia University School
    of Medicine
Morgantown, West Virginia

**Rosa Berga-Bolanos**
National Institute on Aging
National Institutes of Health
Baltimore, Maryland

**Suzanne Bertera**
Institute of Cellular Therapeutics
Allegheny Health Network
Pittsburgh, Pennsylvania

**Mahmood Y. Bilal**
Interdisciplinary Graduate Program
    in Immunology
University of Iowa
Iowa City, Iowa

**Louis-Marie Charbonnier**
Department of Pediatrics
Harvard Medical School
Boston, Massachusetts

**Talal A. Chatila**
Department of Pediatrics
Harvard Medical School
Boston, Massachusetts

**Sijo V. Chemmannur**
Fox Chase Cancer Center
Philadelphia, Pennsylvania

**Jonathan J. Cho**
College of Medicine
University of Florida
Gainesville, Florida

**Mengqi Dong**
Maisonneuve-Rosemont Hospital
    Research Center
University of Montreal
Montreal, Canada

**Shawn P. Fahl**
Fox Chase Cancer Center
Philadelphia, Pennsylvania

**Yong Fan**
Institute of Cellular Therapeutics
Allegheny Health Network
and
Department of Biological Sciences
Carnegie Mellon University
Pittsburgh, Pennsylvania

**Christina Go**
Fels Institute for Cancer Research and
    Molecular Biology
and
Department of Medical Genetics and
    Molecular Biochemistry
Temple University School of Medicine
Philadelphia, Pennsylvania

**Nicola M. Heller**
Department of Anesthesiology
    and Critical Care Medicine
Johns Hopkins University School
    of Medicine
Baltimore, Maryland

**Jon C. D. Houtman**
Interdisciplinary Graduate Program
    in Immunology
and
Carver College of Medicine
University of Iowa
Iowa City, Iowa

**Hui-Chen Hsu**
Department of Medicine
University of Alabama at Birmingham
Birmingham, Alabama

**Dietmar J. Kappes**
Fox Chase Cancer Center
Philadelphia, Pennsylvania

**Nathalie Labrecque**
Maisonneuve-Rosemont Hospital
    Research Center
University of Montreal
Montreal, Canada

**Hao Li**
Internal Medicine
Beth Israel Deaconess Medical Center
Boston, Massachusetts

**Kyle J. Lorentsen**
College of Medicine
University of Florida
Gainesville, Florida

**Robert M. Lowe**
Pediatric Rheumatology
Children's Specialty Center of Nevada
Las Vegas, Nevada

**Heather J. Melichar**
Maisonneuve-Rosemont Hospital
    Research Center
University of Montreal
Montreal, Canada

**Jayati Mookerjee-Basu**
Fox Chase Cancer Center
Philadelphia, Pennsylvania

**John D. Mountz**
Department of Medicine
University of Alabama at Birmingham
and
Birmingham VA Medical Center
Birmingham, Alabama

**Lynette Naler**
National Institute on Aging
National Institutes of Health
Baltimore, Maryland

**Isha Pradhan**
Institute of Cellular Therapeutics
Allegheny Health Network
Pittsburgh, Pennsylvania

**Janaki Purushe**
Department of Microbiology and
    Immunology
Temple University
Philadelphia, Pennsylvania

**Li Qin**
Fox Chase Cancer Center
Philadelphia, Pennsylvania

**Elsie Samakai**
Fels Institute for Cancer Research and
    Molecular Biology
and
Department of Medical Genetics and
    Molecular Biochemistry
Temple University School of Medicine
Philadelphia, Pennsylvania

**Jyoti Misra Sen**
National Institute on Aging
National Institutes of Health
and
Department of Medicine
Johns Hopkins University School of
    Medicine
Baltimore, Maryland

**Jonathan Soboloff**
Fels Institute for Cancer Research and
    Molecular Biology
and
Department of Medical Genetics and
    Molecular Biochemistry
Temple University School of Medicine
Philadelphia, Pennsylvania

**Aditi Sood**
Maisonneuve-Rosemont Hospital
    Research Center
University of Montreal
Montreal, Canada

**Asako Tajima**
Institute of Cellular Therapeutics
Allegheny Health Network
Pittsburgh, Pennsylvania

**Massimo Trucco**
Institute of Cellular Therapeutics
Allegheny Health Network
and
Department of Biological Sciences
Carnegie Mellon University
Pittsburgh, Pennsylvania

**David L. Wiest**
Fox Chase Center
Philadelphia, Pennsylvania

**Yi Zhang**
Fels Institute for Cancer Research and
    Molecular Biology
Temple University
Philadelphia, Pennsylvania

# Contributors

Fred Marcus
National Institute on Aging
National Institute of Health
and
Department of Medicine
Johns Hopkins University School of
Medicine
Baltimore, Maryland

Jonathan Schooler
Fox Institute for Cancer Research and
Molecular Biology
and
Department of Medical Genetics and
Molecular Biochemistry
Temple University School of Medicine
Philadelphia, Pennsylvania

Kiki Sigel
Shriners Hospital
Research Center
University of Montreal
Montreal, Canada

Akiko Tajima
Institute of Cellular Therapeutics
Allegheny Health Network
Pittsburgh, Pennsylvania

Massimo Trucco
Institute of Cellular Therapeutics
Allegheny Health Network
and
Department of Biological Sciences
Carnegie Mellon University
Pittsburgh, Pennsylvania

David L. West
Fox Chase Center
Philadelphia, Pennsylvania

Yi Zhang
Institute for Cancer Research and
Molecular Biology
Temple University
Philadelphia, Pennsylvania

# 1 Consequences of Blocking the Choreography of Double Negative Thymocyte Maturation

*John B. Barnett*

## CONTENTS

## ABSTRACT

It is well recognized that thymocytes must undergo a progression of maturation and differentiation stages on their journey to becoming a mature CD4$^+$ or
CD8$^+$ T cell. Early is this progression are four double-negative (DN) stages, so
named because they express neither CD4 or CD8 on their surface. These DN
stages have been further classified as DN1, DN2, DN3, or DN4 based on the
presence or absence of the specific cell membrane markers, CD25 and CD44.
The process of a cell progressing from one DN stage to another (and beyond) is
a complex "choreography" involving the action of numerous transcription factors, morphogens, and cytokines. The first critical component in this sequence
is the activation of the transmembrane receptor Notch1. Failure to engage the
Notch1 receptor at this stage shunts the cells into an entirely different pathway
that results in production of NK cells, B cells, or myeloid cells. However, if the
cell passes this critical step, it then begins the process of becoming a mature T
cell by the sequential action of the aforementioned transcription factors, morphogens, and cytokines. In this chapter we discuss how these factors contribute
to the maturation of T cells by detailing the consequences of the loss of the
action of each of the factors involved in DN maturation.

## 1.1 INTRODUCTION

Figure 1.1 provides a schematic of the normal progression of thymocytes from the
arrival of the multipotent stem cell (MSC) until the single positive (SP) CD4$^+$ and
CD8$^+$ T cell emerge. At each step in this schematic there are growth factors and morphogens that are required to induce the cell(s) to move to the next stage of development. In this chapter, the goal is to discuss the consequences of blocking one or more

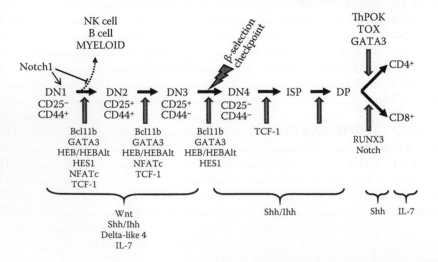

**FIGURE 1.1** Normal progression of thymocyte development and the factors that influence
each stage of development.

of these important factors on the progression of the cells. This chapter will focus on those thymocytes that lack CD4+ and CD8+, and are thus referred to as double negative (DN) cells. Further, we will not cover the development of γδ T cells, which appear to take a simpler pathway to maturation, that is, primarily dependent on Id3 to promote the γδ T cells' fate.[1]

The β-selection checkpoint is also shown in Figure 1.1. β-selection refers to a step when T cell receptor (TCR) rearrangement occurs. At the DN3 step, the TCRβ chain is rearranged and is then paired with the pre-Tα chain to create the pre-TCR. Cells must successfully rearrange the β-chain to produce a functional pre-TCR before progressing beyond the DN3 stage.

Methods to knock out or knock down specific genes as well as induce gain-of-function and overexpression of specific genes in vivo and in vitro have provided the tools to dissect this pathway. However, a given factor may be active at more than one stage and that may not always be clearly shown by the phenotypic result. The action of a factor may be very concentration dependent, as is the case with morphogens, and a knock out of the gene expressing a factor, usually an all-or-none result, which may not allow for concentration-dependent actions to be discerned. Nonetheless, use of these molecular and genetic techniques has contributed a tremendous amount to our understanding of the normal progression of thymocyte development, and much of what we know about T cell development has come from these studies.

A recent editorial by Moon and Gough discusses the challenges of representing pathways in a nonlinear manner, using the Wnt and β-catenin story to illustrate their point.[2] Figure 1.1 presents the thymocyte development in a very linear manner, yet as Moon and Gough point out, there are crosstalk, codependencies and concentration-dependent effects between different elements that drive developmental pathways.[2] Illustrating these cross- and codependencies in this diagram would make it difficult to decipher. Case in point in Figure 4 provided by Rothenberg and Anderson[3] to illustrate four different physiological conditions that would govern lymphopoiesis. Therefore, the pathway is still illustrated in a linear fashion. However, partially to the point presented by Moon and Gough,[2] the action of a particular transcriptional-activator often affects more than a single stage in the process, and there will be references to multiple stages under any given heading.

## 1.2 DN1 STAGE

In addition to the lack of CD4+ and CD8+ (hence double negative), the DN1 stage of thymocyte development is defined by positive cell surface expression of CD44 without the expression of CD25 (CD44+CD25−). Unlike the other cell stages, the DN1 stage cells are heterogeneous and includes cells that can be as easily programmed to become B cells, NK cells, myeloid cells, and dendritic cells. Some workers acknowledge this by referring to this population of cells as thymus seeding progenitors (TSPs)[4] rather than DN1 cells. To designate this fact, however, we will use the convention of DN1 in this chapter. Two transitions occur in this early T cell development that are of particular interest: the onset of T lineage gene expression (specification), and the final exclusion of any fate other than a T cell fate (commitment).[5]

## 1.2.1 Notch1

Mammals have four Notch receptors (Notch1–4). However, Notch1 is the factor that provides the signal for the DN1 cell to continue on the path to becoming a mature T cell. Without Notch1 signaling, the DN1 cell is shunted toward becoming a non-T cell such as a B cell, myeloid cell, or NK cell.[6–9] Notch1 has been extensively studied in hematopoiesis and has a well-studied role in marginal zone B cell development and peripheral T cell development in addition to its critical role in thymic αβ T cell development. The ligand for Notch1 in thymic T cell development is Delta-like 4 (DLL4), as inactivation of Dll4 but not Dll1 in thymic epithelial cells (TECs) resulted in a complete block in T cell development.[7,10] Once Notch1 is bound to DLL4, after a succession of proteolytic cleavages, the intracellular portion of Notch1 is translocated to the nucleus, heterodimerizes with a DNA binding transcription cofactor called CSL (CBF-1 [RBP-J in mouse], Suppressor of Hairless, Lag-1) and it becomes part of a transcription complex.[11] This transcription complex also includes the Mastermind proteins (MAML1-3) and MED8-mediator.[11] One target gene of this transcription complex is Hairy/enhancer of split (HES),[11] the role of which will be addressed later. Notch continues to play a role in thymocyte development at several stages. Active Notch1 transcription continues until the β-selection checkpoint between the DN3 to DN4 stages; however, Notch1 protein levels continue to remain high until the ISP stage.[12]

In addition to directly knocking out Notch1, it is possible to interfere with Notch1 signaling by inducing transgenic expression of Notch modulators. Given the complexity of Notch1 signaling,[11] there are several targets that can be exploited to block Notch1 signaling. Thymocyte signaling requires a "tuneable" approach, that is, the influence of any signaling molecule is often not binary (all on or all off), but more analog (concentration dependent) to provide a more precise regulation of its action. The regulation of Notch1 is controlled at several levels and there are several examples of negative regulators. The sensitivity (hence analog control) of Notch to ligand-mediated activation is decreased by Fringe glycosyl transferases; and Numb antagonizes intracellular signals of intracellular Notch.[13] Conversely, the nuclear factors SKIP and Mastermind enhance the activation of the transcription factor of Notch, which is CSL.[13]

The Fringe glycosyl transferases act by adding $N$-acetylglucosamine to $O$-fucose on the extracellular domain of Notch.[14] There are three known variants of Fringe called Lunatic (Lfng), Manic, and Radical.[15] Mechanistically, the Fringe modification of the Notch receptors alters the binding and response of Notch to its ligands.[16] Lunatic fringe was first discovered to affect T cell development and Lfng is normally expressed in DN progenitors.[17,18] Increased expression of Lfng in DP thymocytes causes increased binding to Notch ligands on stromal cells, blocking access of DN T cell progenitors to the thymic stromal, thus, blocking DN progression. A distinct class of Notch1 modulators are coded by the Deltex1 genes.[13] In the *Drosophila*, Deltex1 is a positive regulator that mediates or augments effects downstream of Notch1.[13] With these past results in mind, Izon et al. reconstituted mice with BM-derived hematopoietic progenitors transduced ex vivo with Deltex1.[13] In contrast to the *Drosophila* results, this treatment resulted in a significant reduction in the number of thymocytes produced and in the small portion of thymocytes shown in the thymus, these cells were all DN.[13] Thus, in mammals, Deltex1 gene products are inhibitory. Deltex1 expression is high in DN1 cells, downregulated in DN2 cells, and then markedly upregulated in DN3 cells.[13] Subsequent mechanistic experiments suggested that Deltex1 antagonized the recruitment of coactivators needed by a Notch1 transcription factor, CSL.[13] Yun and Bevan extended these results by investigating the effects of overexpression of Deltex1 as well as Notch-regulated ankyrin-repeat protein (Nrarp).[19] Nrarp is also an inhibitor of Notch1 that is expressed in all thymocyte subsets.[19] Yun and Bevan saw a significant increase in DN1 and DN2 cells when Nrarp was overexpressed but not when Deltrex1 overexpression blocked Notch1.[19] The block due to Nrarp overexpression was primarily in the DN1 to DN2 transition.[19] Mechanistically, Nrarp inhibits Notch induction of CBF-1, a member of the CSL transcription complex.[19] Yun and Bevan also showed that although Deltrex1 and Nrarp block T versus B cell commitment, only Nrarp was capable of blocking early thymocyte development.[19] An interesting contrast comes from research on Numb and Numblike, which have shown to be Notch inhibitors for neuronal stem cell development,[20] but had no effect on thymic cell development in a double knockout system.[21]

The process of β-selection was described earlier. Wolfer et al. showed that when Notch1 was knocked out, thymocytes show aberrant VDJβ rearrangement and impaired pre-TCR-independent survival occurred.[22] Further studies on the role of Notch1 during the β-selection using a Mastermind-like 1 dominant negative construct to inhibit transcription revealed that the "β-selection checkpoint in vivo is absolute and independent of the pre-TCR."[23] Although there is a sharp downregulation of Notch1 transcription after β-selection, the question of Notch1 expression during thymopoiesis was addressed by Fiorini et al. using monoclonal antibodies (mAbs) specific for Notch1 and Notch2.[24] Their results show that Notch1 is expressed at high levels on all DN cells and is downregulated at the DP and SP stages.[24] They also noted that the Notch1 positive DN2/3 cells localized adjacent to the thymus capsule.[24] This could seemingly facilitate the process of "thymus crosstalk" between TEC and thymocytes raised much earlier by van Ewijk et al.[25]

In summary, activation by Notch1 is essential, first for the decision to proceed to develop from a DN1 cell through the thymocyte differentiation pathway. Second, it is essential for the progression from DN1 to DN2. Third, its influence continues at least through the β-selection stage (DN3 to DN4) and possibly at the DN4 to ISP stage.

## 1.2.2 HES1

HES1 is a transcription factor regulated by Notch1. Although earlier described as a *cd4* silencer, analysis of thymocyte differentiation using thymic organ cultures did not support this conclusion.[26] More recent studies assign a positive role for HES1 in development of thymic granulocytes.[27] Thus, continuation of DN1 differentiation into thymocytes requires transcriptional downregulation of HES1.

## 1.2.3 T Cell Factor-1 (TCF-1)

*Tcf7* (TCF-1 gene[28,29]) is a direct target gene of Notch1. TCF-1 is a signal-dependent transducer of the Wnt pathway; that is, when Wnt activates β-catenin, TCF-1 acts as an activator of target gene expression.[30,31] But the direct targets of TCF-1 in thymopoiesis are *gata3* and *bcl11b* (discussed later).[28] TCF-1 loss of function experiments indicate that it plays an essential role in thymopoiesis, probably by acting on Wnt signaling.[32] TCF-1 can bypass the need for Notch1 when expressed in high levels, which are typically achieved during the DN3 phase. Normally, TCF-1 is at physiologically low levels at the initiation of thymopoiesis, however, TCF-1 can synergize with Notch1 at early stages.[28] The exact role of TCF-1 in thymopoiesis is a subject of intense interest and there are many unanswered questions.

### 1.2.4 WNT

The canonical (linear) Wnt-β-catenin pathway has been beautifully illustrated by Staal and Clevers.[33] In the absence of Wnt, β-catenin binds to a "destruction complex" consisting of AXIN (axis inhibitor), adenomatous polyposis coli (APC), and the serine/threonine kinases casein kinase 1 (CK1) and glycogen-synthase kinase 3β (GSK3β). CK phosphorylates β-catenin, which allows β-transducin-repeat-containing protein (βTRCP) to ubiquitylate β-catenin, which is then destroyed by the proteasome.[33] In the presence of Wnt, Wnt binds to its receptor Frizzled (Fzd) and coreceptor, low-density-lipoprotein-receptor-related protein 5 or 6 (LRP5/6) leading to the inactivation of GSK3β by disheveled (DVL).[33] β-catenin now translocates to the nucleus to participate in a transcription complex consisting of β-catenin, pygopus homologue (PYGO), and legless homologue (LGS).[33] "Wnt" is not a single product but a family consisting of 19 members. Wnt proteins are produced by the thymic epithelium cells (TECs), and Brunk et al. performed a comprehensive assessment of Wnt expression in the TEC.[34] The amount of expression varied depending on the source of the TEC (medullary [mTEC] or cortical [cTEC]) and whether the mTEC expressed high (mTEC^hi) or low (mTEC^low) levels of MHC-II. The Wnts that were most abundant were Wnt4, Wnt5a, Wnt7a, Wnt7b, Wnt8b, Wnt9a, Wnt9b, Wnt10a, and Wnt10b.[34] The expression levels of Fzd was also determined and TECs showed the highest level of expression with Fzd6 showing the highest level and was expressed at approximately the same level in all TEC cells.[34] These workers concluded that TECs were both the source and target of Wnt.[34] The Wnt pathway is also controlled by the action of several naturally occurring decoys—secreted frizzled-related protein-1 (sFRP), Wnt inhibitory factor-1 (WIF-1), and Dickkopf (DKK)—that bind to and block LRP5/6.[34]

There are three different Wnt pathways recognized: the canonical pathway, the planar cell polarity pathway, and the Wnt-Ca^{2+} pathway. The role of the canonical Wnt pathway in thymocyte development has been extensively researched. Primarily, blocking the Wnt signal by overexpressing known natural inhibitors, such as DKK,[33] results in a significant reduction in the number of DN2-4 thymocyte progenitors. However, directly blocking β-catenin did not affect hemopoiesis.[32] Thus, it is generally accepted that a major role for Wnt is to increase DN1 proliferation.[35]

Osada et al. knocked out *kremen1* and examined the thymic epithelial architecture.[36] Kremen1 (Krm) inhibits Wnt by binding to DKK/LRP6 complex, triggering internalization and clearance from the cell surface.[36] Loss of Krm results in excessive Wnt signaling.[36] Loss of Krm caused aberrant development of the thymus epithelium, demonstrated by a loss of separation of the cortical and medullary regions of the developing thymus; however, there was no evidence of changes in thymocyte number or frequency.[36]

## 1.2.5 SHH/IHH

There are three hedgehog (Hh) proteins: sonic hedgehog (Shh), Indian hedgehog (Ihh), and desert hedgehog (Dhh). These proteins share a common signaling pathway, binding to a cell surface receptor Patched. The binding of Hh to Patched releases the signal transduction molecule Smoothened to signal into the cell. This signaling activates the Gli family of transcription factors, which consists of Gli1, Gli2, and Gli3.[37] Gli1 acts only as an activator of transcription, whereas Gli2 and Gli3 can function as an activator or repressor of transcription.[37] Whether Gli2 and Gli3 act as activators or repressors depends on the balance, in both strength and duration, of Gli Repressor (Gli2R and Gli3R) and Gli Activator (Gli2A, Gli1) in the cell.[37] Smoothened-dependent signaling via Gli2 and Gli3 is required for DN1 to DN2 transition and proliferation, but Gli1 is not.[37] Shh and Ihh pathway activation negatively regulate the DN3 to DP in a Gli2-dependent manner.[37] The Gli proteins are not expressed equally in the thymocyte populations. Gli3 is expressed exclusively in fetal thymocytes with the highest expression in the DN1 population.[38] Gli1 and Gli2 are expressed equally in fetal and adult thymocytes.[37] Gli2 expression is highest in the DN1 and DN2 populations, whereas Gli1 expression is highest in DN2 and DN3 cells and downregulated in DN4 and DP cells.[37] These data have resulted in a model of Hh signaling in thymocyte differentiation where Hh is required for DN1 to DN2 differentiation but must be repressed after the pre-TCR signal transduction and has a negative regulatory function after pre-TCR signaling.[37]

Hh also plays a role in thymocyte proliferation. In vitro studies by Ichim et al. using OP9-DL1 cultures overexpressing orphan nuclear receptor v-erb-A related-2 (Ear-2, Nr2f6), which cause decreased expression of Gli1 and Gli2 among other important targets, show decreased proliferation of DN1 cells during days 1 to 5.[39] Ear-2 downregulation also caused reduced survival of DN4-DP cells.[39]

## 1.2.6 DELTA-LIKE 4

See Section 1.2.1.

## 1.2.7 NFATC/NF-κB

Given the large number of gene targets for NFATc and NF-κB, it is not surprising that they have a role in the process of thymocyte differentiation. Activation of NFAT is calcium ($Ca^{2+}$) dependent and the requirements for extracellular $Ca^{2+}$ influx via the calcium release-activated calcium (CRAC) channel for T cell activation has been intensely studied. Because of their involvement in numerous gene activations, experiments to embryonically knockout functional versions of these transcription factors by targeting various substructures are often lethal.

Aifantis et al. found low levels of nuclear NF-κB activity in DN1 cells and much higher levels in DN4 cells in mice.[40] These data are consistent with the conclusion that NF-κB is required either during or prior to β-selection. Similar results were obtained when assaying for nuclear NFAT levels. DN1 cells showed low nuclear levels of NFAT whereas DN4 cells had much higher nuclear levels of NFAT. Use of inhibitors for components of the canonical activation pathways of these transcription factors, such as U73122, which inhibits PLCγ, prevented $Ca^{2+}$ influx and thus NFAT activation, but they did not study its effect on any DN cell types other than DN1 and DN4. These studies[40] preceded the structural identification of the components of the CRAC channel and more recent work using STIM1$^{-/-}$ plus STIM2$^{-/-}$ knockout animals showed no decrease in DP or peripheral T cells.[41] Similarly, neither Orai1- nor Orai2-deficient animals showed any effect on T cell development.[42,43] It is not clear from these studies whether there is a differential role for canonical NF-κB or NFAT in any of the DN populations, and it is possible that the changes noted are due to upregulation of transcription of other factors needed during the differentiation process. Further discussion on the role of NFATc1 follows.

## 1.2.8    IL-7

Patra et al. reported higher expression levels of NFATc1 in DN2 and DN3 cells than in the DN4 population.[44] Similar results were found when measuring nuclear translocation of NFATc2 and NFATc3; NFATc2 and NFATc3 are the other forms of NFAT found in immune cells.[44] Although IL-2 signaling played no role in NFATc1 activation, IL-7 did influence NFATc1 activation.[44] Patra et al. measured IL-7Rα expression and found that the expression levels paralleled the NFATc1 expression pattern, that is, low in DN1, abundant in DN2 and DN3 cells, and absent in DN4 cells.[44] They also found that IL-7 signals induced activation of NFATc1, and further analysis using cyclosporin A (CsA) to attempt to block IL-7 activation of NFATc1 showed no inhibition suggesting that this was calcineurin independent.[44] Addition studies using the Jak3 inhibitor WHI-P131 demonstrated that Jak3 inhibition levels resulted in much less DN1-to-DN3 differentiation and in vitro kinase assays showed direct activation of NFATc1 by Jak3.[44] Bcl2 expression is induced by IL-7 in pre-T cells[45,46] in cooperation with STAT5.[44] Again, inhibition of Jak3 by WHI-P131 or another inhibitor (e.g., PF-956980) prevented activation of NFATc1 and STAT5 as well as reduced Bcl2 expression.[44] NFATc1 knockout animals had arrested thymocyte differentiation at DN1 similar to that of *IL7*$^{-/-}$ and *IL7r*$^{-/-}$ animals.[44] These workers ascribe their results to an alternative NFAT-activation pathway in pre-TCR-negative thymocytes.[44]

## 1.3    DN2 STAGE

Blockage of the DN2 stage can result in the lymphopoietic pathway proceeding toward the production of non-T cells, such as myeloid cells and NK cells. Such is the case with inhibition of Notch1 as described earlier. However, there are factors that are required to drive thymopoiesis forward from DN2, and these will be described next. We have already described numerous factors that are involved in driving differentiation from DN1 through to later stages and these will not be repeated here.

### 1.3.1   BCL11B

BCL11b is a zinc finger protein that functions to ensure the commitment to T cell production and *bcl11b⁻ᐟ⁻* mice die shortly after birth due to a loss of thymocytes.[47] *Bcl11b* is a direct target of TCF-1 as discussed earlier.[28] Expression of Bcl11b has a sharp onset in the early DN2 stage.[48] Expression then continues throughout T cell development and because of this continuous expression, it is available to contribute to T cell development as it progresses.

Blocking Bcl11b has several effects. First, loss of Bcl11b expression results in the progenitor cells being shunted into NK cell production.[48–50] Second, there is a block in the development of the αβ T cells at the first TCR-dependent selection event.[51] Third, there are impaired survival and abnormities of CD4⁺CD8⁺ TCRαβ cells.[47] Fourth, Bcl11b downregulates a number of stem cell and progenitor cell genes that inhibit the progression of DN2 at this pivotal stage.[48] Finally, Bcl11b has tumor suppressor activity and *bcl11b⁺ᐟ⁻* animals subjected to γ-irradiation develop thymic lymphomas.[52]

### 1.3.2   GATA3

GATA3 is necessary for T cell development and peaks during the DN2 stage.[53] *Gata3* is also a direct target of TCF-1.[28] Reduction of GATA3 by RNA interference and conditional deletion resulted in death of DN1 cells, delayed progression to the DN2 stage, and blocked the appearance of DN3 stage cells. Also noted was a skewed DN2 gene regulation and gene expression analyses by quantitative PCR and RNA sequencing showed that GATA-3-deficient DN2 cells quickly upregulated genes, including Spi1 (PU.1) and Bcl11a, and downregulated genes, including *Cpa3*, *Ets1*, *Zfpm1*, *Bcl11b*, *Il9r*, and *Il17rb* with gene-specific kinetics and dose dependencies.[53] GATA3 was able to block B cell development without Bcl11b help.[53] As indicated earlier, thymocyte signaling requires a tuneable concentration-dependent analog approach, that is, not binary (all on or all off), to provide a more precise regulation of its action. This was demonstrated by titration of GATA-3 activity using tamoxifen-inducible GATA-3 in prethymic multipotent precursors, which showed that GATA-3 inhibits B and myeloid developmental alternatives at different threshold doses.[53] Overexpression of GATA3 inhibited Notch1-induced specification to the T cell lineage.[54]

### 1.3.3   IL-7

Based on a global microarray meta-analysis that predicted that IL-18 would have a role in thymopoiesis, Gandhapudi et al. investigated the action of IL-18.[55] They found that IL-18 alone significantly enhanced the expansion of early thymic progenitor cells at a level comparable to IL-7 alone.[55] Further, they showed the combination of high concentrations of IL-18 with IL-7 induced a "modest" effect on the expansion of DN2 cells.[55]

## 1.4 DN3 STAGE

A critical event that occurs during the DN3 to DN4 stage is β-selection where TCR rearrangement occurs. Reports of the role of key mediators not previously discussed that are necessary for this essential step are provided next.

### 1.4.1 IL-7

The role of IL-7 and IL-7R in earlier DN populations has been previously discussed. The role of IL-7 at earlier stages of thymocyte development (DN2 to DN3a) is primarily related to survival of TCRβ+ cells by inducing the expression of Bcl2.[56,57] Boudil et al. recently described a role for IL-7 signaling in TCRβ+ cells DN3 and DN4 cells.[57] Early DN3 cells become quiescent and rearrange gene segments in loci encoding the γ-chain, δ-chain, and β-chain of the TCR.[57] Initiation of β-selection requires successful rearrangement of the β-chain and expression of intracellular TCRβ (iTCRβ) to form the pre-TCR signaling complex.[57] TCR α-chain (*Tcra*) is rearranged in DP cells.[57] IL-7R expression persists through the early stages of β-selection, and Boudil et al. addressed the importance of IL-7 signaling beyond the induction of the pro-survival molecular Bcl-2 in this process.[57] IL-7 signaled TCRβ+ DN3 and DN4 thymocytes to upregulate genes that encode molecules involved in cell growth and repressed Bcl6, a transcriptional repressor.[57] Genes representing molecules involved in the Jak-STAT, GTP binding, Ras-MAPK, and PI(3)K-mTOR signaling pathways; as well as genes encoding receptors and involved in transcription were assayed in DN3a (CD25+iTCRβ−), DN3b (CD25+iTCRβ+), and DN4 (CD25−iTCRβ+) cells in response to IL-7 stimulation of *il7−/−* thymocytes.[57] Genes from all categories were changed in response to IL-7 stimulation, however, the genes most markedly upregulated in the transition from DN3a to DN3b were *Socs3* (Jak-STAT), *Igtp* (GTP binding), *Sgk1* (PI(3)K-mTOR), *Ctla4* (receptors), *Gpr83* (receptors), and *Fos* (transcription), although several genes from each category were upregulated during this transition.[57] As stated earlier, Bcl2 is a survival factor, and transgenic expression of *bcl2* did not reverse the effect of *il7−/−*,[57] however, deletion of *bcl6* partially restored the self-renewal of DN4 cells.[57] The conclusion of these workers was that IL-7 signaling acted "cooperatively with signaling via the pre-TCR and Notch1 to coordinate proliferation, differentiation, and *tcra* recombination during β-selection."[57]

### 1.4.2 HEB/HEBALT/HEBCAN

HEB is an E-protein transcription factor that is required for thymocyte development.[58] The HEB gene encodes for two transcription factors: HEBCan, which is a long form, and a shorter form called HEBAlt.[58] HEBCan (not shown in Figure 1.1) is expressed throughout T cell development, but HEBAlt is only expressed in the DN stages.[58,59] Using HEB−/− transgenic embryos (HEB−/− is embryonically lethal), Braunstein and Anderson determined that HEB−/− precursors were blocked at the β-selection step,[58] however, transgenic expression of HEBAlt restored the ability of HEB[−/−] precursors to pass through the β-selection step. Earlier reports indicated that another E-protein transcription factor, E2A, although known to prevent thymic precursors from passing through the β-selection checkpoint, was not able to block the progression of Rag-1[−/−] precursors from forming DP cells.[58,60] It was determined that HEBAlt did not push cells beyond the β-selection stage in the absence of Rag-1.[58] HEBAlt enabled T-cell precursors to respond to CD3ε stimulation; CD3ε is required for progression beyond the DN3 stage.[61] Taken together, these studies showed that HEB factors are heavily involved in β-selection; HEBAlt was able to partially restore the ability of HEB[−/−] precursors to pass through β-selection and enable T cell precursors to respond to CD3ε stimulation. On the other hand, HEBCan is required to fully restore production of DP and SP cells, implying that HEBAlt and HEBCan act as nonredundant, critical, stage-specific components during T cell development.

## 1.5  DN4 STAGE

DN4 is the final DN stage, and blocking this step results in few DP or more mature forms of thymocytes. Two critical factors involved in this step are discussed next. Of course, as diagrammed in Figure 1.1, there are several factors that are involved in this step, however, their role in hematopoiesis have been discussed in previous sections.

### 1.5.1  CD3ε

CD3 chain-specific transcripts are expressed at the DN1 stage, and by the DN3 stage, most thymocytes express high levels of CD3γε- and CD3δε-complex (CIC) transcripts, although surface expression of CIC is negligible at this stage.[62] There is evidence that the various chains are translated and stored within the ER/Golgi, but the fully assembled structure is not expressed until the CD3ζζ homodimer is inserted into the complex.[62] The effect of CD3 deficiency on thymocyte development depends on the chain affected, that is, CD3δ deficiency impairs thymocyte development at the DP stage, whereas, CD3γ, ζ, or ε deficiency prevents the pre-TCR-mediated DN to DP transition.[62] Reconstituting each of the transcriptionally inactivated subunit chains individually suggests that the intracytoplasmic signaling motifs (ITAMs) are functionally equivalent.[63] Brodeur et al. approached the question of the role of CD3γ, δ, and ε intracytoplasmic structures in thymocyte differentiation.[62] Their data suggest that CD3γ and CD3δ do not affect thymic differentiation except for reducing thymocyte numbers and increasing TCR expression on DP cells, however, the intracytoplasmic domain of CD3ε is required for DN3 to DP thymocyte development.[62] Brodeur et al. were also able to assign distinct roles for various individual motifs within the intracytoplasmic domain of CD3ε.[62]

Membrane surface expression of CD3ε is required for appropriate transition from DN3 to DN4.[64] To accomplish the membrane expression of CD3ε, a CD3ε membrane-proximal basic-rich stretch (BRS) is necessary for membrane binding to the CD3ε cytoplasmic tail. Using mice with a CD3ε-BRS mutation, which reduced or abolished membrane expression, Bettini et al. found that DN3 to DN4 transition was affected.[64] They discovered a significant increase in DN3 thymocytes with a loss of DN4 thymocytes.[64] These CD3ε-BRS mutant mice also have defects in TCR signaling, namely, TCR signaling in DN4 cells was significantly higher than wild-type (WT) mice.[64] These mutant mice had several other downstream defects, most notably, defective TCR signaling.[64,65]

### 1.5.2 SHH

Using a substantially different approach, Barnett and coworkers[66–68] investigated the effect of prenatal exposure to cadmium (Cd) on the adult offspring. Offspring (mixed sexes) less than 1 day old showed significantly more DN4 thymocytes and a trend in fewer DP thymocytes.[68] Shh activity was significantly decreased in the thymic extracts[68] and given that Shh is necessary for DN to DP transition,[69] this suggests a possible block in the transition from DN3 to DN4 due to reduced Shh. Additional studies were conducted on 20-week-old offspring and there was a significant increase in the DN1 populations in both males and females.[67] Shah et al. also showed that Shh activity was necessary for DN1 to DN2 transition.[69] Females showed an increase in DN3 cells with no difference in the DN4 population, however, male offspring did not show any difference in the DN3 or DN4 population.[67] It would be reasonable to conclude that this sex difference is due to a differential effect of Cd on the sexes and not a difference in thymocyte maturation requirements between the sexes.

## 1.6 CONCLUSION

As discussed, the maturation of thymocytes from the time an early thymic progenitor cell arrives at the thymus until it becomes a DP cell and beyond is a tightly controlled choreography of factors acting on the cells. The myriad of transcription factors that participate in the expression of these key factors are not mentioned, with the exception of a few key transcription factors (e.g., NF-κB and NFATc). As mentioned in the Introduction, it is important but difficult to discuss the many instances of cross-talk between factors and concentration-dependencies of some factors in this process. Nonetheless, this chapter provides a framework of this comprehensive process.

## REFERENCES

1. Lauritsen, J. P. et al. Marked induction of the helix-loop-helix protein Id3 promotes the gammadelta T cell fate and renders their functional maturation Notch independent. *Immunity* **31**, 565–575, doi:10.1016/j.immuni.2009.07.010 (2009).
2. Moon, R. T., and Gough, N. R. Beyond canonical: The Wnt and β-catenin story. *Science Signaling* **9**, eg5, doi:10.1126/scisignal.aaf6192 (2016).
3. Rothenberg, E. V., and Anderson, M. K. Elements of transcription factor network design for T-lineage specification. *Dev Biol* **246**, 29–44, doi:10.1006/dbio.2002.0667 (2002).
4. Shah, D. K., and Zuniga-Pflucker, J. C. An overview of the intrathymic intricacies of T cell development. *J Immunol* **192**, 4017–4023, doi:10.4049/jimmunol.1302259 (2014).

5. Tydell, C. C. et al. Molecular dissection of prethymic progenitor entry into the T lymphocyte developmental pathway. *J Immunol* **179**, 421–438 (2007).

6. Hozumi, K. et al. Delta-like 4 is indispensable in thymic environment specific for T cell development. *J Exp Med* **205**, 2507–2513 (2008).

7. Koch, U. et al. Delta-like 4 is the essential, nonredundant ligand for Notch1 during thymic T cell lineage commitment. *J Exp Med* **205**, 2515–2523 (2008).

8. Radtke, F. et al. Deficient T cell fate specification in mice with an induced inactivation of Notch1. *Immunity* **10**, 547–558 (1999).

9. Wilson, A., MacDonald, H. R., and Radtke, F. Notch 1-deficient common lymphoid precursors adopt a B cell fate in the thymus. *J Exp Med* **194**, 1003–1012 (2001).

10. Hozumi, K. et al. Delta-like 1 is necessary for the generation of marginal zone B cells but not T cells in vivo. *Nat Immunol* **5**, 638–644, doi:10.1038/ni1075 (2004).

11. Radtke, F., Fasnacht, N., and MacDonald, H. R. Notch signaling in the immune system. *Immunity* **32**, 14–27 (2010).

12. Fiorini, E. et al. Cutting edge: Thymic crosstalk regulates Delta-like 4 expression on cortical epithelial cells. *J Immunol* **181**, 8199–8203 (2008).

13. Izon, D. J. et al. Deltex1 redirects lymphoid progenitors to the B cell lineage by antagonizing Notch1. *Immunity* **16**, 231–243, doi:10.1016/S1074-7613(02)00271-6 (2002).

14. Haltiwanger, R. S. Regulation of signal transduction pathways in development by glycosylation. *Curr Opin Struct Biol* **12**, 593–598 (2002).

15. Rampal, R. et al. Lunatic fringe, Manic fringe, and Radical fringe recognize similar specificity determinants in O-fucosylated epidermal growth factor-like repeats. *J Biol Chem* **280**, 42454–42463, doi:10.1074/jbc.M509552200 (2005).

16. Moloney, D. J. et al. Fringe is a glycosyltransferase that modifies Notch. *Nature* **406**, 369–375, doi:10.1038/35019000 (2000).

17. Visan, I., Yuan, J. S., Tan, J. B., Cretegny, K., and Guidos, C. J. Regulation of intrathymic T-cell development by Lunatic Fringe-Notch1 interactions. *Immunol Rev* **209**, 76–94, doi:10.1111/j.0105-2896.2006.00360.x (2006).

18. Tsukumo, S., Hirose, K., Maekawa, Y., Kishihara, K., and Yasutomo, K. Lunatic fringe controls T cell differentiation through modulating notch signaling. *J Immunol* **177**, 8365–8371 (2006).

19. Yun, T. J., and Bevan, M. J. Notch-regulated ankyrin-repeat protein inhibits Notch1 signaling: Multiple Notch1 signaling pathways involved in T cell development. *J Immunol* **170**, 5834–5841 (2003).

20. Petersen, P. H., Zou, K., Hwang, J. K., Jan, Y. N., and Zhong, W. Progenitor cell maintenance requires numb and numblike during mouse neurogenesis. *Nature* **419**, 929–934, doi:10.1038/nature01124 (2002).

21. Wilson, A. et al. Normal hemopoiesis and lymphopoiesis in the combined absence of numb and numblike. *J Immunol* **178**, 6746–6751 (2007).

22. Wolfer, A., Wilson, A., Nemir, M., MacDonald, H. R., and Radtke, F. Inactivation of Notch1 impairs VDJβ rearrangement and allows pre-TCR-independent survival of early αβ lineage thymocytes. *Immunity* **16**, 869–879, doi:http://dx.doi.org/10.1016/S1074-7613(02)00330-8 (2002).

23. Maillard, I. et al. The requirement for Notch signaling at the beta-selection checkpoint in vivo is absolute and independent of the pre-T cell receptor. *J Exp Med* **203**, 2239–2245, doi:10.1084/jem.20061020 (2006).

24. Fiorini, E. et al. Dynamic regulation of Notch 1 and Notch 2 surface expression during T cell development and activation revealed by novel monoclonal antibodies. *J Immunol* **183**, 7212–7222, doi:10.4049/jimmunol.0902432 (2009).

25. van Ewijk, W., Shores, E. W., and Singer, A. Crosstalk in the mouse thymus. *Immunol Today* **15**, 214–217, doi:http://dx.doi.org/10.1016/0167-5699(94)90246-1 (1994).

26. Bosselut, R. CD4/CD8-lineage differentiation in the thymus: From nuclear effectors to membrane signals. *Nat Rev Immunol* **4**, 529–540, doi:10.1038/nri1392 (2004).
27. De Obaldia, M. E., Bell, J. J., and Bhandoola, A. Early T-cell progenitors are the major granulocyte precursors in the adult mouse thymus. *Blood* **121**, 64–71, doi:10.1182 /blood-2012-08-451773 (2013).
28. Rothenberg, E. V. Transcriptional drivers of the T-cell lineage program. *Curr Opin Immunol* **24**, 132–138, doi:10.1016/j.coi.2011.12.012 (2012).
29. Yu, S. et al. The TCF-1 and LEF-1 transcription factors have cooperative and opposing roles in T cell development and malignancy. *Immunity* **37**, 813–826, doi:10.1016/j .immuni.2012.08.009 (2012).
30. Gordon, M. D., and Nusse, R. Wnt signaling: Multiple pathways, multiple receptors, and multiple transcription factors. *J Biol Chem* **281**, 22429–22433, doi:10.1074/jbc .R600015200 (2006).
31. Staal, F. J., and Sen, J. M. The canonical Wnt signaling pathway plays an important role in lymphopoiesis and hematopoiesis. *Eur J Immunol* **38**, 1788–1794, doi:10.1002 /eji.200738118 (2008).
32. Jeannet, G. et al. Long-term, multilineage hematopoiesis occurs in the combined absence of beta-catenin and gamma-catenin. *Blood* **111**, 142–149, doi:10.1182/blood -2007-07-102558 (2008).
33. Staal, F. J., and Clevers, H. C. WNT signalling and haematopoiesis: A WNT-WNT situation. *Nature Rev Immunol* **5**, 21–30 (2005).
34. Brunk, F., Augustin, I., Meister, M., Boutros, M., and Kyewski, B. Thymic epithelial cells are a nonredundant source of Wnt ligands for thymus development. *J Immunol* **195**, 5261–5271, doi:10.4049/jimmunol.1501265 (2015).
35. Staal, F. J., and Luis, T. C. Wnt signaling in hematopoiesis: Crucial factors for self-renewal, proliferation, and cell fate decisions. *J Cell Biochem* **109**, 844–849, doi:10.1002/jcb.22467 (2010).
36. Osada, M. et al. The Wnt signaling antagonist Kremen1 is required for development of thymic architecture. *Clinical Develop Immunol* **13**, 299–319, doi:10.1080 /17402520600935097 (2006).
37. Drakopoulou, E. et al. Non-redundant role for the transcription factor Gli1 at multiple stages of thymocyte development. *Cell Cycle* **9**, 4144–4152 (2010).
38. Hager-Theodorides, A. L. et al. The Gli3 transcription factor expressed in the thymus stroma controls thymocyte negative selection via hedgehog-dependent and -independent mechanisms. *J Immunol* **183**, 3023–3032 (2009).
39. Ichim, C. V., Dervović, D. D., Zúñiga-Pflücker, J. C., and Wells, R. A. The orphan nuclear receptor Ear-2 (Nr2f6) is a novel negative regulator of T cell development. *Exp Hematol* **42**, 46–58, doi:10.1016/j.exphem.2013.09.010 (2014).
40. Aifantis, I., Gounari, F., Scorrano, L., Borowski, C., and von Boehmer, H. Constitutive pre-TCR signaling promotes differentiation through $Ca^{2+}$ mobilization and activation of NF-kappaB and NFAT. *Nat Immunol* **2**, 403–409, doi:10.1038/87704 (2001).
41. Oh-hora, M. et al. Dual functions for the endoplasmic reticulum calcium sensors STIM1 and STIM2 in T cell activation and tolerance. *Nat Immunol* **9**, 432–443 (2008).
42. Gwack, Y. et al. Hair loss and defective T- and B-cell function in mice lacking ORAI1. *Mol Cell Biol* **28**, 5209–5222 (2008).
43. Vig, M. et al. Defective mast cell effector functions in mice lacking the CRACM1 pore subunit of store-operated calcium release-activated calcium channels. *Nat Immunol* **9**, 89–96 (2008).
44. Patra, A. K. et al. An alternative NFAT-activation pathway mediated by IL-7 is critical for early thymocyte development. *Nat Immunol* **14**, 127–135, doi:10.1038/ni.2507 (2013).

45. von Freeden-Jeffry, U., Solvason, N., Howard, M., and Murray, R. The earliest T lineage-committed cells depend on IL-7 for Bcl-2 expression and normal cell cycle progression. *Immunity* **7**, 147–154 (1997).

46. Kim, K., Lee, C. K., Sayers, T. J., Muegge, K., and Durum, S. K. The trophic action of IL-7 on pro-T cells: Inhibition of apoptosis of pro-T1, -T2, and -T3 cells correlates with Bcl-2 and Bax levels and is independent of Fas and p53 pathways. *J Immunol* **160**, 5735–5741 (1998).

47. Albu, D. I. et al. BCL11B is required for positive selection and survival of double-positive thymocytes. *J Exp Med* **204**, 3003–3015, doi:10.1084/jem.20070863 (2007).

48. Li, L., Leid, M., and Rothenberg, E. V. An early T cell lineage commitment checkpoint dependent on the transcription factor Bcl11b. *Science* **329**, 89–93, doi:10.1126/science.1188989 (2010).

49. Ikawa, T. et al. An essential developmental checkpoint for production of the T cell lineage. *Science* **329**, 93–96, doi:10.1126/science.1188995 (2010).

50. Li, P. et al. Reprogramming of T cells to natural killer-like cells upon Bcl11b deletion. *Science* **329**, 85–89, doi:10.1126/science.1188063 (2010).

51. Wakabayashi, Y. et al. Bcl11b is required for differentiation and survival of alphabeta T lymphocytes. *Nat Immunol* **4**, 533–539, doi:10.1038/ni927 (2003).

52. Kamimura, K. et al. Haploinsufficiency of Bcl11b for suppression of lymphomagenesis and thymocyte development. *Biochem Biophys Res Commun* **355**, 538–542, doi:http://dx.doi.org/10.1016/j.bbrc.2007.02.003 (2007).

53. Scripture-Adams, D. D et al. GATA-3 dose-dependent checkpoints in early T cell commitment. *J Immunol* **193**, 3470–3491, doi:10.4049/jimmunol.1301663 (2014).

54. Van de Walle, I. et al. GATA3 induces human T-cell commitment by restraining Notch activity and repressing NK-cell fate. *Nature Commun* **7**, 11171, doi:10.1038/ncomms11171 (2016).

55. Gandhapudi, S. K. et al. IL-18 acts in synergy with IL-7 to promote ex vivo expansion of T lymphoid progenitor cells. *J Immunol* **194**, 3820–3828, doi:10.4049/jimmunol.1301542 (2015).

56. Seddon, B. Thymic IL-7 signaling goes beyond survival. *Nat Immunol* **16**, 337–338, doi:10.1038/ni.3128 (2015).

57. Boudil, A. et al. IL-7 coordinates proliferation, differentiation and Tcra recombination during thymocyte beta-selection. *Nat Immunol* **16**, 397–405, doi:10.1038/ni.3122 (2015).

58. Braunstein, M., and Anderson, M. K. Developmental progression of fetal HEB(–/–) precursors to the pre-T-cell stage is restored by HEBAlt. *Eur J Immunol* **40**, 3173–3182, doi:10.1002/eji.201040360 (2010).

59. Wang, D. et al. The basic helix-loop-helix transcription factor HEBAlt is expressed in pro-T cells and enhances the generation of T cell precursors. *J Immunol* **177**, 109–119 (2006).

60. Engel, I., Johns, C., Bain, G., Rivera, R. R., and Murre, C. Early thymocyte development is regulated by modulation of E2A protein activity. *J Exp Med* **194**, 733–745 (2001).

61. Malissen, M. et al. Altered T cell development in mice with a targeted mutation of the CD3-epsilon gene. *EMBO J* **14**, 4641–4653 (1995).

62. Brodeur, J. F., Li, S., Martins, M. S., Larose, L., and Dave, V. P. Critical and multiple roles for the CD3epsilon intracytoplasmic tail in double negative to double positive thymocyte differentiation. *J Immunol* **182**, 4844–4853 (2009).

63. Sommers, C. L. et al. Function of CD3 epsilon-mediated signals in T cell development. *J Exp Med* **192**, 913–919 (2000).

64. Bettini, M. L. et al. Membrane association of the CD3ε signaling domain is required for optimal T cell development and function. *J Immunol* **193**, 258–267, doi:10.4049/jimmunol.1400322 (2014).

65. Deford-Watts, L. M. et al. The cytoplasmic tail of the T cell receptor CD3 epsilon subunit contains a phospholipid-binding motif that regulates T cell functions. *J Immunol* **183**, 1055–1064, doi:10.4049/jimmunol.0900404 (2009).

66. Hanson, M. L. et al. Prenatal cadmium exposure alters postnatal immune cell development and function. *Toxicol Appl Pharmacol* **261**, 196–203 (2012).

67. Holásková, I., Elliott, M., Hanson, M. L., Schafer, R., and Barnett, J. B. Prenatal cadmium exposure produces persistent changes to thymus and spleen cell phenotypic repertoire as well as the acquired immune response. *Toxicol Appl Pharmacol* **265**, 181–189 (2012).

68. Hanson, M. L., Brundage, K. M., Schafer, R., Tou, J. C., and Barnett, J. B. Prenatal cadmium exposure dysregulates sonic hedgehog and Wnt/beta-catenin signaling in the thymus resulting in altered thymocyte development. *Toxicol Appl Pharmacol* **242**, 136–145 (2010).

69. Shah, D. K. et al. Reduced thymocyte development in sonic hedgehog knockout embryos. *J Immunol* **172**, 2296–2306 (2004).

# 2 In Situ Analysis of T Cell Receptor Signals during Positive Selection

*Nathalie Labrecque, Mengqi Dong,\**
*Aditi Sood,\* and Heather J. Melichar*

## CONTENTS

### ABSTRACT

There are many similarities in the T cell receptor (TCR) signaling pathways that support positive selection, deletion of autoreactive T cells in the thymus, and activation of mature T cells in secondary lymphoid organs. However,

---

\* Indicates equal contribution.

several unique aspects of this process allow immature T cells endowed with an antigen receptor capable of recognizing foreign peptides in the context of major histocompatibility molecules to survive and differentiate upon low-affinity encounters with self-antigens during development. We review recent advances toward understanding the molecular and cellular mechanisms contributing to the unique TCR signals that support positive selection. In addition, given the dynamism involved in thymocyte scanning of the thymic stroma for appropriate signals, this process has proven difficult to recapitulate *in vitro*. Accordingly, we discuss the benefits and limitations of current and emerging models of positive selection as well as technologies that are improving our ability to characterize the behavior and TCR signals that accompany positive selection in the complex, three-dimensional environment in which this process occurs.

## 2.1   THYMIC T CELL DIFFERENTIATION

T cells play an essential role in fighting infection via T cell receptor (TCR) recognition of a peptide fragment of foreign antigen presented by major histocompatibility complex (MHC) molecules. Unfortunately, T cells can also contribute to the development of several autoimmune diseases. The precarious balance between their protective and pathologic functions stems from the need for T cells to possess an incredibly diverse TCR repertoire to cover the recognition of a vast array of foreign antigens while not reacting to self-antigens. This TCR diversity is generated during recombination of germ-line encoded gene segments and is bolstered by nontemplated nucleotide additions to form the TCR during development in the thymus. As this is a random process, not all the TCRs that are generated will be useful, that is, endowed with the ability to recognize a peptide derived from the foreign antigen embedded within a host MHC molecule. Thus, there are stringent selection processes that accompany thymopoiesis that underlie the development of a functional (MHC-restricted) and self-tolerant T cell repertoire. In particular, the process of positive selection supports the survival and differentiation of immature thymocytes into functionally mature T cells that express TCRs fit for the MHC molecules of the host. The distinct TCR signals in the thymus that mediate positive selection occur in response to low to moderate affinity interactions with self-peptide presented by MHC proteins (self-pMHC) on specialized cell types embedded within a complex three-dimensional environment and are not completely understood. Here, we highlight a few of the molecular and cellular mechanisms that shape the TCR signals associated with positive selection and the methods for addressing outstanding questions relevant to thymic selection.

### 2.1.1   T CELL DEVELOPMENTAL INTERMEDIATES

T cell developmental intermediates in the thymus are broadly characterized based on cell surface expression of CD4 and CD8 proteins. The early thymic progenitors that seed the thymus arrive from the bone marrow via the vasculature at the cortico-medullary region and maintain multilineage potential. These cells lack CD4 and

CD8 expression as well as lineage markers and comprise one part of a larger, heterogeneous population of CD4⁻CD8⁻ (double negative, DN) cells. During the DN stage, cells become committed to the T lineage and can rearrange and express their β, γ, and δ TCR genes. αβ and γδ lineage T cells diverge at this stage, and the development of conventional αβ lineage T cells proceeds to the CD4⁺CD8⁺ (double positive, DP) stage after a functional TCRβ chain paired with the pre-Tα chain signals cell proliferation, *Tcra* gene rearrangement and expression, and differentiation. At the DP stage, additional mechanisms are in place to trim the highly diverse TCR repertoire to those that are functionally able to recognize peptide when presented by MHC (positive selection), yet tolerant to self-antigen (negative selection). The process of positive selection occurs over 1 to 4 days,[1-6] results in the migration of thymocytes from the cortex to the medulla, and culminates with their differentiation into mature, CD4⁺ and CD8⁺ (single positive, SP) T cells.

### 2.1.2 MODE OF TCR RECOGNITION AND EFFICACY OF POSITIVE SELECTION

In general, the αβ TCR recognizes peptide only when presented in the context of MHC molecules. The α and β chains of the TCR contain hypervariable complementary determining regions (CDRs). The CDR1 and CDR2 loops, for which the diversity is limited by the number of germ-line V regions encoded within the *Tcra* and *Tcrb* locus, primarily interact with the MHC molecules, whereas the highly diverse CDR3 loop, generated by the flexible junction of V-(D)-J gene segments and addition of nontemplate nucleotides, primarily recognizes the embedded peptide.[7,8] Despite a potential bias of germ-line encoded CDR1 and CDR2 loops to interact with MHC molecules, it is estimated that only 3% to 5% of thymocytes survive thymic selection and are exported from the thymus as mature T cells.[7-9] Furthermore, a recent study using an innovative approach to study the preselection TCR repertoire has shown that 15% of the preselection repertoire is reactive to self-pMHC. Among these reactive TCRs, 50% will undergo positive selection while the other 50% will be negatively selected.[10]

### 2.1.3 AFFINITY/AVIDITY THRESHOLD THAT DEFINES POSITIVE VERSUS NEGATIVE SELECTION

The widely accepted affinity model of thymic selection avers that the relative TCR affinity for self-pMHC dictates the fate of a developing T cell. A minimum affinity for self-pMHC is required to induce the expression of Bcl-2 and other antiapoptotic molecules to promote survival.[11,12] TCR interactions with self-pMHC below this threshold result in death by neglect. Only those TCR signals in the low to mid range induce survival signals and promote differentiation of DP thymocytes to mature SP T cells (positive selection), which is accompanied with migration to the medulla. Recognition of self-pMHC with high affinity, however, induces apoptosis (or agonist selection), eliminating or redirecting self-reactive T cells prior to exiting the thymus. Importantly, successful positive selection does not preclude a mature SP thymocyte from undergoing negative selection when it encounters a new repertoire of antigens in the thymic medulla.

Experiments using the fetal thymic organ culture (FTOC) system containing a monoclonal population of preselection DP thymocytes and altered peptide ligands (APLs) in conjunction with measurements of the affinity of the peptide-MHC complex for the TCR have suggested that a broad range of TCR-pMHC affinities appear to support positive selection, whereas the boundary between positive and negative selection is surprisingly narrow.[13–16] Indeed, there was only a twofold difference in affinity between the lowest affinity APL that induces negative selection and the highest affinity APL that promotes positive selection.[13,15] High-affinity peptides induce cell death, whereas a peptide that straddles the boundary between positive and negative selection can push the selection outcome one way or the other based on concentration.[13] Unfortunately, this sharp boundary of affinity between positively and negatively selecting ligands has only been described for three class I-restricted TCRs, including two that have the same antigenic specificity,[13,15] and we lack information on the range of affinity required for positive versus negative selection of thymocytes expressing MHC class II-restricted TCRs.

### 2.1.4 Role of Cortical Thymic Epithelial Cells in Positive Selection

Cortical thymic epithelial cells (cTECs) are specialized to support positive selection.[17–19] They express unique proteolytic components that appear to play important and unique roles in expanding the breadth of positive selection. Incorporation of the β5t subunit into the proteasome of cTECs changes the proteolytic cleavage site of the proteasome to generate a distinct set of self-peptides that are loaded on MHC class I molecules.[20,21] β5t expression in cTECs is essential for positive selection and the absence of this proteasome subunit leads to loss of the thymocytes expressing TCRs at the lower end of the positive selection spectrum in a polyclonal repertoire.[20,22] Indeed, a recent study suggests that β5t generates peptide-MHC class I complexes with a lower affinity for the TCR,[21] and it is possible that β5t generates a specialized subset of peptides rather than simply expanding the diversity of the peptide repertoire.[22] A similar mechanism is at play for the selection of MHC class II-restricted CD4+ T cells. Indeed, cTECs express a different set of proteases involved in the generation of MHC class II binding peptides than peripheral APCs and medullary thymic epithelial cells (mTECs). cTECs specifically express cathepsin L and thymus-specific serine protease (TSSP). Mice deficient for cathepsin L show a severe deficiency in the generation of the polyclonal repertoire of CD4+ SP thymocytes,[23,24] whereas mice deficient for TSSP show only specific effects on the positive selection of certain TCR transgenic T cells.[25,26]

The generation of a unique repertoire of peptides is not the only aspect of cTECs that biases their ability to support positive rather than negative selection. cTECs do not express high levels of costimulatory molecules necessary for negative selection as do mTECs or dendritic cells (DCs).[27–31] In addition, cTECs presenting high-affinity, cognate antigen are unable to sustain stable interactions with thymocytes expressing antigen-specific TCRs to the same extent as their hematopoietic counterparts that can support negative selection.[32] This does not appear to be attributable solely to the differential expression of conventional costimulatory molecules, as CD80/CD86[−/−] DCs maintain stable interactions with thymocytes similar to WT DCs.[32] Given the idea that positive selection requires repeated brief interactions with epithelial cells without synapse formation,[32–34] these intrinsic qualities of cTECs may make them suitable to support positive selection, but the mechanisms remain unknown.

### 2.1.5 SPATIOTEMPORAL DYNAMICS OF THYMIC SELECTION

The cortical and medullary thymic environments are composed of unique stromal components that support distinct aspects of T cell development and selection.[18] Despite this, it is increasingly appreciated that there is significant overlap of the thymic selection processes in space and time, the consequences of which are only beginning to be explored. Just as it is now appreciated that a large portion of negative selection occurs in the cortex to ubiquitously expressed antigen at the DP stage of thymic differentiation concomitant with positive selection,[35–37] positive selection may continue beyond the cortex. The transition from DP to CD4+ and CD8+ SP cells occurs over several days, and a minimum of 48 hours of TCR signals is required for significant positive selection and CD8+ T cell differentiation.[1,4–6,38–40] However, between 12 and 24 hours after initiation of positive selection, a majority of MHC class I-restricted DP cells undergoing positive selection have upregulated CCR7 (and downregulated CXCR4) and migrated to the medulla.[5] These data suggest that developing T cells will acquire additional TCR signals via interaction with distinct tissue-restricted peptides and cell types in the medulla that may shape the mature T cell repertoire. Thus, despite the fact that cTECs are essential for positive selection of developing T cells, how the medullary microenvironment influences positive selection has not yet been addressed.

In summary, it is clear that TCR affinity for self-pMHC plays an important role in the positive selection of T cells and that a special set of peptides is required to promote the positive selection of thymocytes by cTECs within a complex thymic milieu. However, several key questions remain. In particular, we are still working to understand the molecular mechanisms that allow a thymocyte to distinguish a positive from negative selecting ligand.

## 2.2 TCR SIGNALING EVENTS CONTROLLING POSITIVE SELECTION

Distinct TCR signals downstream of low versus high affinity ligand during the thymic selection process are necessary to decode the strength of signaling and determine their fate (Figure 2.1). TCR signaling is initiated following coengagement of the TCR and the coreceptor (CD4 or CD8) with peptide-MHC. This brings the TCR/CD3 complex into proximity with the tyrosine kinase Lck, which is associated with the cytoplasmic tail of the coreceptor allowing for phosphorylation of the ITAM motifs within the intracellular domain of the CD3 complex. The phosphorylated ITAM of the CD3ζ chain then recruits the tyrosine kinase ZAP70 allowing for its phosphorylation and activation by Lck. ZAP70 then propagates the signal, via LAT and other adaptor molecules, leading to the activation of classical intracellular signaling pathways such as mitogen-activated protein kinases (MAPKs) and calcium flux (Figure 2.1).

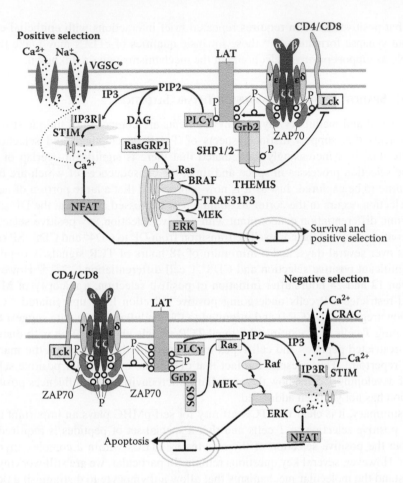

**FIGURE 2.1** Schematic of TCR signaling during positive and negative selection. During thymic selection, the TCR binds self-ligands triggering the activation of a signaling cascade. Themis attenuates TCR proximal signaling during positive selection by interacting with the phosphatases SHP1/2. TCR signals associated with positive selection are accompanied by the localization of RasGRP1 and the activation of the MAPK pathway at the Golgi apparatus whereas these events occur at the plasma membrane in response to high-affinity negative selecting signals. TCR signaling also triggers the release of intracellular calcium from the endoplasmic reticulum (ER) leading to translocation of nuclear factor of activated T cell (NFAT) to the nucleus. [Ca²⁺]i are sustained via CRAC channels in negative selection, yet VSGC channels are necessary to sustain [Ca²⁺]i during positive selection. * VGSC appears to be necessary for MHC class II-restricted positive selection, not MHC class I-restricted positive selection. Together, these events support the survival and differentiation of functional DP thymocytes. Lck, lymphocyte-specific protein tyrosine kinase; ZAP70, zeta-chain-associated protein kinase of 70 kDa; LAT, linker of activated T cells; Grb2, growth factor receptor-bound protein 2; THEMIS, thymocyte-expressed molecule involved in selection; SHP1/2, Src homology region 2 domain-containing phosphatase-1 and -2; PLCγ, phospholipase C gamma; DAG, diacylglycerol; PIP2, phosphatidylinositol 4,5-bisphosphate; IP3, inositol 1,4,5-trisphosphate; IP3R, IP3 receptor; VGSC, voltage-gated sodium channel; SOS, Son of Sevenless; CRAC, Ca²⁺ release-activated Ca²⁺; NFAT, nuclear factor of activated T cells.

## 2.2.1   MAPK Signaling Pathway

The identification of peptides at the affinity threshold between positive and negative selection has provided tools to further define how preselection DP thymocytes
decode TCR affinity for self-pMHC. TCR signaling during positive and negative
selection involves the activation of similar TCR signal cascades, and one of the major
outstanding questions in the field is how an analogue input (i.e., a rather broad range
of peptide affinity) results in a digital output (i.e., positive versus negative selection). This question has been elegantly addressed using fixed, MHC class I-restricted
TCR transgenic thymocytes stimulated with tetramers loaded with APLs of varying affinities.[13] Stimulation of preselection thymocytes with tetramers loaded with
low-affinity ligands resulted in differential compartmentalization of TCR signaling
intermediates in comparison to stimulation of the same cells with high affinity negatively selecting ligands. In particular, activation of the MAPKs ERK1 and ERK2 was
reported to occur in the Golgi in response to low-affinity ligands, whereas ERK1/2
activation occurs at the plasma membrane in response to high-affinity ligands
(Figure 2.1).[13] Furthermore, low but sustained ERK1/2 activation was reported for
positively selecting ligands, whereas strong and transient activation of ERK1/2 was
observed with negatively selecting ligands.[13] Interestingly, these differences in compartmentalization of ERK activation were observed even with the negatively selecting peptide of the lowest affinity.[13] These findings are particularly interesting given
that ERK1/2 are necessary for positive but not negative selection.[41-47]

Analysis of the differential requirements for the guanine exchange factors (GEFs)
that activate Ras—RasGRP1 and Sos1—during positive and negative selection,
respectively, begins to address how, at the molecular level, the strength of the TCR
signal leads to differences in the site at which ERK1/2 are activated. A weaker TCR
signal is believed to activate Ras (a necessary upstream event required to activate the
ERK/MAPK signaling cascade) via RasGRP1,[43,48] while both RasGRP1 and Sos1
will activate Ras during negative selection (Figure 2.1).[49,50] However, RasGRP1 is
only recruited to the Golgi during positive selection, whereas both RasGRP1 and Sos1
are recruited to the membrane upon interaction with negatively selecting ligands.[13]
How can ligands with different affinity lead to differential compartmentalization of
ERK activation? This is likely the result of differential phosphorylation of LAT by
the different ligands. Negatively selecting ligands induce stronger phosphorylation
of LAT that then allows for a stable interaction with Grb2/Sos (which is dependent
on fully phosphorylated LAT) and therefore activation of the Ras-MAPK pathway at
the membrane.[13,51,52] Weaker ligands, such as those inducing positive selection, may
only partially phosphorylate LAT, but this may be sufficient to recruit PLCγ and the
generation of DAG to activate RasGRP1 (Figure 2.1).

Recently, a key role for the subcellular localization of ERK1/2 activation for positive selection *in vivo* has been bolstered via the identification of TRAF31P3 as an
interacting partner of MEK, the upstream kinase phosphorylating ERK1/2, which
leads to its localization at the Golgi.[53] TRAF31P3-deficient mice have a significant
block in the development of both CD4+ and CD8+ T cells, but no significant impact
on negative selection is observed in these mice.[53] Therefore, activation of ERK1/2
at the Golgi during positive selection is mediated by the selective recruitment of

MEK by TRAF3IP3. Thus, differential compartmentalization of TCR signaling intermediates is likely necessary to distinguish positive and negative selection for both CD4+- and CD8+-committed lineages. Further studies are required to understand how positively selecting ligands allow for MEK recruitment by TRAF3IP3 to the Golgi.

### 2.2.1.1    Role of the Atypical MAPK ERK3

The important role of the classical MAPK signaling pathway during thymic T cell development is now quite well established, with a prominent and specific role for the ERK1/2 pathway during positive selection.[44,54] However, very little is known about the contribution of the atypical MAPK family (ERK3, ERK4, ERK7, and NLK) during thymocyte differentiation. In contrast to the classical MAPKs (ERK1/2, ERK5, p38, and JNK), these atypical MAPKs are not phosphorylated and activated by MAPK kinases (or MEK).[55] Unlike classical MAPKs, ERK3 lacks a phosphoacceptor site within its activation loop and possesses a long C-terminal extension. Moreover, ERK3 is constitutively phosphorylated in its activation loop by group I p21-activated kinases.[56,57] The demonstration that the classical ERK1/2 pathway induces expression of ERK3 in T cells raises the possibility that ERK3 is a downstream effector of ERK1/2 that may contribute to positive selection during thymic differentiation.[58] Indeed, a recent study showed a complete abrogation of positive selection of thymocytes in ERK3-deficient mice expressing MHC class I- or II-restricted TCRs,[59] a phenotype similar to the one obtained in the absence of ERK1/2.[54] Moreover, this study has also demonstrated that ERK3-deficient thymocytes are less responsive to TCR signaling *in vitro* suggesting that ERK3 contributes to positive selection by affecting the response of thymocytes to TCR engagement.[59] Further studies are required to determine whether ERK3 acts as a downstream effector of ERK1/2 and how the ERK3 signaling pathway controls positive selection. As ERK3 is a unique signaling molecule and not well characterized within the TCR signaling pathway, deciphering its role during thymopoiesis may bring new and important insights on the regulation of positive selection.

### 2.2.2    THEMIS

The recent discovery of the essential role of Themis for positive selection has added a layer of complexity to the signaling events that regulate positive selection.[60-62] Analysis of TCR signaling events in OT-I preselection DP thymocytes using MHC class I tetramers loaded with the OVA peptide or APLs of different affinities for the OT-I TCR have demonstrated that Themis promotes positive selection of thymocytes by attenuating TCR signaling in response to low-affinity ligands. Indeed, Themis-deficient DP thymocytes showed enhanced TCR signaling in response to low-affinity ligands, whereas there were no changes in the signaling events associated with negatively selecting ligands.[63] Furthermore, the attenuation of TCR signaling to low-affinity ligand by Themis is mediated by its ability to interact with the phosphatases SHP1 and SHP2, which will likely lead to the dephosphorylation of Lck (Figure 2.1).[63,64] Given that DP thymocytes are predisposed to react with high sensitivity to TCR stimulation,[65,66] it appears counterintuitive that Themis-mediated dampening of TCR signaling is essential for positive selection. However, further studies to determine how Themis specifically regulates TCR signaling to low-affinity ligands and if this is essential to promote ERK1/2 activation at the Golgi promise to shed light on this process.

### 2.2.3 CALCIUM FLUX

Another important mediator of TCR signals is intracellular free $Ca^{2+}$ ($[Ca^{2+}]i$). Increases in $[Ca^{2+}]i$ lead to activation of calcineurin and downstream relocalization of nuclear factor of activated T cell (NFAT) to the nucleus. Thus, differences in $[Ca^{2+}]i$ can lead to distinct transcriptional outcomes.[67–70] TCR signals trigger a cascade of signaling events that lead to a $Ca^{2+}$ release from the intracellular stores in the endoplasmic reticulum (Figure 2.1). In mature T cells, $[Ca^{2+}]i$ levels are sustained via $Ca^{2+}$ release-activated $Ca^{2+}$ (CRAC) channels to import extracellular $Ca^{2+}$.[71–73] Though also important for negative selection,[74] CRAC channels are dispensable in thymocytes undergoing positive selection.[75–77] Instead, a voltage-gated $Na^+$ channel (VGSC) has been implicated in sustaining $[Ca^{2+}]i$ required for the positive selection of MHC class II-restricted, but not MHC class I-restricted, thymocytes, though the molecular mechanisms have not yet been determined.[78] Whether there is a unique mechanism to sustain $[Ca^{2+}]i$ in MHC class I-restricted positive selection is unknown. Despite the observation that *in vitro* stimulation of preselection MHC class I TCR transgenic thymocytes with tetramers presenting low-affinity ligand suggested that there is a low, slow accumulation of $[Ca^{2+}]i$ in response to positive selecting ligands,[13] two-photon imaging of $[Ca^{2+}]i$ oscillations in MHC class I-restricted thymocytes undergoing positive selection in thymic slices (see later) suggest that these events are, in fact, heterogeneous, but generally quite short (3–5 minutes) and infrequent (1–2 per hour).[32] This is in contrast to their MHC class II-restricted counterparts that show average increases in $[Ca^{2+}]i$ for 15 to 30 minutes *in situ* suggesting that the unique mechanisms to sustain TCR signals during positive selection of CD4 lineage T cells may not be necessary for MHC class I-restricted CD8 lineage T cells.[33]

Intermittent TCR signals are required over the several days it takes to complete positive selection, and interruption of the TCR signals for rather short segments significantly impedes this process.[1,5,32] Recent data suggests that these transient events may contribute to a slow increase in basal $[Ca^{2+}]i$; in postselection DP cells that must reach a predetermined "threshold" for terminal differentiation into mature $CD4^+$ and $CD8^+$ T cells.[1,5] In mature T cells, there is evidence of signal "memory" or "summation" during sequential, serial TCR signals for c-fos and NFAT.[79,80] Thus, it is possible that the serial TCR signals necessary for positive selection lead to an accumulation of signaling intermediates that gradually results in the drastic changes in transcriptional programs that accompany the DP to SP thymocyte transition.

## 2.3 EMERGING TOOLS TO STUDY TCR SIGNALS DURING POSITIVE SELECTION *IN SITU*

In addition to TCR affinity for self-pMHC, the quantity, quality, and frequency of TCR signals provided by the varied stromal cell subsets that comprise the complex three-dimensional environment in which T cells develop modulate thymic selection. As such, discrepancies in the characterization between *in vitro* and *in situ* observations describing the characteristics of TCR signals associated with positive selection exist. Despite the utility of reductionist *in vitro* systems to provide a synchronous population of cells to dissect the molecular mechanisms associated with stimulation of preselection TCR transgenic thymocytes with low-affinity peptides, the contribution of cellular interactions, thymocyte migration, and the thymic milieu to modulating the TCR signal cannot be addressed. In particular, distinct differences in TCR signals relayed via peptide presentation by varied cell types, the density and diversity of ligands, and thymocyte motility play important roles in modulating TCR signals.[32,33] In addition, TCR signaling during positive selection becomes increasingly complicated when one begins to factor in changes in sensitivity to TCR signals throughout differentiation, that multiple interactions with self-pMHC are required for positive selection, and that there can exist multiple selecting ligands for a single TCR. Thus, the long-standing affinity model of T cell selection must be revised in order to take into account the spatiotemporal context of thymic selection, and more recent experiments combining thymic slice organotypic culture systems with two-photon microscopy have begun to address these topics.

### 2.3.1 ANALYSIS OF TCR SIGNALING *IN SITU*

As they develop, T cells go from slow moving progenitors in the thymic cortex sampling thymic epithelial cells for low-affinity interactions with their TCR to signal differentiation and migration to the medulla where they become among the fastest cells in the body and are exposed to a unique environment expressing tissue-restricted antigens.[81–85] Thus, T cell development is deeply intertwined and reliant on migration, cell–cell interactions, and TCR signaling. Two-photon microscopy allows us to monitor these dynamic events, processes that cannot easily and efficiently be replicated in a tissue culture dish, in real-time and *in situ*, with minimal phototoxic effects. Using intact tissue or thymic organ culture models, we can now monitor thymocytes and antigen presenting cells labeled with fluorescent dyes, genetic fluorescent labels, fusion proteins, and/or calcium indicators.

### 2.3.1.1 Thymic Slice Organotypic Culture

Several *in vitro* and *in situ* models of positive selection exist (Table 2.1). Tetramers have been used to stimulate thymocytes with low-affinity peptides in order to generate a synchronous population of signaling cells.[13,63] It does not appear, however, that tetramer stimulation alone supports positive selection, and tetramer stimulation of thymocytes with low-affinity peptides can, in fact, promote cell death.[86] In addition, it has been notoriously difficult to recapitulate positive selection in the absence of a cellular component. Coculture systems, of which the OP9-DL1/4 system is the mostly widely used, support limited positive selection.[87–89] However, these *in vitro* systems lack the complex three-dimensional environment of intact tissue and the unique peptide processing machinery of cortical thymic epithelial cells. Of several *in situ* models of positive selection, thymic slice organ cultures are particularly conducive to imaging TCR signals in real-time using two-photon microscopy (Figure 2.2a).

**TABLE 2.1**

**Comparison of Experimental Models of Positive Selection**

| Model | Features | Advantages | Disadvantages | Refs |
|---|---|---|---|---|
| | | *In Vitro* | | |
| OP9-DL1/4 | Coculture of stem cells or T cell precursors with bone marrow stromal cells expressing a Notch ligand. | Support T lineage commitment and limited positive selection. | Lack MHC class II expression. Lack the thymoproteasome. Genetic background not compatible with many available TCR transgenic models. | 88,89 |
| Tetramer stimulation | Stimulation of TCR transgenic DP thymocytes with tetramers loaded with defined antigens. | Synchronous TCR signals allow dissection of the molecular events associated with positive versus negative selection. | Does not support positive selection. | 13,86 |
| | | *In Situ* | | |
| FTOC | Isolated fetal thymic lobes maintained in culture. | Fully support positive selection of CD4+ and CD8+ T cells. Analysis of one wave of T cell differentiation. | Underdeveloped medulla. Non synchronized TCR signals. | 90,91 |
| RTOC | Reaggregates of thymocytes and thymic stromal cells. | Fully support development of CD4+ and CD8+ T cells. Defined cell populations. | Lack well-defined thymic cortical and medullary regions. Nonsynchronized TCR signals. Technically difficult. | 17 |
| Thymic slices | Thymocytes at a defined developmental stage can be overlaid onto thymic slices. | Fully support positive selection of CD4+ and CD8+ T cells. Intact thymic cortex and medulla. More physiological as overlaid TCR Tg cells make up only 1% of the total cells. Provide access to the medulla for imaging by two-photon microscopy. | Lack thymic capsule. Can only be maintained in culture for 3–4 days without significant cell loss and death. Nonsynchronized TCR signals. | 32,33,81,84 |

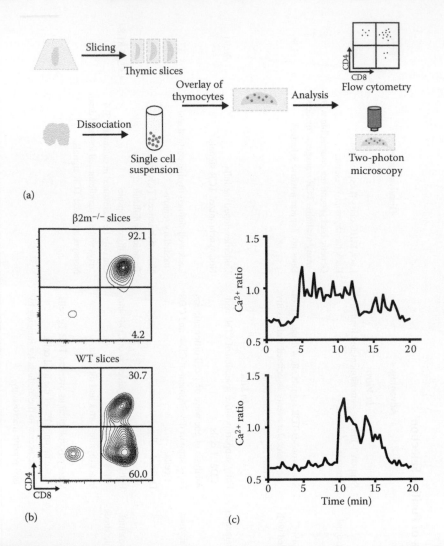

**FIGURE 2.2** The thymic slice model for analysis of positive selection. (a) Schematic representation of the thymic slice model. Preselection thymocytes overlaid onto thymic slices migrate into the tissue and their development can be followed overtime by flow cytometry and two-photon microscopy. (b) Representative flow cytometric analysis of positive selection in thymic slices. CFSE labeled preselection OT-I TCR transgenic Rag1$^{-/-}$ β2m$^{-/-}$ thymocytes were overlaid onto nonselecting β2m$^{-/-}$ or selecting wild type (WT) slices. After 72 hours of incubation at 37°C, thymocytes were analyzed by flow cytometry. WT, but not β2m$^{-/-}$, slices support the development of live, CFSE$^+$ TCRβ$^{high}$ CD8$^+$ T cells. (c) Representative two-photon analysis of calcium flux in individual cells. Indo-1LR labeled preselection OT-I TCR transgenic thymocytes were overlaid onto WT slices and imaged for 20 minutes (60 time points of 20 second intervals) via two-photon microscopy. Note that TCR signaling events are heterogeneous but last, on average, 3 to 5 minutes.

Similar to reaggregate thymic organ cultures (RTOCs) and FTOCs, thymic slices support positive selection of both MHC class I and II-restricted thymocytes (Figure 2.2b; data not shown).[1,5,17,32,33,90,91] In contrast to other *in situ* models, however, thymic slices maintain intact and well-defined cortical and medullary regions.[81,82,84] Further, the ease with which one can add well-defined, labeled cell subsets to thymic slices for imaging offers an advantage over the use of intact thymic tissue.[92,93] Additionally, intact thymic lobes are not compatible with imaging thymocyte behavior in the medulla that is usually more than 200 microns below the capsule, generally too deep within the tissue for successful imaging by two-photon microscopy. Again, the thymic slice overcomes this hurdle, and one can use landmarks such as the density of CD11c-YFP+ dendritic cell from reporter mice or stromal cell morphology using EGFP transgenic mice to identify cortical and medullary regions in live tissue.[81,84,94,95]

To generate thymic slices, individual thymic lobes are embedded in low-melt agarose and generally sliced with a vibratome to a thickness of ~400 to 1000 microns. The thymic slices can be maintained in culture for several days on tissue culture inserts. These slices maintain distinct cortical and medullary compartments as well as chemokine gradients to direct the appropriate localization of distinct thymocyte populations that can be overlaid atop the thymic slices.[81,82,84] Typically, ~1 to 3 × 10^6 thymocytes in 10 to 20 μL of complete media are overlaid on individual thymic slices and maintained at 37°C for 2 hours to allow a portion of the cells to migrate into the tissue. At this time, the thymic slices are washed by indirect pipetting of media over the slices to remove any excess cells that did not enter the tissue and results in ~0.5% to 1% of total cells in each thymic slice derived from the overlaid thymocytes. This is particularly important in that, though still significantly overrepresented, the low proportion of antigen-specific thymocytes recapitulates the low frequency of such cells in a physiological thymus and limits competition for selecting peptides. Overlaid cells can be prelabeled with proliferation dyes, genetically encoded fluorescent markers, or identified via congenic markers to distinguish them from thymic slice endogenous cells. Genetically encoded reporters and dyes useful for tracking TCR signals in real-time are described in Section 2.3.2.

For analysis of positive selection in particular, TCR transgenic thymocytes from nonselecting mice arrested at the preselection DP stage are overlaid onto thymic slices, and in the presence of endogenous, selecting ligands, will develop into CD4+ or CD8+ SP cells over several days (Figure 2.2b; data not shown).[1,5,32,33] Their differentiation can be tracked via flow cytometry, and their migration, cell–cell interactions, and TCR signaling by two-photon microscopy (Figure 2.2c; data not shown). Importantly, however, due to the thickness of the slice and the fact that thymic slices lack a capsule, there is both an accumulation of necrotic cells as well as cell loss over time. Thus, there is a limited timeline for maintenance of thymic slices in culture (generally <4 days).

### 2.3.1.2  Two-Photon Microscopy

Advances in microscopy technology now allow the measurement of the behavior of individual thymocytes in four dimensions: $x$, $y$, $z$, and time. Two-photon microscopy, in particular, has several advantages for imaging the dynamic behavior of immune cells due to the ability to image deeper in tissue.[96] Excitation of fluorescent molecules with two photons at near-infrared wavelengths (~700 to 1000 nm) leads to decreased tissue damage and photobleaching, and allows imaging to >200 microns in biological samples. Though these advantages come at a cost in terms of spatial resolution, particularly in the $z$ dimension, the technology is sufficient to analyze large-scale subcellular protein relocalization in individual cells.

The health of the tissue is paramount for successful imaging experiments. Thus, tissue is maintained in warm, oxygenated phenol red-free media throughout the imaging session either via an incubated chamber or perfused media.[93] If careful consideration to prepare the tissue under conditions to limit damage is undertaken and the tissue is maintained appropriately during the imaging session, tissue imaging can continue for 2 to 4 hours. Evidence of tissue damage is generally quite obvious, as thymocytes (or the dendrites of dendritic cells) stop moving.

Choosing fluorochrome combinations with compatible excitation and emission wavelengths is important. Excitation of fluorescent molecules with two-photon lasers is generally twice that of a single-photon laser, though this should be optimized for each fluorochrome used. Thus, when conducting imaging experiments with multiple fluorescent parameters, it is necessary to determine if the fluorochromes can be excited using similar wavelengths. Although some two-photon microscopes allow for adjustments to the laser wavelength during imaging, the time required for these changes is not conducive with tracking individual cells in time-lapse experiments. Newer technologies, however, are beginning to support simultaneous excitation with two-wavelengths. Second, it is important to consider the emission spectra of the fluorescent molecules and the available filter sets to limit overlap and successfully distinguish labeled cells. Partially overlapping emission spectra from different fluorophores can be further separated using custom spectral unmixing programs. Indeed, two-cell populations labeled with GFP and YFP with significantly overlapping emission spectra can be imaged concurrently in the same volume with careful application of filter sets and unmixing programs.

Additional considerations for two-photon imaging of thymocyte behavior are the density of labeled cells and the intervals between time points. The importance of these parameters becomes evident in the analysis of two-photon imaging data sets. There are several commercially available software programs for automated tracking of cells over time in three dimensions. Analysis of the data sets is most effective when the density of labeled cells and the interval between time points is appropriate. Although this must be optimized on an experiment-to-experiment basis, generally a density of ~0.5% of labeled cells in a tissue is ideal. This is generally the proportion obtained when labeled cells are overlaid on thymic slices as described in Section 2.3.1.1. In addition, it is important to maximize the interval between time points in order to limit residual phototoxicity and photobleaching, while ensuring that it is not so long that it is difficult to identify the trajectory of individual cells between time points. Depending on the speed of the population of cells, 20 to 30 s intervals between time points is generally appropriate for these experiments. Despite optimization of these parameters, it is paramount that automated tracking results are manually verified. These various programs also allow measurements of fluorescence intensity, fluorescence volumes, and cellular interactions among other parameters.

## 2.3.2 Reporters of TCR Signals during Positive Selection

TCR affinity/avidity for self-pMHC plays a dominant role in discriminating cell fate during thymic selection. As such, markers of TCR affinity/avidity are extremely useful to measure how perturbations in factors influencing positive selection affect the TCR affinity for self-pMHC of the developing T cell repertoire. Cell-surface levels of CD5 are useful indicators of TCR avidity during positive selection.[97] In addition, it has been shown that Nur77, a protein upregulated via antigen receptor stimulation in lymphocytes, is an immediate early gene downstream of TCR signals in mature T cells and thymocytes that is not influenced by other inflammatory stimuli.[98–101] Intracellular staining for Nur77 is possible, but precludes further analysis of live cells. However, Nur77[GFP] transgenic mice, in which GFP is under the control of the *Nr4a1* regulatory elements in a bacterial artificial chromosome, have emerged as a useful reporter of antigen receptor signal strength.[99,102] Nur77[GFP] is particularly valuable for *ex vivo* analysis of accumulated TCR signals during positive selection, but not for real-time TCR signal events, as the transient increases in Nur77[GFP] peak several hours after the TCR stimulus.[99] Instead, the calcium indicators and fusion proteins described later are preferred for use in conjunction with two-photon analysis of TCR signals associated with positive selection *in situ*.

### 2.3.2.1 Real-Time Relocalization of Signaling Intermediates

TCR signaling downstream of positive selection is a cascade of signals that leads to differential localization of signaling intermediates as well as relocalization of transcription factors from the cytoplasm to the nucleus. Fusion proteins of TCR signaling intermediates are an underutilized tool to study TCR signals associated with positive selection in real time. Unlike the TCR synapses that are associated with mature, naive T cell activation, the TCR signals associated with positive selection appear to be more transient in nature.[34] Thus, whether redistribution of TCR proximal signaling components to an interaction interface (such as has been reported for LAT-GFP in mature, naïve T cells) is useful remains unknown.[103] However, translocation of NFAT fusion proteins from the cytoplasm to the nucleus has been demonstrated *in situ* for both mature T cells and thymocyte undergoing negative selection as well as during MHC class II-restricted positive selection.[32,34,80] Similar results have not yet been reported for TCR signals associated with positive selection of MHC class I-restricted thymocytes, and it will be interesting to determine to what extent TCR induced differences in [Ca$^{2+}$]i between MHC class I- and II-restricted thymocytes undergoing positive selection manifests in differences in nuclear NFAT and downstream transcriptional outcomes. Localization of the fusion proteins to discrete intracellular locations will require higher-resolution imaging modalities such as confocal microscopy as compared with two-photon imaging. It is also important to consider that these fusion proteins have generally been introduced via retroviral vectors and that overexpression may perturb the TCR signals themselves or downstream developmental outcomes. If possible, endogenous knock-ins for such molecules could be generated, though depending on the imaging modality, the level of fluorescence may be a limiting factor.

### 2.3.2.2 Calcium Flux

Increases in [$Ca^{2+}$]i indicate productive TCR signals, and differences in [$Ca^{2+}$]i induce unique transcriptional programs. [$Ca^{2+}$]i can be measured in real time using a number of different tools that have been used extensively *in vitro* via flow cytometry and fluorescence/confocal microscopy as well as *in situ* via two-photon microscopy. Calcium indicator dyes such as Indo-1 among many others, can distinguish minor fluctuations in [$Ca^{2+}$]i and are sensitive enough to detect basal calcium levels in resting thymocytes as well as TCR signals during thymic selection by two-photon microscopy.[1,5,32,33,104,105] Ratiometric calcium indicator dyes have the advantage that one can normalize for differences in labeling and cell size, and between experiments. For ratiometric calcium indicator dyes, either the excitation or emission wavelength of the dye changes in the $Ca^{2+}$-bound state.[104] Particularly useful for two-photon microscopy due to its single-excitation/dual emission characteristics for unbound and $Ca^{2+}$ bound dye, is Indo-1.[104] Despite the utility of these dyes and ease with which cells can be labeled, consideration must be taken when choosing the best reagent for measuring [$Ca^{2+}$]i. Calcium indicator dyes are useful in the relatively short term, as they can become compartmentalized and extruded from the cells.[104] Leak-resistant alternatives are being developed, but, relatively speaking, cells must be labeled shortly before analysis and are not suitable for experiments lasting over several days or weeks. In contrast, genetically encoded calcium indicators such as the cameleons, G-CaMP/pericams, and their newer generation counterparts enable calcium monitoring in longer-term experiments.[106] Cells can be isolated from transgenic mice ubiquitously expressing the genetically encoded calcium indicator, or genetic indicators can be introduced into the cell via retroviral infection. Newer generation genetically encoded calcium indicators have demonstrated success *in vivo* during mature T cell activation,[107–110] but it has not yet been demonstrated whether their sensitivity and dynamic range of signal change is sufficient to detect the $Ca^{2+}$ fluctuations associated with positive selection. As these constructs evolve, new versions will likely overcome the sensitivity hurdle. This will be particularly important in order to demonstrate how TCR signals and basal calcium levels evolve over the several days that it can take to complete positive selection.

### 2.3.3 EXPERIMENTAL CHALLENGES

At present it is difficult to determine if the differences in subcellular localization of TCR signaling intermediates that have been observed after *in vitro* stimulation with low- and high-affinity peptides is the mechanism by which thymocytes distinguish positive from negative selection signals *in situ*. Analysis of the dynamic relocalization of fusion proteins in immune cells has been accomplished with two-photon microscopy, but this has at present been studied with overexpression of fluorescent fusion proteins and quantification of robust relocalization.[32,34,80,111] In particular, NFAT and nuclear marker H2b fusion proteins allow the dissection of NFAT relocalization from the cytoplasm to the nucleus after TCR activation.[32,34,80] However, visualization of more subtle changes in protein movement (LAT or ERK1/2, for example) to discrete subcellular compartments (TCR-self-pMHC interface or membrane versus

Golgi) during positive selection has not yet been achieved. To do this, it will likely be necessary to further develop two-photon excitation fluorescence resonance energy transfer (2P-FRET) and 2P-FRET fluorescence lifetime imaging (FLIM) methods suitable for imaging positive selection in thymic tissue.

## 2.4 CONCLUSION

Positive selection has been studied at length, yet several fundamental questions remain. We have only just begun to understand the processes of thymic selection as they occur *in situ* in a tightly packed, three-dimensional structure composed of unique cell subsets, and, in fact, *in situ* studies using two-photon microscopy to track dynamic TCR signals and migration during thymic selection in intact tissue paints a different picture of T cell development than *in vitro* studies. These technologies, however, are currently limited by their sensitivity and resolution, and the best approach to address outstanding questions in the field will be through a combination of complementary methods, recognizing their limitations, and pushing the boundaries of the current technology.

## ACKNOWLEDGMENTS

Support for this research is provided by a grant from NSERC (RGPIN-2015-06645) to Nathalie Labrecque, as well as grants from the SickKids Foundation and CIHR-IHDCYN (NI15-002), an operating grant from the CIHR-III (MOP-142254), and start-up funds from the FRQS (Établissement de jeunes chercheurs) and Hôpital Maisonneuve-Rosemont Foundation to Heather J. Melichar. Melichar is a junior 1 scholar of the FRQS, a CIHR new investigator (MSH-141967), and a Cole Foundation Early Career Transition award recipient.

## REFERENCES

1. Au-Yeung BB, Melichar HJ, Ross JO, Cheng DA, Zikherman J, Shokat KM et al. Quantitative and temporal requirements revealed for Zap70 catalytic activity during T cell development. *Nat Immunol* 2014; **15**(7): 687–694.
2. Egerton M, Scollay R, Shortman K. Kinetics of mature T-cell development in the thymus. *Proc Natl Acad Sci USA* 1990; **87**(7): 2579–2582.
3. Huesmann M, Scott B, Kisielow P, von Boehmer H. Kinetics and efficacy of positive selection in the thymus of normal and T cell receptor transgenic mice. *Cell* 1991; **66**(3): 533–540.
4. Lucas B, Vasseur F, Penit C. Production, selection, and maturation of thymocytes with high surface density of TCR. *J Immunol* 1994; **153**(1): 53–62.
5. Ross JO, Melichar HJ, Au-Yeung BB, Herzmark P, Weiss A, Robey EA. Distinct phases in the positive selection of CD8+ T cells distinguished by intrathymic migration and T-cell receptor signaling patterns. *Proc Natl Acad Sci USA* 2014; **111**(25): E2550–E2558.
6. Saini M, Sinclair C, Marshall D, Tolaini M, Sakaguchi S, Seddon B. Regulation of Zap70 expression during thymocyte development enables temporal separation of CD4 and CD8 repertoire selection at different signaling thresholds. *Sci Signal* 2010; **3**(114): ra23.

7. Garcia KC. Reconciling views on T cell receptor germline bias for MHC. *Trends Immunol* 2012; **33**(9): 429–436.
8. Garcia KC, Adams JJ, Feng D, Ely LK. The molecular basis of TCR germline bias for MHC is surprisingly simple. *Nat Immunol* 2009; **10**(2): 143–147.
9. Scollay RG, Butcher EC, Weissman IL. Thymus cell migration: Quantitative aspects of cellular traffic from the thymus to the periphery in mice. *Eur J Immunol* 1980; **10**(3): 210–218.
10. McDonald BD, Bunker JJ, Erickson SA, Oh-Hora M, Bendelac A. Crossreactive alpha-beta T cell receptors are the predominant targets of thymocyte negative selection. *Immunity* 2015; **43**(5): 859–869.
11. Linette GP, Grusby MJ, Hedrick SM, Hansen TH, Glimcher LH, Korsmeyer SJ. Bcl-2 is upregulated at the CD4+ CD8+ stage during positive selection and promotes thymocyte differentiation at several control points. *Immunity* 1994; **1**(3): 197–205.
12. Punt JA, Suzuki H, Granger LG, Sharrow SO, Singer A. Lineage commitment in the thymus: Only the most differentiated (TCRhibcl-2hi) subset of CD4+CD8+ thymocytes has selectively terminated CD4 or CD8 synthesis. *J Exp Med* 1996; **184**(6): 2091–2099.
13. Daniels MA, Teixeiro E, Gill J, Hausmann B, Roubaty D, Holmberg K et al. Thymic selection threshold defined by compartmentalization of Ras/MAPK signalling. *Nature* 2006; **444**(7120): 724–729.
14. Juang J, Ebert PJ, Feng D, Garcia KC, Krogsgaard M, Davis MM. Peptide-MHC heterodimers show that thymic positive selection requires a more restricted set of self-peptides than negative selection. *J Exp Med* 2010; **207**(6): 1223–1234.
15. Naeher D, Daniels MA, Hausmann B, Guillaume P, Luescher I, Palmer E. A constant affinity threshold for T cell tolerance. *J Exp Med* 2007; **204**(11): 2553–2559.
16. Alam SM, Travers PJ, Wung JL, Nasholds W, Redpath S, Jameson SC et al. T-cell-receptor affinity and thymocyte positive selection. *Nature* 1996; **381**(6583): 616–620.
17. Anderson G, Owen JJ, Moore NC, Jenkinson EJ. Thymic epithelial cells provide unique signals for positive selection of CD4+CD8+ thymocytes in vitro. *J Exp Med* 1994; **179**(6): 2027–2031.
18. Klein L, Kyewski B, Allen PM, Hogquist KA. Positive and negative selection of the T cell repertoire: What thymocytes see (and don't see). *Nat Rev Immunol* 2014; **14**(6): 377–391.
19. Laufer TM, DeKoning J, Markowitz JS, Lo D, Glimcher LH. Unopposed positive selection and autoreactivity in mice expressing class II MHC only on thymic cortex. *Nature* 1996; **383**(6595): 81–85.
20. Murata S, Sasaki K, Kishimoto T, Niwa S, Hayashi H, Takahama Y et al. Regulation of CD8+ T cell development by thymus-specific proteasomes. *Science* 2007; **316**(5829): 1349–1353.
21. Sasaki K, Takada K, Ohte Y, Kondo H, Sorimachi H, Tanaka K et al. Thymoproteasomes produce unique peptide motifs for positive selection of CD8(+) T cells. *Nat Commun* 2015; **6**: 7484.
22. Xing Y, Jameson SC, Hogquist KA. Thymoproteasome subunit-beta5T generates peptide-MHC complexes specialized for positive selection. *Proc Natl Acad Sci USA* 2013; **110**(17): 6979–6984.
23. Nakagawa T, Roth W, Wong P, Nelson A, Farr A, Deussing J et al. Cathepsin L: Critical role in Ii degradation and CD4 T cell selection in the thymus. *Science* 1998; **280**(5362): 450–453.
24. Honey K, Nakagawa T, Peters C, Rudensky A. Cathepsin L regulates CD4+ T cell selection independently of its effect on invariant chain: A role in the generation of positively selecting peptide ligands. *J Exp Med* 2002; **195**(10): 1349–1358.

25. Gommeaux J, Gregoire C, Nguessan P, Richelme M, Malissen M, Guerder S et al. Thymus-specific serine protease regulates positive selection of a subset of CD4+ thymocytes. *Eur J Immunol* 2009; **39**(4): 956–964.

26. Viret C, Lamare C, Guiraud M, Fazilleau N, Bour A, Malissen B et al. Thymus-specific serine protease contributes to the diversification of the functional endogenous CD4 T cell receptor repertoire. *J Exp Med* 2011; **208**(1): 3–11.

27. Buhlmann JE, Elkin SK, Sharpe AH. A role for the B7-1/B7-2:CD28/CTLA-4 pathway during negative selection. *J Immunol* 2003; **170**(11): 5421–5428.

28. Foy TM, Page DM, Waldschmidt TJ, Schoneveld A, Laman JD, Masters SR et al. An essential role for gp39, the ligand for CD40, in thymic selection. *J Exp Med* 1995; **182**(5): 1377–1388.

29. Kishimoto H, Sprent J. Several different cell surface molecules control negative selection of medullary thymocytes. *J Exp Med* 1999; **190**(1): 65–73.

30. Li R, Page DM. Requirement for a complex array of costimulators in the negative selection of autoreactive thymocytes in vivo. *J Immunol* 2001; **166**(10): 6050–6056.

31. Pobezinsky LA, Angelov GS, Tai X, Jeurling S, Van Laethem F, Feigenbaum L et al. Clonal deletion and the fate of autoreactive thymocytes that survive negative selection. *Nat Immunol* 2012; **13**(6): 569–578.

32. Melichar HJ, Ross JO, Herzmark P, Hogquist KA, Robey EA. Distinct temporal patterns of T cell receptor signaling during positive versus negative selection in situ. *Sci Signal* 2013; **6**(297): ra92.

33. Bhakta NR, Oh DY, Lewis RS. Calcium oscillations regulate thymocyte motility during positive selection in the three-dimensional thymic environment. *Nat Immunol* 2005; **6**(2): 143–151.

34. Ebert PJ, Ehrlich LI, Davis MM. Low ligand requirement for deletion and lack of synapses in positive selection enforce the gauntlet of thymic T cell maturation. *Immunity* 2008; **29**(5): 734–745.

35. McCaughtry TM, Baldwin TA, Wilken MS, Hogquist KA. Clonal deletion of thymocytes can occur in the cortex with no involvement of the medulla. *J Exp Med* 2008; **205**(11): 2575–2584.

36. Stritesky GL, Xing Y, Erickson JR, Kalekar LA, Wang X, Mueller DL et al. Murine thymic selection quantified using a unique method to capture deleted T cells. *Proc Natl Acad Sci USA* 2013; **110**(12): 4679–4684.

37. Daley SR, Hu DY, Goodnow CC. Helios marks strongly autoreactive CD4+ T cells in two major waves of thymic deletion distinguished by induction of PD-1 or NF-kappaB. *J Exp Med* 2013; **210**(2): 269–285.

38. Kisielow P, Miazek A. Positive selection of T cells: Rescue from programmed cell death and differentiation require continual engagement of the T cell receptor. *J Exp Med* 1995; **181**(6): 1975–1984.

39. McNeil LK, Starr TK, Hogquist KA. A requirement for sustained ERK signaling during thymocyte positive selection in vivo. *Proc Natl Acad Sci USA* 2005; **102**(38): 13574–13579.

40. Wilkinson RW, Anderson G, Owen JJ, Jenkinson EJ. Positive selection of thymocytes involves sustained interactions with the thymic microenvironment. *J Immunol* 1995; **155**(11): 5234–5240.

41. Alberola-Ila J, Forbush KA, Seger R, Krebs EG, Perlmutter RM. Selective requirement for MAP kinase activation in thymocyte differentiation. *Nature* 1995; **373**(6515): 620–623.

42. Alberola-Ila J, Hogquist KA, Swan KA, Bevan MJ, Perlmutter RM. Positive and negative selection invoke distinct signaling pathways. *J Exp Med* 1996; **184**(1): 9–18.

43. Dower NA, Stang SL, Bottorff DA, Ebinu JO, Dickie P, Ostergaard HL et al. RasGRP is essential for mouse thymocyte differentiation and TCR signaling. *Nat Immunol* 2000; **1**(4): 317–321.

44. McGargill MA, Ch'en IL, Katayama CD, Pages G, Pouyssegur J, Hedrick SM. Cutting edge: Extracellular signal-related kinase is not required for negative selection of developing T cells. *J Immunol* 2009; **183**(8): 4838–4842.

45. Neilson JR, Winslow MM, Hur EM, Crabtree GR. Calcineurin B1 is essential for positive but not negative selection during thymocyte development. *Immunity* 2004; **20**(3): 255–266.

46. Sugawara T, Moriguchi T, Nishida E, Takahama Y. Differential roles of ERK and p38 MAP kinase pathways in positive and negative selection of T lymphocytes. *Immunity* 1998; **9**(4): 565–574.

47. Werlen G, Hausmann B, Palmer E. A motif in the alphabeta T-cell receptor controls positive selection by modulating ERK activity. *Nature* 2000; **406**(6794): 422–426.

48. Priatel JJ, Teh SJ, Dower NA, Stone JC, Teh HS. RasGRP1 transduces low-grade TCR signals which are critical for T cell development, homeostasis, and differentiation. *Immunity* 2002; **17**(5): 617–627.

49. Gong Q, Cheng AM, Akk AM, Alberola-Ila J, Gong G, Pawson T et al. Disruption of T cell signaling networks and development by Grb2 haploid insufficiency. *Nat Immunol* 2001; **2**(1): 29–36.

50. Kortum RL, Sommers CL, Pinski JM, Alexander CP, Merrill RK, Li W et al. Deconstructing Ras signaling in the thymus. *Mol Cell Biol* 2012; **32**(14): 2748–2759.

51. Houtman JC, Higashimoto Y, Dimasi N, Cho S, Yamaguchi H, Bowden B et al. Binding specificity of multiprotein signaling complexes is determined by both cooperative interactions and affinity preferences. *Biochemistry* 2004; **43**(14): 4170–4178.

52. Zhu M, Janssen E, Zhang W. Minimal requirement of tyrosine residues of linker for activation of T cells in TCR signaling and thymocyte development. *J Immunol* 2003; **170**(1): 325–333.

53. Zou Q, Jin J, Xiao Y, Hu H, Zhou X, Jie Z et al. T cell development involves TRAF3IP3-mediated ERK signaling in the Golgi. *J Exp Med* 2015; **212**(8): 1323–1336.

54. Fischer AM, Katayama CD, Pages G, Pouyssegur J, Hedrick SM. The role of Erk1 and Erk2 in multiple stages of T cell development. *Immunity* 2005; **23**(4): 431–443.

55. Coulombe P, Meloche S. Atypical mitogen-activated protein kinases: Structure, regulation and functions. *Biochim Biophys Acta* 2007; **1773**(8): 1376–1387.

56. De la Mota-Peynado A, Chernoff J, Beeser A. Identification of the atypical MAPK Erk3 as a novel substrate for p21-activated kinase (Pak) activity. *J Biol Chem* 2011; **286**(15): 13603–13611.

57. Deleris P, Trost M, Topisirovic I, Tanguay PL, Borden KL, Thibault P et al. Activation loop phosphorylation of ERK3/ERK4 by group I p21-activated kinases (PAKs) defines a novel PAK-ERK3/4-MAPK-activated protein kinase 5 signaling pathway. *J Biol Chem* 2011; **286**(8): 6470–6478.

58. Marquis M, Boulet S, Mathien S, Rousseau J, Thebault P, Daudelin JF et al. The non-classical MAP kinase ERK3 controls T cell activation. *PLoS One* 2014; **9**(1): e86681.

59. Sirois J, Daudelin JF, Boulet S, Marquis M, Meloche S, Labrecque N. The atypical MAPK ERK3 controls positive selection of thymocytes. *Immunology* 2015; **145**(1): 161–169.

60. Johnson AL, Aravind L, Shulzhenko N, Morgun A, Choi SY, Crockford TL et al. Themis is a member of a new metazoan gene family and is required for the completion of thymocyte positive selection. *Nat Immunol* 2009; **10**(8): 831–839.

61. Lesourne R, Uehara S, Lee J, Song KD, Li L, Pinkhasov J et al. Themis, a T cell-specific protein important for late thymocyte development. *Nat Immunol* 2009; **10**(8): 840–847.

62. Patrick MS, Oda H, Hayakawa K, Sato Y, Eshima K, Kirikae T et al. Gasp, a Grb2-associating protein, is critical for positive selection of thymocytes. *Proc Natl Acad Sci USA* 2009; **106**(38): 16345–16350.
63. Fu G, Casas J, Rigaud S, Rybakin V, Lambolez F, Brzostek J et al. Themis sets the signal threshold for positive and negative selection in T-cell development. *Nature* 2013; **504**(7480): 441–445.
64. Paster W, Bruger AM, Katsch K, Gregoire C, Roncagalli R, Fu G et al. A THEMIS:SHP1 complex promotes T-cell survival. *EMBO J* 2015; **34**(3): 393–409.
65. Davey GM, Schober SL, Endrizzi BT, Dutcher AK, Jameson SC, Hogquist KA. Preselection thymocytes are more sensitive to T cell receptor stimulation than mature T cells. *J Exp Med* 1998; **188**(10): 1867–1874.
66. Lucas B, Stefanova I, Yasutomo K, Dautigny N, Germain RN. Divergent changes in the sensitivity of maturing T cells to structurally related ligands underlies formation of a useful T cell repertoire. *Immunity* 1999; **10**(3): 367–376.
67. Dolmetsch RE, Lewis RS, Goodnow CC, Healy JI. Differential activation of transcription factors induced by Ca2+ response amplitude and duration. *Nature* 1997; **386**(6627): 855–858.
68. Dolmetsch RE, Xu K, Lewis RS. Calcium oscillations increase the efficiency and specificity of gene expression. *Nature* 1998; **392**(6679): 933–936.
69. Gu X, Spitzer NC. Distinct aspects of neuronal differentiation encoded by frequency of spontaneous Ca2+ transients. *Nature* 1995; **375**(6534): 784–787.
70. Timmerman LA, Clipstone NA, Ho SN, Northrop JP, Crabtree GR. Rapid shuttling of NF-AT in discrimination of Ca2+ signals and immunosuppression. *Nature* 1996; **383**(6603): 837–840.
71. Oh-hora M. Calcium signaling in the development and function of T-lineage cells. *Immunol Rev* 2009; **231**(1): 210–224.
72. Cahalan MD, Chandy KG. The functional network of ion channels in T lymphocytes. *Immunol Rev* 2009; **231**(1): 59–87.
73. Feske S. ORAI1 and STIM1 deficiency in human and mice: Roles of store-operated Ca2+ entry in the immune system and beyond. *Immunol Rev* 2009; **231**(1): 189–209.
74. Kim KD, Srikanth S, Yee MK, Mock DC, Lawson GW, Gwack Y. ORAI1 deficiency impairs activated T cell death and enhances T cell survival. *J Immunol* 2011; **187**(7): 3620–3630.
75. Beyersdorf N, Braun A, Vogtle T, Varga-Szabo D, Galdos RR, Kissler S et al. STIM1-independent T cell development and effector function in vivo. *J Immunol* 2009; **182**(6): 3390–3397.
76. Gwack Y, Srikanth S, Oh-Hora M, Hogan PG, Lamperti ED, Yamashita M et al. Hair loss and defective T- and B-cell function in mice lacking ORAI1. *Mol Cell Biol* 2008; **28**(17): 5209–5222.
77. Vig M, DeHaven WI, Bird GS, Billingsley JM, Wang H, Rao PE et al. Defective mast cell effector functions in mice lacking the CRACM1 pore subunit of store-operated calcium release-activated calcium channels. *Nat Immunol* 2008; **9**(1): 89–96.
78. Lo WL, Donermeyer DL, Allen PM. A voltage-gated sodium channel is essential for the positive selection of CD4(+) T cells. *Nat Immunol* 2012; **13**(9): 880–887.
79. Clark CE, Hasan M, Bousso P. A role for the immediate early gene product c-fos in imprinting T cells with short-term memory for signal summation. *PLoS One* 2011; **6**(4): e18916.
80. Marangoni F, Murooka TT, Manzo T, Kim EY, Carrizosa E, Elpek NM et al. The transcription factor NFAT exhibits signal memory during serial T cell interactions with antigen-presenting cells. *Immunity* 2013; **38**(2): 237–249.
81. Ehrlich LI, Oh DY, Weissman IL, Lewis RS. Differential contribution of chemotaxis and substrate restriction to segregation of immature and mature thymocytes. *Immunity* 2009; **31**(6): 986–998.

82. Halkias J, Melichar HJ, Taylor KT, Ross JO, Yen B, Cooper SB et al. Opposing chemokine gradients control human thymocyte migration in situ. *J Clin Invest* 2013; **123**(5): 2131–2142.

83. Halkias J, Yen B, Taylor KT, Reinhartz O, Winoto A, Robey EA et al. Conserved and divergent aspects of human T-cell development and migration in humanized mice. *Immunol Cell Biol* 2015; **93**(8): 716–726.

84. Le Borgne M, Ladi E, Dzhagalov I, Herzmark P, Liao YF, Chakraborty AK et al. The impact of negative selection on thymocyte migration in the medulla. *Nat Immunol* 2009; **10**(8): 823–830.

85. Witt CM, Raychaudhuri S, Schaefer B, Chakraborty AK, Robey EA. Directed migration of positively selected thymocytes visualized in real time. *PLoS Biol* 2005; **3**(6): e160.

86. Rybakin V, Gascoigne NR. Negative selection assay based on stimulation of T cell receptor transgenic thymocytes with peptide-MHC tetramers. *PLoS One* 2012; **7**(8): e43191.

87. de Pooter RF, Schmitt TM, Zuniga-Pflucker JC. In vitro generation of T lymphocytes from embryonic stem cells. *Methods Mol Biol* 2006; **330**: 113–121.

88. Dervovic DD, Ciofani M, Kianizad K, Zuniga-Pflucker JC. Comparative and functional evaluation of in vitro generated to ex vivo CD8 T cells. *J Immunol* 2012; **189**(7): 3411–3420.

89. Schmitt TM, Zuniga-Pflucker JC. Induction of T cell development from hematopoietic progenitor cells by delta-like-1 in vitro. *Immunity* 2002; **17**(6): 749–756.

90. Ceredig R, Jenkinson EJ, MacDonald HR, Owen JJ. Development of cytolytic T lymphocyte precursors in organ-cultured mouse embryonic thymus rudiments. *J Exp Med* 1982; **155**(2): 617–622.

91. Fairchild PJ, Austyn JM. Developmental changes predispose the fetal thymus to positive selection of CD4+CD8- T cells. *Immunology* 1995; **85**(2): 292–298.

92. Ross JO, Melichar HJ, Halkias J, Robey EA. Studying T cell development in thymic slices. *Methods Mol Biol* 2016; **1323**: 131–140.

93. Dzhagalov IL, Melichar HJ, Ross JO, Herzmark P, Robey EA. Two-photon imaging of the immune system. *Curr Protoc Cytom* 2012; **Chapter 12**: Unit 12.26.

94. Lindquist RL, Shakhar G, Dudziak D, Wardemann H, Eisenreich T, Dustin ML et al. Visualizing dendritic cell networks in vivo. *Nat Immunol* 2004; **5**(12): 1243–1250.

95. Wright DE, Cheshier SH, Wagers AJ, Randall TD, Christensen JL, Weissman IL. Cyclophosphamide/granulocyte colony-stimulating factor causes selective mobilization of bone marrow hematopoietic stem cells into the blood after M phase of the cell cycle. *Blood* 2001; **97**(8): 2278–2285.

96. Helmchen F, Denk W. Deep tissue two-photon microscopy. *Nat Methods* 2005; **2**(12): 932–940.

97. Azzam HS, Grinberg A, Lui K, Shen H, Shores EW, Love PE. CD5 expression is developmentally regulated by T cell receptor (TCR) signals and TCR avidity. *J Exp Med* 1998; **188**(12): 2301–2311.

98. Baldwin TA, Hogquist KA. Transcriptional analysis of clonal deletion in vivo. *J Immunol* 2007; **179**(2): 837–844.

99. Moran AE, Holzapfel KL, Xing Y, Cunningham NR, Maltzman JS, Punt J et al. T cell receptor signal strength in Treg and iNKT cell development demonstrated by a novel fluorescent reporter mouse. *J Exp Med* 2011; **208**(6): 1279–1289.

100. Osborne BA, Smith SW, Liu ZG, McLaughlin KA, Grimm L, Schwartz LM. Identification of genes induced during apoptosis in T lymphocytes. *Immunol Rev* 1994; **142**: 301–320.

101. Liu ZG, Smith SW, McLaughlin KA, Schwartz LM, Osborne BA. Apoptotic signals delivered through the T-cell receptor of a T-cell hybrid require the immediate-early gene nur77. *Nature* 1994; **367**(6460): 281–284.

102. Zikherman J, Parameswaran R, Weiss A. Endogenous antigen tunes the responsiveness of naive B cells but not T cells. *Nature* 2012; **489**(7414): 160–164.

103. Azar GA, Lemaitre F, Robey EA, Bousso P. Subcellular dynamics of T cell immunological synapses and kinapses in lymph nodes. *Proc Natl Acad Sci USA* 2010; **107**(8): 3675–3680.

104. Paredes RM, Etzler JC, Watts LT, Zheng W, Lechleiter JD. Chemical calcium indicators. *Methods* 2008; **46**(3): 143–151.

105. Dzhagalov IL, Chen KG, Herzmark P, Robey EA. Elimination of self-reactive T cells in the thymus: A timeline for negative selection. *PLoS Biol* 2013; **11**(5): e1001566.

106. Perez Koldenkova V, Nagai T. Genetically encoded Ca(2+) indicators: Properties and evaluation. *Biochim Biophys Acta* 2013; **1833**(7): 1787–1797.

107. Mues M, Bartholomaus I, Thestrup T, Griesbeck O, Wekerle H, Kawakami N et al. Real-time in vivo analysis of T cell activation in the central nervous system using a genetically encoded calcium indicator. *Nat Med* 2013; **19**(6): 778–783.

108. Le Borgne M, Raju S, Zinselmeyer BH, Le VT, Li J, Wang Y et al. Real-time analysis of calcium signals during the early phase of T cell activation using a genetically encoded calcium biosensor. *J Immunol* 2016; **196**(4): 1471–1479.

109. Shulman Z, Gitlin AD, Weinstein JS, Lainez B, Esplugues E, Flavell RA et al. Dynamic signaling by T follicular helper cells during germinal center B cell selection. *Science* 2014; **345**(6200): 1058–1062.

110. Thestrup T, Litzlbauer J, Bartholomaus I, Mues M, Russo L, Dana H et al. Optimized ratiometric calcium sensors for functional in vivo imaging of neurons and T lymphocytes. *Nat Methods* 2014; **11**(2): 175–182.

111. Melichar HJ, Li O, Herzmark P, Padmanabhan RK, Oliaro J, Ludford-Menting MJ et al. Quantifying subcellular distribution of fluorescent fusion proteins in cells migrating within tissues. *Immunol Cell Biol* 2011; **89**(4): 549–557.

# 3 Regulation of Negative Selection in the Thymus by Cytokines
## Novel Role of IL-23 to Regulate RORγt

*Robert M. Lowe, Hao Li, Hui-Chen Hsu,
and John D. Mountz*

## CONTENTS

### ABSTRACT

The thymus is a complex organ that performs its critical function through the action of several key groups of cells that are located in spatially distinct areas of the thymus: cortical thymic epithelial cells, medullary thymic epithelial cells, and medullary dendritic cells. Negative selection occurs in the thymus when dendritic cells (DCs) presenting self-antigens interact with $CD4^+CD8^+$ double-positive thymocytes that express a self-reactive T-cell receptor. Expression of major histocompatibility and costimulatory molecules on the DCs are considered the major factors determining the fate of thymocytes. Accumulating evidence suggests that additional factors including cell-to-cell interactions between thymocytes and permanent thymic resident cells as well as the presence of local cytokines and soluble mediators that act on thymocytes through the JAK-STAT pathway, or through Nur77 or RORγt signaling pathways, can also play an important role to refine the selection process. This chapter will describe key cytokines and nuclear hormone receptors that act together at different stages of thymocyte development to mediate thymocyte survival and apoptosis. The connection of IL-23 in influencing T-cell diversity and regulation in the periphery through a central mechanism is also described.

## 3.1 INTRODUCTION

Cytokines, including interleukin-2 (IL-2), IL-7, IL-12, IL-17, IL-22, and IL-23, along with interferon gamma (IFN-γ) actively participate in the regulation of thymocyte development and negative selection. In addition to cytokine signaling, which is mediated through the Janus kinase-signal transducer and activation of transcription (JAK-STAT) pathway, steroid hormone receptors, especially the orphan steroid hormone receptors including retinoic acid-related orphan receptor gamma t (RORγt) and Nur77, play a key role in thymocyte development, selection, and survival (Table 3.1).

**TABLE 3.1**

**Effects of Cytokines on Thymocyte Apoptosis or Survival**

| Cytokine | Overexpression Effect(s)[a] | Underexpression Effect(s)[b] |
|---|---|---|
| IL-1 | Promotes IL-2 production[1] | |
| IL-2 | Promotes survival and induces IL-3 production[2] | No change in thymocyte numbers[3] |
| | | Lowered antigen-mediated thymocyte negative selection[4] |
| IL-3 | Rescues thymocyte proliferation in Jak3[(-/-)] knockout mice[5] | |
| IL-4 | | No differences in thymocyte count at young age (4 weeks old) but increased thymocyte count at 6 weeks old |
| IL-6 | | Lowered thymocyte count[7] |
| IL-7 | Protection of thymocytes from apoptosis in a developmental stage-dependent manner[8,9] | Decreased cellularity with normal distribution of CD4 and CD8[10] |
| IL-9 | | No effect on thymocyte cell population[11] |
| IL-10 | | No effect on thymocyte cell population[12] |
| IL-12 | Enhanced IL-7 or IL-2-induced thymocytes proliferation[13] | Required for normal negative selection/apoptosis[14,15] |
| | | Accelerated age-related thymic involution in aged *Il12b* (p40)[-/-] mice[13] |
| IFN-γ | | No effect on thymocyte cell population[16] |
| IL-12Rβ2 | | Accelerated maturation[17] |
| IL-22 | Enhancement of thymic recovery following total body irradiation[18] | |
| IL-23 | Enhancement of TCR-mediated late DP thymocytes deletion[19] | Lowered TCR-mediated late DP thymocyte apoptosis[19] |

[a]   Overexpression effects of cytokines on thymocyte development.
[b]   Underexpression effects of cytokines (including knockout animal models) on developing thymocytes.

The thymus is a complex organ that performs its critical function through the action of several key groups of cells that are located in spatially distinct areas of the thymus: cortical thymic epithelial cells (cTECs), medullary thymic epithelial cells (mTECs), and medullary dendritic cells (mDCs). In addition to direct cell-to-cell interactions with these permanent thymic resident cells, certain cytokines act on thymocytes through the JAK-STAT pathway, or through Nur77 or RORγt signaling pathways (Figure 3.1). This chapter will describe key cytokines and nuclear hormone receptors that act together at different stages of thymocyte development to mediate thymocyte selection. Defective regulation of these pathways leads to the development of autoreactive thymocytes that can play an important role in some forms of autoimmune disease in humans and animal models of autoimmune disease.[14,17,20–27] However, a full discussion of abnormalities of thymic negative selection in these conditions is beyond the scope of this review.

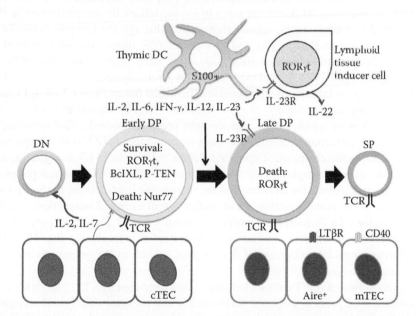

**FIGURE 3.1** Cytokines and nuclear hormone receptors help regulate thymocyte selection. T-cell progenitor cells enter the thymus via postcapillary venules in the cortico-medullary junction area of the thymus. Once in the thymus, these progenitor cells undergo a series of differentiation steps as they pass through three distinct microenvironment regions: the outer cortex, the inner cortex, and the medulla. As thymocytes proceed through these regions, their interactions with cortical thymic epithelial cells (cTECs), medullary thymic epithelial cells (mTECs), and thymic dendritic cells (DCs) each play a key role in determining the selection fate of thymocytes. Although TCR signaling strength is a key factor in determining the fate of thymocyte selection, the multiple cytokines produced by cTECs, mTECs, and thymic DCs modulate this process. Furthermore, the nuclear hormone receptor retinoic acid-related orphan receptor gamma t (RORγt) exhibits a uniquely dichotomous effect based on its temporal expression by either promoting early DP thymocyte survival or inducing late thymocyte apoptosis.

## 3.2   CELLS IN THE THYMUS

In the thymic cortex, positive selection occurs by interaction of CD4$^+$CD8$^+$ (DP) thymocytes with cortical-thymic epithelial cells (cTECs, ERTR4$^+$CD205$^+$Ly51$^+$IL-7$^+$DL4$^+$),[28] which trigger a differentiation program involving expression of CCR7, enabling these positively selected cells to enter the medulla.[29,30] Once they have entered the medulla, the DP thymocytes interact with medullary-thymic epithelial cells (mTECs) including the autoimmune regulator positive (Aire$^+$) subset. This results in negative selection in the medulla that eliminates autoreactive thymocytes by programmed cell death (apoptosis) through interactions of the thymocytes with both the medullary dendritic cells (mDCs) and mTECs.[31] Interestingly, the formation of the thymic medulla depends on interactions between medullary resident innate and adaptive immune cells.[32] Among the cell surface molecules expressed on mTECs in the thymic medulla are the lymphotoxin-beta receptor (LTβR) and the cell surface protein CD40. Matsumoto and colleagues have proposed a model for the establishment and maintenance of Aire expression in the medullary thymic environment.[33–35] In both fetal and adult stages, signals transmitted through the receptor activator of nuclear factor kappa-B (RANK) and probably LTβR induce the development of medullary-thymic epithelial progenitor cells at the Aire$^+$ stage.[36] After initial development, combined RANK-ligand expression and positively selected CD4$^+$CD8$^-$ thymocytes, as well as intrathymically generated invariant natural killer T (iNKT) cells, ensure the continued production of Aire$^+$ mTECs.[37,38] LTβR signaling is a key event in the formation of lymph nodes, and we have recently showed LTβR signaling is important for maintenance of marginal zone macrophages (MZMs) in the spleen to promote tolerogenic clearance of apoptotic debris.[39] This places LTβR signaling and its signaling event through nuclear factor-kappa B (NF-kB) as essential pathways in initially establishing the environment for tolerogenic interactions in the thymus, spleen, and lymph node. Together, these studies reveal a critical interaction of tumor necrosis factor (TNF) receptor superfamily members and T cells to promote the proper thymic environment. As such, thymic epithelial cells clearly play a key role in nurturing normal development of thymocytes, as well as the proper environment for subsequent negative selection.

Another subset of αβ$^+$ T cells include the CD1d-restricted iNKT.[40] The iNKT cells predominately express a restricted Vα14-Jα18 T cell receptor (TCR) and recognize CD1d/glycolipid complexes, which are expressed by cortical-resident CD4$^+$CD8$^+$ thymocytes. iNKT cells have been identified to be both IL-17$^+$ and RORγt$^+$ that develop intrathymically by CD1d-mediated interactions.[41,42] White and colleagues[31,32,43] showed that normal development of both RORγt$^-$ and RORγt$^+$ iNKT17 thymocytes depends on the presence of mTECs and can be modulated by IL-15.[43]

## 3.3   IL-7 AND THYMIC T CELL DEVELOPMENT

IL-7 signaling has been extensively reviewed and shown to play an important survival role at all stages of thymocyte development.[44–46] Therefore, regulation of IL-7 is known to be a key mechanism for regulating thymocyte development and survival. IL-7Rα expression is initiated during the early CD4$^-$CD8$^-$ double negative stage 2

(DN2) of thymocyte development and is essential for IL-7-dependent survival and proliferation of TCRβ-selected thymocytes.[46] However, IL-7 becomes dispensable as thymocytes develop to the double negative stage 4 (DN4) and DP stage where IL-7α expression is low.[47] The DP thymocytes are then maintained in an IL-7[lo] environment, which enables their programming for possible subsequent cell death. However, after TCR-mediated positive selection, IL-7Rα expression is upregulated and these cells again become IL-7-signaling competent. Persistent TCR signaling desensitizes this IL-7 signaling and permits CD4 lineage differentiation.[48] When TCR signaling is terminated, this allows IL-7Rα signaling in intermediate cells and leads to CD8 lineage development.[44]

The mechanism of action of IL-7 includes prosurvival Bcl2 family members including Bcl-2, Bcl-xL, Bcl-w, and Mcl-1[49-51] and redistribution of the proapoptotic proteins, Bax and BAD.[52] IL-7 also induces activation of phosphoinositide 3 (PI3) kinase and phosphorylation of the serine/threonine kinase Akt, which upregulates cell metabolism and induces the Glut1 glucose transporter. Therefore, as a prosurvival mechanism, it is likely that the geographic distribution of IL-7 into areas of the thymus of earliest and late-stage proselection thymocyte development enables the precise survival role of IL-7. This has been investigated initially using an IL-7 bacterial artificial chromosome (BAC) reporter mouse to drive expression of yellow fluorescent protein (YFP) downstream of the translational start codon of the transgenic IL-7 locus.[53,54] IL-7 expression was limited to thymic epithelial cells.[47] Thus, both temporal and spatial regulation of IL-7, as well as IL-7 receptor (IL-7R, CD127), modulates IL-7-mediated thymocyte survival. IL-7 signaling, while essential, does not apparently involve steroid receptors but does require JAK1,[55] JAK3,[56-58] and STAT5.[59]

## 3.4 IL-12 ROLE IN DELAYING THYMIC INVOLUTION

IL-12 was initially studied in IL-12β knockout (KO) mice, which lack the IL-12 p40 subunit. At the time of these initial studies,[13] IL-12 was known to be a heterodimer composed of a disulfide bond between the p35 and p40 subunits. Both the p35 and p40 subunits of IL-12 are secreted by thymic epithelial cells and were shown previously to act by a negative selection process.[15] IL-12 is capable of upregulating expression of CD28, thus potentially promoting negative selection. Importantly, at the time of these studies, IL-23 (p19/p40 heterodimer) was not yet fully identified and therefore was not assayed for expression during these initial studies of the p40-deficient mice. It is now known that these p40-deficient mice lack both IL-12 and IL-23.

Interestingly, the initial studies with the IL-12β KO mouse showed that thymic involution occurred at a later age that was after 12 months of age.[13] The results of these studies suggest that increased thymic involution in the IL-12β KO mice may not be due to only a primary defect in IL-12 expression. Accelerated thymic involution may also be occurring due to the loss of IL-12 signaling that normally helps compensate for a decline in the level of signaling capacity of other cytokines necessary to sustain thymocytes of animals that are of older age. It was known that with aging, compensatory mechanisms often occurred to allow for survival that may not be evident in younger mice. In support of this, IL-12 was found to promote thymocyte development by interactions with both IL-7 and IL-2. These findings were consistent with other studies that showed that expression of IL-7 did not decline in BALB/c mice until 7 months of age.[60] Given the critical role of IL-7, compensatory mechanisms for IL-7 signaling, such as through IL-12, can be identified as a potential mechanism. A synergistic effect of IL-12 with IL-2 and IL-18 was observed to modulate thymocyte apoptosis.[61] IL-12 plus IL-18 was shown to upregulate class I and class II MHC expression on cortical and medullary thymic epithelial cell lines in an IFN-γ-dependent manner. The results of in vitro experiments indicated that thymocytes produce IFN-γ in response to IL-18 in combination with IL-12 and IL-2. Thus, the combinations of IL-2, IL-12, and IL-18 were found to induce phenotypic and functional changes of thymocytes that altered the migration, differentiation, and apoptosis of these intrathymic T cells. Together, these results provide evidence for a model in which IL-12 in animals of a younger age plays a proapoptotic role in combination with other cytokines; whereas at a later age, IL-12 may play more of a compensatory role and inhibit late-stage thymic involution.

## 3.5   IL-22 IN THYMIC REGENERATION

IL-22 is primarily associated with maintenance of a barrier function and induction of innate antimicrobial molecules at mucosal surfaces[62,63] and is primarily produced by Th17 T cells and innate lymphoid cell (ILC) subsets. As such, IL-22 has an established role in potentially either promoting autoimmune disease or helping prevent it by acting on epithelial cells that maintain developing thymocytes.

IL-22 was similarly identified for its role in maintaining thymic epithelial cells after injury.[18] Although there was no difference in total thymic cellularity or thymocyte subpopulations in WT or *Il22*[-/-] mice, there was impaired thymic regeneration after total body radiation in *Il22*[-/-] mice.[18] IL-22 was shown to signal through thymic epithelial cells, effectively promoting their proliferation and survival. This prosurvival mechanism was upregulated in radio-resistant RORγt[+] CCR6[+] NKp46[-] lymphoid tissue inducer cells after thymic injury. Importantly, this induction was dependent on IL-23, which could signal through RORγt in these lymphoid tissue inducer cells. These results underscore again the importance of signaling interactions between cytokines on the extrathymic compartment, as well as highlighting the complex role of orphan steroid hormone receptors, such as RORγt, in affecting thymocyte development, apoptosis, and regeneration after injury.

## 3.6   IL-23 ACTS ON LATE DP THYMOCYTES
##         TO ENABLE NEGATIVE SELECTION

In addition to acting on lymphoid tissue inducer cells, IL-23 is known, especially in the periphery, to act on T cells to maintain and promote development of Th17 peripheral T cells by signaling through RORγt. We have recently identified a unique effect of IL-23 to regulate effector T-cell development through thymocyte selection.[19] We have found that in the *IL-23 p19*[-/-] × D[b]/H-Y mouse, there was a decrease in negative selection most obviously observed in D$^b$/H-Y male mice.[19] Compared to IL-23 p19 wild-type (WT) male mice, the *IL-23 p19*$^{-/-}$ mice exhibited an increased cortical area and an increase in both DP thymocytes, as well as CD8$^+$ single-positive (SP) thymocytes. Negative selection was observed at the inner cortex of the thymus and in the medullary area in the vicinity of thymic S100$^+$ dendritic cells in the IL-23 WT D$^b$/H-Y mice. These apoptotic cells were dramatically reduced in the male *IL-23 p19*$^{-/-}$ × D$^b$/H-Y mouse.

One possible mechanism for this finding could be that TCR signaling could be upregulating the IL-23R, as evidenced by the observation that apoptosis was synergistically mediated by the addition of IL-23. This finding was detected on a subpopulation of DP thymocytes that were found to be late-stage CD4$^{hi}$CD8$^{hi}$ DP thymocytes. We have further demonstrated that IL-23 preferentially eliminated natural regulatory T cells (nTregs) in the thymus, leading to diminished thymic regulatory T cell output.[19] This work suggests that inflammatory cytokines may act through both central and peripheral mechanisms to promote autoimmune disease.

## 3.7 RORγT DIFFERENTIALLY REGULATES EARLY VERSUS LATE THYMOCYTES

Through recent studies, we have identified that signaling downstream of IL23R in late DP thymocyte that regulates apoptosis was induced by RORγt signaling and subsequent downregulation of the transcriptional activator, cRel (an NF-kB family member).[19] This finding is consistent with previous results from studies conducted by He and Bevan, where it was previously shown that downregulation of RORγt at the late stage of thymocyte development is essential for thymocyte survival.[64] However, forced expression of RORγt during the late DP stage resulted in apoptosis and the loss of thymocyte survival.[65] In contrast, RORγt activates the gene encoding the antiapoptotic protein Bcl-xL and has been shown to be required for the survival of early-stage DP thymocytes.[66] Together, these results help confirm and extend the model of the dual contrasting functions for RORγt depending on the stage of thymocyte development by both promoting early-stage thymocyte survival as well as facilitating late-stage thymocyte apoptosis.

## 3.8 NUR77 IN THYMOCYTE APOPTOSIS

In addition to RORγt, the Nur77 family of orphan steroid hormone receptors, including Nur77, Nor1, and Nurr1, can be induced by strong TCR signaling that results in negative selection. We have previously shown that the expression of a Nur77 dominant-negative protein resulted in an inefficient clonal deletion of self-reactive T cells.[67] This was also investigated using the D$^b$/H-Y TCR-Tg mouse, in which there was a fivefold increase in the total number of thymocytes expressing the D$^b$/H-Y transgenic TCR on T cells in the dominant-negative Nur77 male mouse. There was also a tenfold increase in DP thymocytes and an eight-fold increase of CD8$^+$D$^b$/H-Y$^+$ T cells in the lymph nodes of dominant-negative Nur77 mice. Interestingly, despite the resulting defect in clonal deletion, the T cells in the male $D^b$/H-Y dominant-negative Nur77 mice were anergic with increased activation of apoptosis in these T cells due to upregulation of Fas and FasL in the lymph node. These results demonstrate the importance of compensatory or backup mechanisms in the periphery that can effectively silence or eliminate autoreactive T cells that may escape negative selection in the thymus. A more comprehensive view of Nur77 signal transduction of apoptosis in the thymus was reviewed by Winoto and Littman.[68] This review proposes the concept that TCR and CD28 signaling acts through both a calcium and calcineurin pathway. It is also suggested that the MEKK2/MEK5 pathway acts through the nuclear factor of activated T cells (NFAT) pathway to upregulate Nur77 and other Nur77 family genes to induce apoptosis. This effect can be modulated by signaling through PTEN (phosphatase and tensin homolog, deleted on chromosome 10), which inactivates Akt, thus promoting survival, as well as through Bcl2 family members in the mitochondria that regulate caspases and downstream apoptosis pathways. Recently, Winoto found that the T cell-specific overexpression of the Bcl2-BH3 mutant transgene can result in T cell-driven multiorgan autoimmunity.[69] This finding indicates that the Bcl2-BH3 mutant can rescue the Nur77 proapoptosis function as a result of defective interactions with the mutant BH3 domain of Bcl2.

## 3.9 CONCLUSION

Regulation of negative selection in the thymus is a complex process that involves cytokine as well as orphan steroid hormone receptor signaling in both thymic epithelial cells and thymic dendritic cells. Thymocytes are further regulated by induction through lymphoid tissue inducer cells.[49] These thymic epithelial cells produce important survival factors, as well as negative selection molecules such as those expressed by the Aire family. These cells also produce prosurvival molecules such as IL-7 that act in specific regions of the thymus to mediate proper negative selection. Thymic dendritic cells in contrast likely produce cytokines such as IL-12 and IL-23, which can influence thymocyte survival, usually by inducing apoptosis. One example of this is that IL-23 can induce apoptosis through the IL-23R on late DP thymocytes by sustaining the expression of RORγt, which must then be downregulated at the later stage to enable thymocyte survival. Thymocytes themselves express receptors to cytokines including those for IL-2, IL-7, IL-12, and IL-23, which can also directly deliver a signal for survival or induce apoptosis through the JAK-STAT pathway. Alternatively, RORγt expression is essential for early thymocyte survival in DP thymocytes. Additionally, TCR/CD3 signaling regulates Nur77, generally during DP thymocyte development. These molecules regulate thymocyte apoptosis through the transcription factor Rel-b (NF-kB family) or Bcl2 family members that can also generally promote thymocyte survival at early-stage development, or induce thymocyte deletion as part of negative selection in a late stage of thymocyte development. These studies support a more complex model in which TCR signaling strength may not be the only factor in determining the fate of thymocyte selection. Also, that certain cytokines may act as important factors to mediating the dialogue between thymocyte and DCs or stromal cells, and thereby may play an important role in mediating thymic selection. The functions of many cytokines in the thymus are still not well established. Further detailed analysis of these intrathymic cytokines will likely reveal additional important central mechanisms in regulating T-cell diversity and differentiation.

## REFERENCES

1. Falk, W., Krammer, P. H., and Mannel, D. N. A new assay for interleukin-1 in the presence of interleukin-2. *J Immunol Methods* **99**, 47–52 (1987).
2. Bellio, M., and Dos Reis, G. A. Triggering of thymocyte function by IL-2 as the only exogenous stimulus; analysis of two distinct modes of IL-2-induced thymocyte proliferation and IL-3 secretion in vitro. *Immunology* **68**, 175–180 (1989).
3. Schorle, H., Holtschke, T., Hunig, T., Schimpl, A., and Horak, I. Development and function of T cells in mice rendered interleukin-2 deficient by gene targeting. *Nature* **352**, 621–624 (1991).
4. Bassiri, H., and Carding, S. R. A requirement for IL-2/IL-2 receptor signaling in intrathymic negative selection. *J Immunol* **166**, 5945–5954 (2001).
5. Brown, M. P. et al. Reconstitution of early lymphoid proliferation and immune function in Jak3-deficient mice by interleukin-3. *Blood* **94**, 1906–1914 (1999).
6. Kuhn, R., Rajewsky, K., and Muller, W. Generation and analysis of interleukin-4 deficient mice. *Science* **254**, 707–710 (1991).
7. Kopf, M. et al. Impaired immune and acute-phase responses in interleukin-6-deficient mice. *Nature* **368**, 339–342 (1994).
8. Kim, K., Lee, C. K., Sayers, T. J., Muegge, K., and Durum, S. K. The trophic action of IL-7 on pro-T cells: Inhibition of apoptosis of pro-T1, -T2, and -T3 cells correlates with Bcl-2 and Bax levels and is independent of Fas and p53 pathways. *J Immunol* **160**, 5735–5741 (1998).
9. Jiang, Q. et al. Cell biology of IL-7, a key lymphotrophin. *Cytokine Growth Factor Rev* **16**, 513–533 (2005).
10. von Freeden-Jeffry, U. et al. Lymphopenia in interleukin (IL)-7 gene-deleted mice identifies IL-7 as a nonredundant cytokine. *J Exp Med* **181**, 1519–1526 (1995).
11. Townsend, J. M. et al. IL-9-deficient mice establish fundamental roles for IL-9 in pulmonary mastocytosis and goblet cell hyperplasia but not T cell development. *Immunity* **13**, 573–583 (2000).
12. Kuhn, R., Lohler, J., Rennick, D., Rajewsky, K., and Muller, W. Interleukin-10-deficient mice develop chronic enterocolitis. *Cell* **75**, 263–274 (1993).
13. Li, L. et al. IL-12 inhibits thymic involution by enhancing IL-7- and IL-2-induced thymocyte proliferation. *J Immunol* **172**, 2909–2916 (2004).
14. Ludviksson, B. R., Gray, B., Strober, W., and Ehrhardt, R. O. Dysregulated intrathymic development in the IL-2-deficient mouse leads to colitis-inducing thymocytes. *J Immunol* **158**, 104–111 (1997).
15. Ludviksson, B. R., Ehrhardt, R. O., and Strober, W. Role of IL-12 in intrathymic negative selection. *J Immunol* **163**, 4349–4359 (1999).
16. Dalton, D. K. et al. Multiple defects of immune cell function in mice with disrupted interferon-gamma genes. *Science* **259**, 1739–1742 (1993).
17. Gran, B., Yu, S., Zhang, G.X., and Rostami, A. Accelerated thymocyte maturation in IL-12Rbeta2-deficient mice contributes to increased susceptibility to autoimmune inflammatory demyelination. *Exp Mol Pathol* **89**, 126–134 (2010).
18. Dudakov, J. A. et al. Interleukin-22 drives endogenous thymic regeneration in mice. *Science* **336**, 91–95 (2012).
19. Li, H. et al. IL-23 promotes TCR-mediated negative selection of thymocytes through the upregulation of IL-23 receptor and RORγt. *Nat Commun* **5**, 4259 (2014).

20. Baxter, A. G., Kinder, S. J., Hammond, K. J., Scollay, R., and Godfrey, D. I. Association between alphabetaTCR+CD4-CD8- T-cell deficiency and IDDM in NOD/Lt mice. *Diabetes* **46**, 572–582 (1997).

21. Bunting, M. D. et al. CCX-CKR deficiency alters thymic stroma impairing thymocyte development and promoting autoimmunity. *Blood* **121**, 118–128 (2013).

22. Cameron, M. J. et al. IL-4 prevents insulitis and insulin-dependent diabetes mellitus in nonobese diabetic mice by potentiation of regulatory T helper-2 cell function. *J Immunol* **159**, 4686–4692 (1997).

23. Casteels, K. M. et al. Sex difference in resistance to dexamethasone-induced apoptosis in NOD mice: Treatment with 1,25(OH)2D3 restores defect. *Diabetes* **47**, 1033–1037 (1998).

24. Chiu, P. P., Jevnikar, A. M., and Danska, J. S. Genetic control of T and B lymphocyte activation in nonobese diabetic mice. *J Immunol* **167**, 7169–7179 (2001).

25. Damjanovic, M., Vidic-Dankovic, B., Kosec, D., and Isakovic, K. Thymus changes in experimentally induced myasthenia gravis. *Autoimmunity* **15**, 201–207 (1993).

26. Gebe, J. A. et al. Autoreactive human T-cell receptor initiates insulitis and impaired glucose tolerance in HLA DR4 transgenic mice. *J Autoimmun* **30**, 197–206 (2008).

27. Nenninger, R. et al. Abnormal thymocyte development and generation of autoreactive T cells in mixed and cortical thymomas. *Lab Invest* **78**, 743–753 (1998).

28. Baik, S., Jenkinson, E. J., Lane, P. J., Anderson, G., and Jenkinson, W. E. Generation of both cortical and Aire(+) medullary thymic epithelial compartments from CD205(+) progenitors. *Eur J Immunol* **43**, 589–594 (2013).

29. Hinterberger, M. et al. Autonomous role of medullary thymic epithelial cells in central CD4(+) T cell tolerance. *Nat Immunol* **11**, 512–519 (2010).

30. Derbinski, J., and Kyewski, B. How thymic antigen presenting cells sample the body's self-antigens. *Curr Opin Immunol* **22**, 592–600 (2010).

31. White, A. J. et al. Sequential phases in the development of Aire-expressing medullary thymic epithelial cells involve distinct cellular input. *Eur J Immunol* **38**, 942–947 (2008).

32. White, A. J. et al. Lymphotoxin signals from positively selected thymocytes regulate the terminal differentiation of medullary thymic epithelial cells. *J Immunol* **185**, 4769–4776 (2010).

33. Matsumoto, M. Transcriptional regulation in thymic epithelial cells for the establishment of self tolerance. *Arch Immunol Ther Exp (Warsz)* **55**, 27–34 (2007).

34. Akiyoshi, H. et al. Subcellular expression of autoimmune regulator is organized in a spatiotemporal manner. *J Biol Chem* **279**, 33984–33991 (2004).

35. Kuroda, N. et al. Development of autoimmunity against transcriptionally unrepressed target antigen in the thymus of Aire-deficient mice. *J Immunol* **174**, 1862–1870 (2005).

36. Mouri, Y. et al. Lymphotoxin signal promotes thymic organogenesis by eliciting RANK expression in the embryonic thymic stroma. *J Immunol* **186**, 5047–5057 (2011).

37. Mi, Q. S. et al. The autoimmune regulator (Aire) controls iNKT cell development and maturation. *Nat Med* **12**, 624–626 (2006).

38. Akiyama, N. et al. Limitation of immune tolerance-inducing thymic epithelial cell development by Spi-B-mediated negative feedback regulation. *J Exp Med* **211**, 2425–2438 (2014).

39. Li, H. et al. Interferon-induced mechanosensing defects impede apoptotic cell clearance in lupus. *J Clin Invest* **125**, 2877–2890 (2015).

40. Bezbradica, J. S., Stanic, A. K., and Joyce, S. Characterization and functional analysis of mouse invariant natural T (iNKT) cells. *Curr Protoc Immunol* **Chapter 14**, Unit 14.13 (2006).

41. Michel, M. L. et al. Critical role of ROR-γt in a new thymic pathway leading to IL-17-producing invariant NKT cell differentiation. *Proc Natl Acad Sci USA* **105**, 19845–19850 (2008).
42. Mars, L. T. et al. Invariant NKT cells inhibit development of the Th17 lineage. *Proc Natl Acad Sci USA* **106**, 6238–6243 (2009).
43. White, A. J. et al. An essential role for medullary thymic epithelial cells during the intrathymic development of invariant NKT cells. *J Immunol* **192**, 2659–2666 (2014).
44. Hong, C., Luckey, M. A., and Park, J. H. Intrathymic IL-7: The where, when, and why of IL-7 signaling during T cell development. *Semin Immunol* **24**, 151–158 (2012).
45. Patra, A. K. et al. An alternative NFAT-activation pathway mediated by IL-7 is critical for early thymocyte development. *Nat Immunol* **14**, 127–135 (2013).
46. Tani-ichi, S. et al. Interleukin-7 receptor controls development and maturation of late stages of thymocyte subpopulations. *Proc Natl Acad Sci USA* **110**, 612–617 (2013).
47. Ribeiro, A. R., Rodrigues, P. M., Meireles, C., Di Santo, J. P., and Alves, N. L. Thymocyte selection regulates the homeostasis of IL-7-expressing thymic cortical epithelial cells in vivo. *J Immunol* **191**, 1200–1209 (2013).
48. Weinreich, M. A., Jameson, S. C., and Hogquist, K. A. Postselection thymocyte maturation and emigration are independent of IL-7 and ERK5. *J Immunol* **186**, 1343–1347 (2011).
49. Dunkle, A., Dzhagalov, I., and He, Y. W. Mcl-1 promotes survival of thymocytes by inhibition of Bak in a pathway separate from Bcl-2. *Cell Death Differ* **17**, 994–1002 (2010).
50. Dunkle, A., Dzhagalov, I., and He, Y. W. Cytokine dependent and cytokine-independent roles for Mcl-1: Genetic evidence for multiple mechanisms by which Mcl-1 promotes survival in primary T lymphocytes. *Cell Death Dis* **2**, e214 (2011).
51. Dzhagalov, I., Dunkle, A., and He, Y. W. The anti-apoptotic Bcl-2 family member Mcl-1 promotes T lymphocyte survival at multiple stages. *J Immunol* **181**, 521–528 (2008).
52. Jiang, Q. et al. Distinct regions of the interleukin-7 receptor regulate different Bcl2 family members. *Mol Cell Biol* **24**, 6501–6513 (2004).
53. Alves, N. L. et al. Characterization of the thymic IL-7 niche in vivo. *Proc Natl Acad Sci USA* **106**, 1512–1517 (2009).
54. Repass, J. F. et al. IL7-hCD25 and IL7-Cre BAC transgenic mouse lines: New tools for analysis of IL-7 expressing cells. *Genesis* **47**, 281–287 (2009).
55. Rodig, S. J. et al. Disruption of the Jak1 gene demonstrates obligatory and nonredundant roles of the Jaks in cytokine-induced biologic responses. *Cell* **93**, 373–383 (1998).
56. Nosaka, T. et al. Defective lymphoid development in mice lacking Jak3. *Science* **270**, 800–802 (1995).
57. Park, S. Y. et al. Developmental defects of lymphoid cells in Jak3 kinase-deficient mice. *Immunity* **3**, 771–782 (1995).
58. Thomis, D. C., Gurniak, C. B., Tivol, E., Sharpe, A. H., and Berg, L. J. Defects in B lymphocyte maturation and T lymphocyte activation in mice lacking Jak3. *Science* **270**, 794–797 (1995).
59. Yao, Z. et al. Stat5a/b are essential for normal lymphoid development and differentiation. *Proc Natl Acad Sci USA* **103**, 1000–1005 (2006).
60. Ortman, C. L., Dittmar, K. A., Witte, P. L., and Le, P. T. Molecular characterization of the mouse involuted thymus: Aberrations in expression of transcription regulators in thymocyte and epithelial compartments. *Int Immunol* **14**, 813–822 (2002).
61. Rodriguez-Galan, M. C., Bream, J. H., Farr, A., and Young, H. A. Synergistic effect of IL-2, IL-12, and IL-18 on thymocyte apoptosis and Th1/Th2 cytokine expression. *J Immunol* **174**, 2796–2804 (2005).

62. Aujla, S. J., and Kolls, J. K. IL-22: A critical mediator in mucosal host defense. *J Mol Med (Berl)* **87**, 451–454 (2009).

63. Sonnenberg, G. F., Fouser, L. A., and Artis, D. Border patrol: Regulation of immunity, inflammation and tissue homeostasis at barrier surfaces by IL-22. *Nat Immunol* **12**, 383–390 (2011).

64. He, Y. W., Deftos, M. L., Ojala, E. W., and Bevan, M. J. RORgamma t, a novel isoform of an orphan receptor, negatively regulates Fas ligand expression and IL-2 production in T cells. *Immunity* **9**, 797–806 (1998).

65. He, Y. W. et al. Down-regulation of the orphan nuclear receptor ROR gamma t is essential for T lymphocyte maturation. *J Immunol* **164**, 5668–5674 (2000).

66. Sun, Z. et al. Requirement for RORgamma in thymocyte survival and lymphoid organ development. *Science* **288**, 2369–2373 (2000).

67. Zhou, T. et al. Inhibition of Nur77/Nurr1 leads to inefficient clonal deletion of self-reactive T cells. *J Exp Med* **183**, 1879–1892 (1996).

68. Winoto, A., and Littman, D. R. Nuclear hormone receptors in T lymphocytes. *Cell* **109 Suppl**, S57–S66 (2002).

69. Burger, M. L., Leung, K. K., Bennett, M. J., and Winoto, A. T cell-specific inhibition of multiple apoptotic pathways blocks negative selection and causes autoimmunity. *Elife* **3** (2014).

# 4 Modulating Thymic Negative Selection with Bioengineered Thymus Organoids

*Isha Pradhan, Asako Tajima, Suzanne Bertera, Massimo Trucco, and Yong Fan*

## CONTENTS

### ABSTRACT

A diverse, yet self-tolerant T-cell repertoire is essential for adaptive immunity. Negative selection is a key checkpoint in T-cell development that is responsible for the elimination of T-cell clones that express T-cell receptors (TCRs) with high affinity to self-peptide, MHC complexes (pMHCs). Both medullary thymic epithelial cells (mTECs) and thymic antigen presenting cells of hematopoietic origin (e.g., thymic dendritic cells, macrophages, and B-cells) contribute to the clonal deletion process. Advances over the past decade have highlighted the critical roles of tissue-restricted antigen (TRA) expression in mTECs in defining "immunological self" and establishing immune tolerance of peripheral tissues and organs. Although numerous genetic tools and mouse models, such as TCR transgenic strains, targeted mutagenic mouse models, and reaggregate thymus organ culture (RTOC), have been developed to study various aspects of negative selection, its underlying mechanism remains not fully understood, primarily due to the complexity of the process. We have recently developed a tissue engineering technique that allows us to reconstruct a functional thymus organoid by repopulating a decellularized thymus scaffold with TECs and other thymic stromal cells. The bioengineered thymus organoid can recapitulate the function of a thymus, support the development of a complex T-cell repertoire, and reestablish T-cell adaptive immunity in athymic nude mice. This bottom-up approach enables us to further delineate the roles of different thymic stromal components in negative selection. Here, using islet cell autoantigen 69 (ICA69), a known pancreatic beta cell auto-antigen in the pathogenesis of type 1 diabetes, as a model antigen, we show that ICA69-reactive T-cells can escape clonal deletion in athymic nude mice transplanted with thymus organoids constructed with ICA69-deficient mTECs. Thymus bioengineering could be a useful approach for further understanding and modulating thymic negative selection.

## 4.1   OVERVIEW OF THYMIC NEGATIVE SELECTION

Recognition of foreign substances and removing them from the body are fundamental to the actions of the immune system. Innate immunity utilizes a limited number of germline-encoded pathogen recognition receptors (PRRs) to recognize conserved molecular structures shared by a wide range of microorganisms.[1] The adaptive immune system, on the contrary, relies on its capability to generate a myriad of antigen-specific receptors on the surface of T or B lymphocytes. The diversity of the antigen receptor repertoire allows the immune system to recognize almost all the possible epitopes that the host might encounter throughout its life. Adaptive immunity empowers the jawed vertebrates to defend against novel pathogens, to effectively direct resources of the immune system to battle specific invading microbes, and to promote quick recall response upon reexposure.[2] Since the microbes and the vertebrate hosts share many key building blocks (e.g., sugar chains, lipids, amino acids), discriminating self from nonself structures becomes critical for the survival of the hosts, preventing them from developing "horror autotoxicus."

The thymus, the primary immune organ responsible for the generation of mature T lymphocytes, is essential for establishing adaptive immune tolerance to self. It is a dynamic organ primarily made of two sets of cells: (1) the transient, migratory, developing thymocytes derived from the hematopoietic stem cells and (2) the resident thymic stromal cells (TSCs), which constitute the three-dimensional (3-D) thymic microenvironment.[3,4] Under normal physiological conditions in young mice, the ratio of the migratory thymocytes and the resident TSCs is above 1000:1. Strikingly, most of the bone marrow-generated, immature thymocytes fail to pass key checkpoints of thymopoiesis (e.g., lineage determination, TCR beta unit selection, positive selection, and negative selection) and undergo either passive or induced cell death.[5,6] Of the more than 50 million CD4+CD8+ double-positive (DP) thymocytes generated daily in a young mouse, more than 90% are eliminated within the cortical region by the mechanism of positive selection, primarily as a consequence of the failure to produce T cell receptors (TCRs) that are able to form stable interactions with the self-peptide presenting major histocompatibility complexes (pMHCs) on the surface of cortical thymic epithelial cells (cTECs).

Thymocytes that pass the positive selection checkpoint, including both DP cells and the subsequently differentiated single-positive (SP; CD4⁻CD8⁺ and CD4⁺CD8⁻) cells, have to undergo negative selection (also termed *clonal deletion*). Thymocytes expressing TCRs with high affinities to self-peptide pMHCs will be actively induced to undergo apoptosis. In contrast to the positive selection, which is mediated by a single population of thymic APCs (aka cTECs), multiple subsets of thymic APCs, including CD11⁺ conventional dendritic cells (cDCs), medullary TECs (mTECs), macrophages, thymic B-cells,[7,8] as well as plasmacytoid DCs (pDCs),[9] are involved in negative selection, among which mTECs and thymic DCs play predominant roles. Moreover, recent studies have shown that negative selection occurs in both the cortical and medullary regions of the thymus, in contrast to the classic view that it is largely restricted to the medulla.[10]

Over the past decades, significant progress has been made to understand the underlying mechanisms of immune self-recognition and tolerance induction in the thymus. Based on their surface markers and function, thymic cDCs can be divided into two subsets: the migratory CD8α–SIRPα⁺ (signal-regulatory protein-α) subset from circulation and the CD8α⁺SIRPα⁻ population originated from the intrathymic differentiation of earliest thymic progenitors (ETPs).[11-13] It is conceivable that migratory thymic cDCs may present peripherally acquired self-antigens in the thymic stroma, while the resident cDCs might play more important roles in sensing and presenting self-antigens of the thymic microenvironment. Consistently, it has been shown that migratory cDCs are more effective in capturing and presenting circulating antigens than resident cDCs.

While the thymic immigration of the migratory DCs and other thymic APCs can establish a direct route for the negative selection machinery to sample peripheral self-antigens and eliminate developing thymocytes with high affinity to self, the underlying mechanisms to establish immune tolerance to antigens, whose expression are subjected to tissue restriction and/or developmental control, were largely unknown. About two decades ago, the somewhat serendipitous discovery that insulin is transcribed in a rare population of thymic stromal cells, known today as mTECs, revealed a critical aspect of the negative selection mechanism. Hanahan and colleagues observed that different lines of RIP-Tag (rat insulin promoter driving SV40 large T-antigen expression) transgenic mice displayed various degrees of immune response against Tag upon immunization.[14,15] Interestingly, ectopic expression of Tag in the thymus was found in some mouse lines. Transgenic lines with high levels of Tag thymic expression display less or no response, whereas those with no thymic Tag expression elicit robust immune response upon immunization. These results suggested the existence of a direct correlation between the thymic expression of neo antigen SV40 Tag and its peripheral immune tolerance. It was later found that mTECs express transcripts, at low levels, of almost all the genes of the hosts, thereby presenting and defining the immunological self of the entire peripheral tissues.[5,16-19]

The underlying molecular mechanism(s) that mTECs use to transcribe a large pool of tissue-restricted antigens (TRAs) remain poorly understood. In the past decade, extensive efforts have been focusing on the autoimmune regulatory (Aire) gene, whose mutation is responsible for human autoimmune polyendocrine syndrome 1 (APS1), a rare autoimmune disorder that affects multiple endocrine and exocrine organs. Anderson et al. showed significant decrease of thymic TRA expression in Aire-deficient mice, suggesting that Aire functions as one of the key regulators of TRA transcription in mTECs.[17] These mice spontaneously develop multiple organ autoimmunity, similar to some extent to human APS1. Thus, ectopic TRA expression in mTECs functions as a "molecular mirror," reflecting the molecular composition of the periphery within the thymus to establish immune self-tolerance. Indeed, mice genetically engineered to have the expression of insulin abolished specifically in mTECs develop full-blown diabetes around 3 to 4 weeks after birth.[20,21] Insulin-producing beta cells in the pancreas are specifically targeted and destroyed by insulin-reactive T cells that escaped thymus censorship. Ectopic expression of insulin in mTECs is critical for clonal deletion of insulin-reactive T cells in the thymus and to establish immune tolerance of pancreatic beta cells.

However, not all the thymic expressing TRAs are affected by Aire knockout. Recently, a second regulator of thymus TRA expression, the forebrain expressed zinc finger 2 (*Fezf2*) gene, was identified.[22] Similar to Aire knockout mice, mice with deletion of Fezf2 specifically in TECs develop autoimmunity affecting multiple organs/tissues. Notably, there is mostly no overlapping between the population of TRAs that is Fezf2-dependent and those that are dependent on Aire, suggesting that the two mechanisms complement each other to broaden the spectrum of TRA expression in the thymus. Interestingly, the mechanisms by which Aire and Fezf2 undertake to promote TRA expression in mTECs are also different. Aire binds to nonmethylated histone 3 lysine 4 (H3K4) of the nucleosome to induce chromatin configuration changes in a large segment of chromosomal region, and, thereby, facilitates the access of the transcriptional factor complexes to the promoter regions of a cluster of TRAs.[23,24] In contrast, Fezf2 directly binds to specific DNA targets in the promoter regions of more than 10,000 genes, functioning as a master regulator of gene transcription. Whether other mechanisms/factors are involved in promoting TRA expression in mTECs remains to be further investigated.

## 4.2 GENERAL APPROACHES TO STUDY THYMUS NEGATIVE SELECTION

Studying negative selection is challenging due to the complexity of the process, such as the involvement of multiple types of thymic stromal cells, the complexities of the ligandomes (the library of self peptides presented by the MHCs) and the large numbers of continuously developing DP or SP thymocytes. Many studies use transgenic mouse lines expressing T cell receptors (TCRs) specific to various self or "model" antigens.[25–28] Although many key aspects of negative selection have been successfully unraveled by this approach, one of the major caveats is that expression of the preassembled transgenic TCR might lead to allelic exclusion bias, which blocks the somatic recombinant events of other TCR alleles. Whether the mechanisms underlying the selection of monoclonal T cells can be applied to normal repertoires remains unclear. Recently, Malhotra et al. developed eGFPp–I-A$^b$ tetramers that can detect, with high specificity and sensitivity, polyclonal populations of CD4$^+$ T cells that recognize the same eGFP peptide presented by MHC II I-A$^b$ molecules. Using eGFP as a model antigen, they were able to trace the fate of the eGFP-reactive CD4$^+$ T cells during and after thymus negative selection in a number of transgenic mouse lines that have various patterns of eGFP-expression. Their data strongly suggest that the self-peptide expression pattern in the thymus (e.g., levels and TSC subsets) determines the fates of peptide-specific polyclonal CD4$^+$ thymocytes during negative selection, ranging from clonal deletion, T-regulatory cell induction and impaired effector function.[29]

Using transgenic and/or gene targeting approaches to dissect the roles of thymic APCs is another commonly used strategy to study negative selection. This approach enables the interrogation of the function of specific factors in defined thymic APC types during negative selection. For example, to specifically target the mTECs, a number of genetically modified mouse lines have been established, which express the Cre recombinase gene under the transcriptional regulation of the promoter/enhancer elements of the Aire gene.[20,30–33] Various genome manipulation methods have been used to generate the Aire-Cre transgenic lines, such as knocking-in the Cre gene in-frame with the starting codon of the Aire gene, manipulating bacteria artificial chromosomes (BACs) to express the Cre gene in-frame with the Aire coding exon, or assembling the Cre transgene with 5' upstream and/or 3' downstream regions of the Aire gene.[20] Most of these lines can recapitulate to some extent the Aire expression pattern and have been successfully used to specifically delete loxP-tagged, targeted genes in postnatal mTECs.

One of the major caveats of this approach is that it is rather difficult to identify the regulatory elements that are responsible for mTEC specific expression. Extrathymic expression of the Aire-Cre transgene is often observed, resulting in deletion of the targeted gene in extrathymic tissues and organs, which might complicate the interpretation of the experimental results. Notably, extrathymic expression might reflect the temporal and spatial expression pattern of the endogenous gene. Aire is also expressed in male and female germ cells, and in early developing embryos in mice. Mice with the Cre gene knocked-in at the endogenous Aire locus displayed high levels of Cre activities in cells of early embryos, resulting in deletion of the targeted gene(s) in a broad range of tissues and organs.[33] To gain better time and tissue control of the Cre recombination events, the tamoxifen-dependent CreET2, a chimeric protein made from fusing the Cre recombinase to a modified fragment of the estrogen receptor (ERT2), is commonly used.[34] In the presence of tamoxifen, CreET2, sequestered in the cytoplasm, is translocated to the nucleus for its action. However, the efficiency of the Cre-mediated recombination events might be affected by the dose and accessibility of the tamoxifen to the target cells, and a high dose of tamoxifen administration might cause unwanted side effects.[35,36] Moreover, accumulation of high levels of Cre in TECs might have adverse impacts on their differentiation and turnover rates, which in turn, affect the negative selection process.[37,38]

Reaggregate thymus organ culture (RTOC) is another approach that can help to dissect the underlying mechanism of negative selection.[39,40] Thymic cells isolated from dissociation of thymus lobes of E14 to E16 mouse embryos are dissociated by enzymatic digestion and cultured on 0.8 μm filters sitting on Gelfoam sponges placed in a petri dish. Under this air–liquid interface culture condition, the thymic cells form aggregates that can support the development of T cells for several weeks.[41,42] When

CD4$^+$CD8$^+$ DP cells and thymic stromal cells isolated from E14–E16 embryos are mixed to form RTOCs, cohorts of CD4$^+$8$^-$ and CD4$^-$CD8$^+$ SP cells can be detected as early as 5 days. When transplanted under the kidney capsules of athymic nude mice, these thymic aggregates can promote T-lymphogenesis in vivo. The RTOC approach allows the preparation of 3-D thymus microenvironments with defined stromal and lymphoid cells to delineate their specific roles in thymic selection. One of the major disadvantages is that only limited numbers of TECs can be harvested from the mouse embryos. Moreover, for reasons that remain largely unknown, post-natal thymic stromal cells lose their capabilities to reaggregate. Therefore, the adult thymus microenvironments cannot be reconstructed with the RTOC approach. It is yet unknown whether thymopoiesis in RTOCs can faithfully recapitulate that of the postnatal thymus.

Recently, thymus organoids have been successfully constructed by repopulating decellularized thymus scaffolds with TECs, as well as other thymic stromal cells (e.g., thymic fibroblasts and endothelial cells), in conjunction with bone marrow progenitors.[43] This bottom-up tissue engineering approach takes advantage of the findings that biological scaffolds prepared from a decellularized solid organ can largely retain the genuine extracellular matrix (ECM) microenvironment suitable for recolonization of cells of the same organ.[44–52] Indeed, the microenvironments of the thymus scaffolds can support the survival and function of adult TECs in vitro, without changing their unique molecular properties. When lineage-negative bone marrow progenitors are incorporated in the reconstructed thymus, both DP and SP thymocytes are detectable within 2 weeks, suggesting that the thymus organoids can support thymopoiesis in vitro. Nude mice reconstructed with the bioengineered thymus organoids develop a complex T cell repertoire that can exert various T cell mediated adaptive immune function (e.g., T-helper function to mediate affinity mat-uration and class switching in B cells, and allogeneic skin graft rejection). Notably, when TECs expressing donor MHCs are incorporated in the thymus organoid recon-struction, thymus-transplanted nude mice display donor-specific tolerance to skin grafts but can promptly reject third-party allogeneic skin grafts. Thus, the bioengi-neered thymus organoids can not only regenerate a functional T cell compartment in the immune compromised hosts but also serve as an immunomodulating tool to negatively select the alloantigen reactive thymocytes, and, thereby, modify the iden-tity of "immunological self."

Here, we will describe the protocol for bioengineering of thymus organoids, and give an example on how we can apply this approach to study thymic islet autoantigen expression in negative selection of islet-reactive T cells, using islet autoantigen 69 (ICA69) as a model antigen.

## 4.3  THYMUS BIOENGINEERING TO MODULATE NEGATIVE SELECTION

All the animals used in the experiments were housed in the animal facility at Allegheny-Singer Research Institute under the protocol reviewed and approved by the Institutional Animal Care and Use Committee of the Allegheny Singer Research Institute.

### 4.3.1 HARVESTING THE THYMUS FROM MICE

Euthanize 3- to 5-week-old C57BL6/J mice with $CO_2$ inhalation and cervical dislocation. Lay the animal on its back and make a horizontal midline incision through the skin. Pull the two loose ends of the cut skin on the opposite direction to expose the inner skin. Cut open the diaphragm from the sides and flip to expose the thymic lobes. Gently remove the pair of thymic lobes using curved serrated forceps. Place the thymus gland in a washing buffer (1x PBS, 2 mM EDTA, 0.5% BSA).

### 4.3.2 THYMUS DECELLULARIZATION

Transfer the harvested thymi in separate cryogenic vials and freeze them in −80°C for 20 minutes. Thaw frozen thymi in a 30°C water bath for 20 minutes. Freeze (−80°C) the thymic samples in a Styrofoam box for 30 minutes. Repeat the freeze-and-thaw process two more times. Thymus glands that have undergone freeze-and-thaw can be preserved at −80°C until further treatment. Transfer the thymus in 5 mL flow tubes with 3 mL of 0.5% sodium dodecyl sulfate (SDS) solution. Place the tubes on a test tube rocker at room temperature. Check the clearness of the thymic samples every hour and change the SDS solution every 1.5 to 2 hours three times. Transfer the thymus in a new flow tube with 0.1% SDS and continue the decellularization step by rocking on the test tube rocker at 4°C overnight. Transfer the thymus to a new flow tube with 3 mL renature buffer (5 mM $MgSO_4$, 5 mM $CaCl_2$, 1% Triton-X 100) and shake on the test tube rocker for 15 minutes at 4°C. Repeat the washing step three more times. Transfer the thymus to a new tube and wash three times with 2 mL of PBS for 15 minutes each at 4°C. Store the thymic scaffolds in the washing solution at 4°C until use.

### 4.3.3 ISOLATION OF THYMIC EPITHELIAL CELLS

Thymic lobes harvested from 3- to 5-week-old mice ($n = 3–4$) are placed in a 60 mm tissue culture dish containing 5 mL RPMI-1640, and dissected with 28G insulin syringe needles into small pieces. Rinse three times with RPMI-1640. Incubate the thymus fragments with 3 mL of collagenase digestion buffer (RPMI-1640 with 0.025 mg/mL purified collagenase, 10 mM HEPES, and 0.2 mg/mL DNase I) at 37°C water for 6 minutes with mild shake. Let the thymic fragments settle to the bottom of the tube via gravity. Collect the supernatant in a 50 mL centrifuge tube with 10 mL of ice-cold washing buffer. Keep on ice. Repeat the collagenase digestion steps two more times. Pool the dissociated thymic cells and pellet with centrifugation. Resuspend in 20 mL of washing buffer. Set up the density gradient centrifugation to enrich thymic stromal cells: Prepare 21% density gradient solution by mixing 2.4 mL density gradient medium (60% w/v of iodixanol in water) and 9.6 mL washing buffer in 50 mL centrifuge tube. Lay 20 mL of thymic cells in RPMI-1640 on top of the density gradient solution. Centrifuge at 600× g for 20 minutes at room temperature in a swinging-bucket rotor. Set the deceleration rate at the slowest level. Collect the TECs retained in the interface. Wash cells three times and resuspend in 2 mL of washing buffer. Adjust the concentration to $1 \times 10^7$ cells/mL. Stain cells with anti-CD45 and anti-EpCAM antibodies. Sort TECs with a FACS sorter instrument (e.g., BD Influx) by gating on the CD45-EpCAM+ TEC population.

### 4.3.4 RECONSTRUCTION OF THYMUS ORGANOIDS IN ATHYMIC NUDE MICE

Isolated TECs are mixed with other accessory cells (e.g., lineage marker negative bone marrow progenitors, endothelial progenitor cells, thymic fibroblasts) and injected into the decellularized thymus scaffolds with 28G insulin syringes under the dissection microscope. The reconstructed thymus organoids are cultured in RPMI-10 medium in the top chambers of six-well Transwell plates at 37°C for 7 to 14 days for in vitro analysis of CD4+CD8− or CD4−CD8+ SP cell development, or for 1 to 3 days before being transplanted underneath the kidney capsules of B6 nude mice for in vivo studies.

Aseptic conditions are observed throughout surgery. The recipient animal is anesthetized and the incision site shaved. A small incision (~1 cm) is made in the flank about one half inch to the right or left of midline. Another incision is made through the abdominal wall. The kidney is externalized and a pocket is made between the capsule and the kidney through a hole made in the capsule at one pole of the kidney. The scaffold is inserted into the pocket being careful not to tear the capsule further. The kidney is then replaced and the abdominal wall is sutured with 6-0 suture before the outer skin is closed with small wound clips. The animal is kept warm and observed throughout recovery. Wound clips are removed in 7 to 10 days.

Starting from 4 weeks post-op, peripheral blood leukocytes (PBLs) will be sampled every 4 weeks from the thymus-reconstructed mice, stained with antibodies specific to surface markers of T cells (e.g., anti-CD45, -CD3, -CD4, and -CD8 antibodies), and analyzed by flow cytometry (FCM) analysis. Circulating T cells can be detected as early as 4 weeks, and gradually reach a plateau around 16 to 20 weeks. Typically, the mice are ready for experimental manipulation and characterization by 20 to 24 weeks post-op.

## 4.4   REPRESENTATIVE DATA

We show here, using Islet autoantigen 69 (ICA69) as an example, that ICA69-specific autoreactive T cells can escape negative selection in bioengineered thymus organoids constructed from ICA69-deficient TECs. ICA69 is one of the known autoantigens in Type 1 diabetes (T1D), potentially involved in the autoimmune destruction of pancreatic β cells.[53–56] In addition to β cells, ICA69 is also expressed in other endocrine and exocrine tissues (e.g., the thyroid and the salivary glands). While its function is not fully known, recent studies suggest that ICA69 regulates the synthesis of secretory vesicles in a wide range of cell types.[57–61] Notably, ICA69 is ectopically expressed in Aire-expressing MHCII^HiCD80^Hi mTECs, but in an Aire-independent fashion. Using a targeted mutagenesis approach, we have previously demonstrated that abrogation of ICA69 expression in mTECs induces autoimmunity not only in islet β cells, but also in extrapancreatic ICA69-expressing tissues and organs (e.g., salivary glands and thyroids).[53]

Thymus lobes were harvested from ICA69-deficient B6 mice. TECs and other CD45- stromal cells were isolated by fluorescence-activated cell sorting (FACS) (Figure 4.1a). Bioengineered thymus organoids were constructed from repopulating the decellularized thymus scaffolds (Figure 4.1b) with ICA69-deficient TECs, stromal cells, and lineage-negative bone marrow progenitor cells of B6.CD45.1 mice. Thymus organoids were transplanted underneath the kidney capsules of athymic B6 nude mice (Figure 4.1c). Twenty weeks post-op, the thymus-reconstructed nude mice were sacrificed and the development of CD4$^+$ and CD8$^+$ T cells were demonstrated by flow cytometry (FCM) (Figure 4.2a). To demonstrate that ICA69-reactive T cells can escape the negative selection in ICA69-deficient thymus organoids, ELISpot (enzyme-linked immune spot) assays were performed on harvested peripheral T cells. As shown in Figure 4.2b, ICA69-reactive T cells were detected in B6 nude mice reconstructed with ICA69-deficient thymic cells, but not in those with wild-type cells ($112 \pm 15$, $n = 3$ versus $9 \pm 5$, $n = 3$, $p < 0.01$). These results highlight

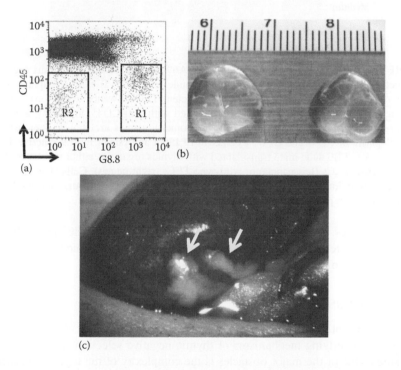

**FIGURE 4.1**    Reconstruction of thymus organoids with decellularized scaffolds. (a) Thymus glands were harvested from 3- to 4-week-old B6 mice ($n = 3$–$4$) and treated with collagenase. The dissociated cells were stained with anti-CD45 and anti-EpCAM antibodies, and were subjected to FACS isolation. Both CD45$^-$EpCAM$^+$ TEC (R1) and CD45$^-$EpCAM$^-$ stromal cell (R2) populations were harvested and mixed with lineage negative bone marrow progenitor cells (not shown) for thymus reconstruction. (b) Representative photographic image of decellularized thymus scaffolds. (c) Representative photographic image showing the transplantation of reconstructed thymus organoids (yellow arrows) underneath the kidney capsule of a nude mouse.

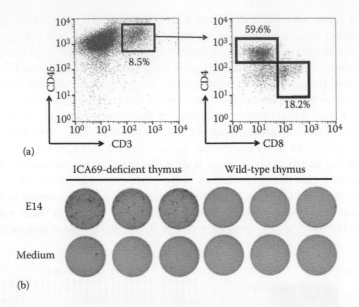

(a)

(b)

**FIGURE 4.2**  Escape of negative selection of ICA69-reactive T cells in thymus organoids constructed with ICA69-deficient TECs. (a) B6 nude mice ($n = 3$) were transplanted with thymus organoids reconstructed with ICA69-deficient TECs. Twenty weeks post-op, cells were harvested from lymph node cells and stained with anti-CD45, anti-CD3, anti-CD4, and anti-CD8 antibodies. Representative flow cytometric panels are shown, demonstrating the development of CD4+ and CD8+ T cells. (b) ELISpot analysis of ICA69-reactive T cells in lymph nodes of B6 nude mice transplanted with thymus organoids either constructed with ICA69-deficient TECs (left panels) or wild-type TECs (right panels). Cells ($1 \times 10^6$/well in triplicate) were cultured with ICA69 C-terminal peptide E14 for 72 hours, and challenged with either E14 (top panels) or medium (lower panels).

the essential roles of ICA69 expression in mTECs in negative selection of ICA69-reactive thymocytes and suggest that reconstructed thymus organoids can be used as an effective tool to study negative selection.

## 4.5   CONCLUSION

Although substantial progresses have been achieved in the past decades in TEC biology, the underlying mechanisms of thymic negative selection remain not fully understood. One of the major obstacles is the complexity of the thymic microenvironments, especially the dynamic nature of the TECs. TECs are a very heterogeneous population, which are dependent on the 3-D configuration of thymic ECMs for their function and survival. Moreover, the precursor populations in adult TECs remain ill defined. With further optimization of the thymus tissue engineering process, such as improvement of vascularization and organization of the thymus microenvironment, this "reductionist" approach can help to dissect the roles of individual subpopulations of TECs and other thymic APCs in mediating negative selection of autoreactive thymocytes.

## ACKNOWLEDGMENTS

The authors thank Drs. Kia Goh and Ipsita Banerjee for their valuable assistance. This work was supported by NIH 1R01 AI123392 (Yong Fan) and Department of Defense grant W81XWH-10-1-1055 (Massimo Trucco).

## REFERENCES

1. Akira, S., Uematsu, S., and Takeuchi, O., Pathogen recognition and innate immunity, *Cell* 124(4), 783–801, 2006.
2. Boehm, T., and Swann, J. B., Origin and evolution of adaptive immunity, *Annu Rev Anim Biosci* 2, 259–283, 2014.
3. Manley, N. R., Richie, E. R., Blackburn, C. C., Condie, B. G., and Sage, J., Structure and function of the thymic microenvironment, *Front Biosci* 17, 2461–2477, 2012.
4. Boehm, T., Thymus development and function, *Curr Opin Immunol* 20(2), 178–184, 2008.
5. Klein, L., Kyewski, B., Allen, P. M., and Hogquist, K. A., Positive and negative selection of the T cell repertoire: What thymocytes see (and don't see), *Nat Rev Immunol* 14(6), 377–391, 2014.
6. Starr, T. K., Jameson, S. C., and Hogquist, K. A., Positive and negative selection of T cells, *Ann Rev Immunol* 21(1), 139–176, 2003.
7. Yamano, T., Nedjic, J., Hinterberger, M., Steinert, M., Koser, S., Pinto, S., Gerdes, N., Lutgens, E., Ishimaru, N., Busslinger, M., Brors, B., Kyewski, B., and Klein, L., Thymic B cells are licensed to present self antigens for central T cell tolerance induction, *Immunity* 42(6), 1048–1061, 2015.
8. Perera, J., Meng, L., Meng, F., and Huang, H., Autoreactive thymic B cells are efficient antigen-presenting cells of cognate self-antigens for T cell negative selection, *Proc Natl Acad Sci USA* 110(42), 17011–17016, 2013.
9. Hadeiba, H., Lahl, K., Edalati, A., Oderup, C., Habtezion, A., Pachynski, R., Nguyen, L., Ghodsi, A., Adler, S., and Butcher, E. C., Plasmacytoid dendritic cells transport peripheral antigens to the thymus to promote central tolerance, *Immunity* 36(3), 438–450, 2012.
10. Sawicka, M., Stritesky, G. L., Reynolds, J., Abourashchi, N., Lythe, G., Molina-Paris, C., and Hogquist, K. A., From pre-DP, post-DP, SP4, and SP8 thymocyte cell counts to a dynamical model of cortical and medullary selection, *Front Immunol* 5, 19, 2014.
11. Ardavin, C., Wu, L., Li, C. L., and Shortman, K., Thymic dendritic cells and T cells develop simultaneously in the thymus from a common precursor population, *Nature* 362(6422), 761–763, 1993.
12. Atibalentja, D. F., Murphy, K. M., and Unanue, E. R., Functional redundancy between thymic CD8alpha+ and Sirpalpha+ conventional dendritic cells in presentation of blood-derived lysozyme by MHC class II proteins, *J Immunol* 186(3), 1421–1431, 2011.
13. Atibalentja, D. F., Byersdorfer, C. A., and Unanue, E. R., Thymus-blood protein interactions are highly effective in negative selection and regulatory T cell induction, *J Immunol* 183(12), 7909–7918, 2009.
14. Smith, K. M., Olson, D. C., Hirose, R., and Hanahan, D., Pancreatic gene expression in rare cells of thymic medulla: Evidence for functional contribution to T cell tolerance, *Int Immunol* 9(9), 1355–1365, 1997.
15. Teitz, T., Chang, J. C., Kan, Y. W., and Yen, T. S., Thymic epithelial neoplasms in transgenic mice expressing SV40 T antigen under the control of an erythroid-specific enhancer, *J Pathol* 177(3), 309–315, 1995.
16. Kyewski, B., and Klein, L., A central role for central tolerance, *Ann Rev Immunol* 24(1), 571–606, 2006.

17. Anderson, M. S., Venanzi, E. S., Klein, L., Chen, Z., Berzins, S. P., Turley, S. J., von Boehmer, H., Bronson, R., Dierich, A., Benoist, C., and Mathis, D., Projection of an immunological self shadow within the thymus by the aire protein, *Science* 298(5597), 1395–1401, 2002.
18. Derbinski, J., Schulte, A., Kyewski, B., and Klein, L., Promiscuous gene expression in medullary thymic epithelial cells mirrors the peripheral self, *Nat Immunol* 2(11), 1032–1039, 2001.
19. Klein, L., and Kyewski, B., "Promiscuous" expression of tissue antigens in the thymus: A key to T-cell tolerance and autoimmunity?, *J Mol Med* 78(9), 483–494, 2000.
20. Fan, Y., Rudert, W. A., Grupillo, M., He, J., Sisino, G., and Trucco, M., Thymus-specific deletion of insulin induces autoimmune diabetes, *EMBO J* 28(18), 2812–2824, 2009.
21. Grupillo, M., Gualtierotti, G., He, J., Sisino, G., Bottino, R., Rudert, W. A., Trucco, M., and Fan, Y., Essential roles of insulin expression in Aire+ tolerogenic dendritic cells in maintaining peripheral self-tolerance of islet beta-cells, *Cell Immunol* 273(2), 115–123, 2012.
22. Takaba, H., Morishita, Y., Tomofuji, Y., Danks, L., Nitta, T., Komatsu, N., Kodama, T., and Takayanagi, H., Fezf2 orchestrates a thymic program of self-antigen expression for immune tolerance, *Cell* 163(4), 975–987, 2015.
23. Kont, V., Murumagi, A., Tykocinski, L. O., Kinkel, S. A., Webster, K. E., Kisand, K., Tserel, L., Pihlap, M., Strobel, P., Scott, H. S., Marx, A., Kyewski, B., and Peterson, P., DNA methylation signatures of the AIRE promoter in thymic epithelial cells, thymomas and normal tissues, *Mol Immunol* 49(3), 518–526, 2011.
24. Meredith, M., Zemmour, D., Mathis, D., and Benoist, C., Aire controls gene expression in the thymic epithelium with ordered stochasticity, *Nat Immunol* 16(9), 942–949, 2015.
25. Sha, W. C., Nelson, C. A., Newberry, R. D., Kranz, D. M., Russell, J. H., and Loh, D. Y., Positive and negative selection of an antigen receptor on T cells in transgenic mice, *Nature* 336(6194), 73–76, 1988.
26. Fukui, Y., Ishimoto, T., Utsuyama, M., Gyotoku, T., Koga, T., Nakao, K., Hirokawa, K., Katsuki, M., and Sasazuki, T., Positive and negative CD4+ thymocyte selection by a single MHC class II/peptide ligand affected by its expression level in the thymus, *Immunity* 6(4), 401–410, 1997.
27. Ghendler, Y., Teng, M. K., Liu, J. H., Witte, T., Liu, J., Kim, K. S., Kern, P., Chang, H. C., Wang, J. H., and Reinherz, E. L., Differential thymic selection outcomes stimulated by focal structural alteration in peptide/major histocompatibility complex ligands, *Proc Natl Acad Sci USA* 95(17), 10061–10066, 1998.
28. Buch, T., Rieux-Laucat, F., Forster, I., and Rajewsky, K., Failure of HY-specific thymocytes to escape negative selection by receptor editing, *Immunity* 16(5), 707–718, 2002.
29. Malhotra, D., Linehan, J. L., Dileepan, T., Lee, Y. J., Purtha, W. E., Lu, J. V., Nelson, R. W., Fife, B. T., Orr, H. T., Anderson, M. S., Hogquist, K. A., and Jenkins, M. K., Tolerance is established in polyclonal CD4(+) T cells by distinct mechanisms, according to self-peptide expression patterns, *Nat Immunol* 17(2), 187–195, 2016.
30. Nishikawa, Y., Hirota, F., Yano, M., Kitajima, H., Miyazaki, J., Kawamoto, H., Mouri, Y., and Matsumoto, M., Biphasic Aire expression in early embryos and in medullary thymic epithelial cells before end-stage terminal differentiation, *J Exp Med* 207(5), 963–971, 2010.
31. Yano, M., Kuroda, N., Han, H., Meguro-Horike, M., Nishikawa, Y., Kiyonari, H., Maemura, K., Yanagawa, Y., Obata, K., Takahashi, S., Ikawa, T., Satoh, R., Kawamoto, H., Mouri, Y., and Matsumoto, M., Aire controls the differentiation program of thymic epithelial cells in the medulla for the establishment of self-tolerance, *J Exp Med* 205(12), 2827–2838, 2008.

32. Gardner, J. M., Devoss, J. J., Friedman, R. S., Wong, D. J., Tan, Y. X., Zhou, X., Johannes, K. P., Su, M. A., Chang, H. Y., Krummel, M. F., and Anderson, M. S., Deletional tolerance mediated by extrathymic Aire-expressing cells, *Science* 321(5890), 843–847, 2008.

33. Dertschnig, S., Nusspaumer, G., Ivanek, R., Hauri-Hohl, M. M., Hollander, G. A., and Krenger, W., Epithelial cytoprotection sustains ectopic expression of tissue-restricted antigens in the thymus during murine acute GVHD, *Blood* 122(5), 837–841, 2013.

34. Hayashi, S., and McMahon, A. P., Efficient recombination in diverse tissues by a tamoxifen-inducible form of Cre: A tool for temporally regulated gene activation/inactivation in the mouse, *Dev Biol* 244(2), 305–318, 2002.

35. Reinert, R. B., Kantz, J., Misfeldt, A. A., Poffenberger, G., Gannon, M., Brissova, M., and Powers, A. C., Tamoxifen-induced Cre-loxP recombination is prolonged in pancreatic islets of adult mice, *PLoS One* 7(3), e33529, 2012.

36. Harno, E., Cottrell, E. C., and White, A., Metabolic pitfalls of CNS Cre-based technology, *Cell Metab* 18(1), 21–28, 2013.

37. Lee, J. Y., Ristow, M., Lin, X., White, M. F., Magnuson, M. A., and Hennighausen, L., RIP-Cre revisited, evidence for impairments of pancreatic beta-cell function, *J Biol Chem* 281(5), 2649–2653, 2006.

38. Lee, K. Y., Russell, S. J., Ussar, S., Boucher, J., Vernochet, C., Mori, M. A., Smyth, G., Rourk, M., Cederquist, C., Rosen, E. D., Kahn, B. B., and Kahn, C. R., Lessons on conditional gene targeting in mouse adipose tissue, *Diabetes* 62(3), 864–874, 2013.

39. Yamaguchi, Y., Kudoh, J., Yoshida, T., and Shimizu, N., In vitro co-culture systems for studying molecular basis of cellular interaction between Aire-expressing medullary thymic epithelial cells and fresh thymocytes, *Biol Open* 3(11), 1071–1082, 2014.

40. Ramsdell, F., Zuniga-Pflucker, J. C., and Takahama, Y., In vitro systems for the study of T cell development: Fetal thymus organ culture and OP9-DL1 cell coculture, *Curr Protoc Immunol* Chapter 3, Unit 3.18, 2006.

41. Anderson, G., and Jenkinson, E. J., Review article: Thymus organ cultures and T-cell receptor repertoire development, *Immunology* 100(4), 405–410, 2000.

42. Anderson, G., and Jenkinson, E. J., Fetal thymus organ culture, *CSH Protoc* 2007.

43. Fan, Y., Tajima, A., Goh, S. K., Geng, X., Gualtierotti, G., Grupillo, M., Coppola, A., Bertera, S., Rudert, W. A., Banerjee, I., Bottino, R., and Trucco, M., Bioengineering thymus organoids to restore thymic function and induce donor-specific immune tolerance to allografts, *Mol Ther* 23(7), 1262–1277, 2015.

44. Gilbert, T. W., Sellaro, T. L., and Badylak, S. F., Decellularization of tissues and organs, *Biomaterials* 27(19), 3675–3683, 2006.

45. Song, J. J., Guyette, J. P., Gilpin, S. E., Gonzalez, G., Vacanti, J. P., and Ott, H. C., Regeneration and experimental orthotopic transplantation of a bioengineered kidney, *Nat Med* 19(5), 646–651, 2013.

46. Orlando, G., Soker, S., and Stratta, R. J., Organ bioengineering and regeneration as the new Holy Grail for organ transplantation, *Ann Surg* 258(2), 221–232, 2013.

47. Goh, S. K., Olsen, P., and Banerjee, I., Extracellular matrix aggregates from differentiating embryoid bodies as a scaffold to support ESC proliferation and differentiation, *PLoS One* 8(4), e61856, 2013.

48. Goh, S. K., Bertera, S., Olsen, P., Candiello, J. E., Halfter, W., Uechi, G., Balasubramani, M., Johnson, S. A., Sicari, B. M., Kollar, E., Badylak, S. F., and Banerjee, I., Perfusion-decellularized pancreas as a natural 3D scaffold for pancreatic tissue and whole organ engineering, *Biomaterials* 34(28), 6760–6772, 2013.

49. Petersen, T. H., Calle, E. A., Zhao, L., Lee, E. J., Gui, L., Raredon, M. B., Gavrilov, K., Yi, T., Zhuang, Z. W., Breuer, C., Herzog, E., and Niklason, L. E., Tissue-engineered lungs for in vivo implantation, *Science* 329(5991), 538–541, 2010.

50. Ott, H. C., Clippinger, B., Conrad, C., Schuetz, C., Pomerantseva, I., Ikonomou, L., Kotton, D., and Vacanti, J. P., Regeneration and orthotopic transplantation of a bioartificial lung, *Nat Med* 16(8), 927–933, 2010.

51. Baptista, P. M., Orlando, G., Mirmalek-Sani, S. H., Siddiqui, M., Atala, A., and Soker, S., Whole organ decellularization: A tool for bioscaffold fabrication and organ bioengineering, *Conf Proc IEEE Eng Med Biol Soc* 6526–6529, 2009.

52. Chan, B. P., and Leong, K. W., Scaffolding in tissue engineering: General approaches and tissue-specific considerations, *Eur Spine J* 17 Suppl 4, 467–479, 2008.

53. Fan, Y., Gualtierotti, G., Tajima, A., Grupillo, M., Coppola, A., He, J., Bertera, S., Owens, G., Pietropaolo, M., Rudert, W. A., and Trucco, M., Compromised central tolerance of ICA69 induces multiple organ autoimmunity, *J Autoimmun* 53, 10–25, 2014.

54. Bonner, S. M., Pietropaolo, S. L., Fan, Y., Chang, Y., Sethupathy, P., Morran, M. P., Beems, M., Giannoukakis, N., Trucco, G., Palumbo, M. O., Solimena, M., Pugliese, A., Polychronakos, C., Trucco, M., and Pietropaolo, M., Sequence variation in promoter of Ica1 gene, which encodes protein implicated in type 1 diabetes, causes transcription factor autoimmune regulator (AIRE) to increase its binding and down-regulate expression, *J Biol Chem* 287(21), 17882–17893, 2012.

55. Winer, S., Gunaratnam, L., Astsatourov, I., Cheung, R. K., Kubiak, V., Karges, W., Hammond-McKibben, D., Gaedigk, R., Graziano, D., Trucco, M., Becker, D. J., and Dosch, H. M., Peptide dose, MHC affinity, and target self-antigen expression are critical for effective immunotherapy of nonobese diabetic mouse prediabetes, *J Immunol* 165(7), 4086–4094, 2000.

56. Pietropaolo, M., Castano, L., Babu, S., Buelow, R., Kuo, Y. L., Martin, S., Martin, A., Powers, A. C., Prochazka, M., Naggert, J. et al., Islet cell autoantigen 69 kD (ICA69). Molecular cloning and characterization of a novel diabetes-associated autoantigen, *J Clin Invest* 92(1), 359–371, 1993.

57. Holst, B., Madsen, K. L., Jansen, A. M., Jin, C., Rickhag, M., Lund, V. K., Jensen, M., Bhatia, V., Sorensen, G., Madsen, A. N., Xue, Z., Moller, S. K., Woldbye, D., Qvortrup, K., Huganir, R., Stamou, D., Kjaerulff, O., and Gether, U., PICK1 deficiency impairs secretory vesicle biogenesis and leads to growth retardation and decreased glucose tolerance, *PLoS Biol* 11(4), e1001542, 2013.

58. Cao, M., Mao, Z., Kam, C., Xiao, N., Cao, X., Shen, C., Cheng, K. K., Xu, A., Lee, K. M., Jiang, L., and Xia, J., PICK1 and ICA69 control insulin granule trafficking and their deficiencies lead to impaired glucose tolerance, *PLoS Biol* 11(4), e1001541, 2013.

59. Buffa, L., Fuchs, E., Pietropaolo, M., Barr, F., and Solimena, M., ICA69 is a novel Rab2 effector regulating ER-Golgi trafficking in insulinoma cells, *Eur J Cell Biol* 87(4), 197–209, 2008.

60. Cao, M., Xu, J., Shen, C., Kam, C., Huganir, R. L., and Xia, J., PICK1 ICA69 heteromeric BAR domain complex regulates synaptic targeting and surface expression of AMPA receptors, *J Neurosci* 27(47), 12945–12956, 2007.

61. Winer, S., Astsaturov, I., Cheung, R., Tsui, H., Song, A., Gaedigk, R., Winer, D., Sampson, A., McKerlie, C., Bookman, A., and Dosch, H. M., Primary Sjögren's syndrome and deficiency of ICA69, *Lancet* 360(9339), 1063–1069, 2002.

# 5 ThPOK, a Key Regulator of T Cell Development and Function

*Jayati Mookerjee-Basu, Sijo V. Chemmannur,
Li Qin, and Dietmar J. Kappes*

## CONTENTS

### ABSTRACT

The molecular basis of differentiation of immature thymocyte precursors to alternate CD4 or CD8 lineage cells remains poorly understood, but appears to be controlled by differences in TCR signaling. Importantly, we have shown that the transcription factor ThPOK acts as a master regulator of this process, such that its presence or absence dictates development to the CD4 or CD8 lineages, respectively. We have also shown that ThPOK expression in thymocytes and mature T cells is regulated primarily at the transcriptional level via several stage- and lineage-specific cis elements. Of particular importance is the ultraconserved ThPOK silencer, whose deletion is sufficient to cause promiscuous expression of ThPOK in thymocytes and diversion of all developing thymocytes to the CD4 lineage. In this review, we provide an overview of the role of ThPOK in cd4 versus cd8 commitment, as well as its transcriptional regulation and additional important roles of ThPOK in development and function of other T cell subsets.

## 5.1    INTRODUCTION

Coordinated function of specialized immune cell types is required to elicit effective immune responses against invading pathogens. The adaptive arms of the immune system, comprised of T and B lymphocytes, play distinct but critical roles in the elimination and memory responses to encountered pathogens. While B lymphocytes trigger an antibody-mediated immune response, T lymphocytes function by recognizing short peptides of antigens presented on the surface of "antigen presenting cells." There are two major T lymphocyte subsets: killer cells, which typically recognize major histocompatibility complex (MHC) class I molecules and express the CD8 coreceptor exclusively; and helper T cells, which recognize MHC class II molecules and express the CD4 coreceptor exclusively. Maintaining the proper number and proportion of CD4 and CD8 T cells in the periphery is critical for an optimal immune response and is enforced in the first instance by stringent and complex T cell development programs in the thymus. Elucidating how this strict correlation between MHC specificity and lineage choice is orchestrated has been a major research area for over three decades. This chapter focuses on the recent advances in the understanding of the molecular basis of CD4/CD8 lineage choice.

## 5.2    T CELL DEVELOPMENT IN A NUTSHELL

Thymic development of T cells is a highly complex and closely controlled program, beginning with early progenitor cells from the bone marrow, which express neither CD4 nor CD8 surface markers, known as double negative (DN) thymocytes, and ends with mature single positive (SP) CD4 and CD8 thymocytes. Within the DN compartment, four sequential stages are marked by differential expression of the CD44 and CD25 markers: DN1 (CD25$^-$CD44$^{hi}$), DN2 (CD25$^+$CD44$^+$), DN3 (CD25$^+$CD44$^-$), and DN4 (CD25$^-$CD44$^-$). Cells belonging to the major $\alpha\beta$ T cell lineage develop further to the double positive CD4$^+$CD8$^+$ (DP) stage, where the $\alpha\beta$ T cell receptor (TCR) complex is first expressed at the cell surface, allowing engagement by intrathymic peptide/MHC ligands (Figure 5.1a). At this stage, cells with high self-affinity undergo death by apoptosis (negative selection), whereas cells with lower self-affinity undergo activation and further differentiation into single positive T cells (positive selection). These processes are extremely stringent leading to the programmed cell death of >90% of DP thymocytes. Defects in these selection processes can result in survival of strongly self-reactive T cells and consequent autoimmune disease. The few thymocytes that are selected for survival undergo alternate commitment to either SP CD4 or CD8 lineages [1–7]. Regardless of the final lineage commitment, DP cells initially downregulate the CD8 coreceptor to give rise to intermediate CD4$^+$CD8$^{lo}$ cells [8–9] (Figure 5.1b). From this common CD4$^+$8$^{lo}$ stage, cells can mature directly into SP CD4 cells or into SP CD8 cells, via additional

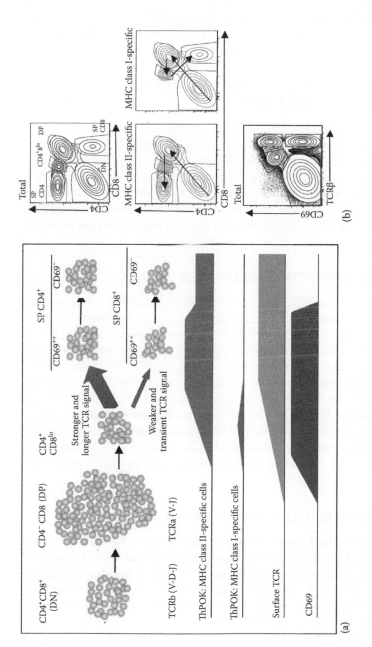

**FIGURE 5.1** T lymphocyte development. (a) Simplified diagram of stages in thymic development, including, in sequential order, double negative (DN), double positive (DP), intermediate CD4+8lo, and finally single-positive (SP) CD4 and CD8. Thymic development is also marked by changes in expression of TCR and CD69 surface markers. TCR and CD69 surface markers are both upmodulated as thymocytes mature from the DP to the CD4+8lo stage and even further during subsequent transition to the SP CD4 and CD8 stages. Finally, just before exiting the thymus, SP thymocytes lose CD69 expression but still maintain high TCR expression. At the CD4+8lo stage, MHC class II- and class I-specific cells express high or low levels of ThPOK, respectively, and diverge into alternate CD4 and CD8 lineages. (b) Schematics of the surface expression pattern of CD4, CD8, CD69, and TCR by total thymocytes, or by MHC class I- or II-specific thymocytes, as indicated.

intermediate stages (CD4$^{lo}$CD8$^{lo}$ and CD4$^{lo}$CD8$^+$) [9–10]. There is an almost perfect correlation between commitment to the CD8 or CD4 lineages, and specificity of a developing thymocyte to MHC class I or II, respectively. The correct match between lineage choice and MHC class specificity is highly important for proper T cell function. How this near perfect correlation between MHC specificity and functional lineage choice is achieved is still not fully understood.

## 5.3 T CELL RECEPTOR (TCR) SIGNALING, THE CRITICAL REGULATOR OF LINEAGE CHOICE

TCR signaling is believed to represent the major driving force underlying alternate commitment to the CD4 and CD8 lineages. Accumulating genetic and biochemical evidence suggests that the strength and duration of the TCR–MHC interaction during thymic T cell development are critical for determining alternate lineage choice, such that relatively long and strong TCR signals give rise to SP CD4 cells, whereas weak and transient signals give rise to CD8 T cells. Two alternative models have been proposed, based on differential TCR signal strength (signal strength model) or duration (kinetic model).

## 5.4 TCR SIGNAL STRENGTH AND LINEAGE CHOICE

During thymic selection the TCR interacts with self-peptide/MHCs expressed on thymic selecting cells. Simultaneously, CD4 and CD8 coreceptors bind directly to nonpolymorphic determinants on MHC class II and I molecules, respectively. Both CD4 and CD8 coreceptors bind the major signaling molecule, p56Lck tyrosine kinase, but the cytoplasmic tail of CD4 does so with significantly higher affinity leading to enhanced phosphorylation of downstream Lck targets [11–12]. In support of such a model, replacing the cytoplasmic tail of CD8α with that of CD4 confers high affinity Lck interaction and causes redirection of class I restricted cells to the CD4 lineage [13], limiting availability of downstream signaling molecules like Lck or ZAP 70 causes quantitative reduction of TCR signaling [14–16], and reducing Lck activity promotes redirection of class I-restricted thymocytes to the CD4 lineage [14].

## 5.5 TCR SIGNAL DURATION AND LINEAGE CHOICE

If signal strength alone could determine lineage choice, there would be a significant risk that some high affinity class I-restricted and low affinity class II-restricted thymocytes could adopt the inappropriate lineage. The kinetic signaling model provides a mechanism to avoid this problem, based on the observation that both class I- and class II-restricted thymocytes pass through the CD4$^+$CD8$^{lo}$ stage during the DP to SP transition. Therefore, class I-restricted thymocytes, which require CD8 for efficient TCR signaling, but not class II-restricted cells undergo an interruption of TCR signaling specifically at the CD4$^+$CD8$^{lo}$ stage. The kinetic signaling model postulates that interrupted or persistent TCR signal transduction leads to commitment of CD4$^+$CD8$^{lo}$ cells to the CD4 or CD8 lineages, respectively. Thus, this model emphasizes the importance of TCR signal duration at the CD4$^+$CD8$^{lo}$ stage for lineage commitment [17]. Interestingly, a complete block of TCR signaling at this stage (by ablation of ZAP70) perturbs CD8 commitment, suggesting the requirement for a minimal level of TCR signaling at the CD4$^+$CD8$^{lo}$ stage even for CD8 commitment [18–20]. Recent studies highlight the importance of IL-7 receptor (IL-7R) mediated signaling for CD8 commitment. Weak and transient TCR engagement permits IL-7R signaling, whereas persistent TCR engagement blocks this signal. Thus, treatment of ex-vivo sorted CD4$^+$CD8$^{lo}$ thymocytes with IL-7 alone or in combination with IL-2 and IL-15 promoted CD8 upregulation and differentiation to the SP CD8 stage [21]. Furthermore, continuous TCR stimulation suppresses IL-4 and IL-7 signaling [22], while forced activation of IL-7R signaling in DP thymocytes induces Runx3 transcription and CD8 differentiation [23]. Moreover, premature CD8 lineage commitment in preselection DP thymocytes is prevented by SOCS1-mediated inhibition of IL-7R signaling [24]. Finally, conditional ablation of IL-7R results in the reduction of CD8 SP thymocytes confirming its importance for SP CD8 development, although it is not absolutely required [25].

## 5.6  APPROACHES TO ELUCIDATE MOLECULAR BASIS OF CD4/CD8 LINEAGE CHOICE

Although TCR signaling is now recognized as the major factor that regulates CD4/CD8 lineage commitment, it remains to be elucidated how TCR-initiated signaling cascades culminate in the differential activation of genes specific to either CD4 or CD8 T-cell commitment. A candidate gene approach using pharmacological inhibitors provided the first evidence of involvement of the Ras–MEK–ERK pathway [26–27]. Another approach was to work backward from transcriptional regulation of the CD4 and CD8 genes, the most obvious genes to be regulated in a lineage-specific manner, the so-called bottom-up strategy [28]. Indeed, the latter approach has been validated, as many of the major transcription factors involved in regulation of CD4 and CD8 expression have also been shown to contribute to the regulatory network that governs helper versus cytotoxic-lineage differentiation programs.

## 5.7  ThPOK PROMOTES CD4 COMMITMENT

Identification by Kappes et al., in the late 1990s, of a spontaneous mutant mouse line (helper deficient, or HD) that lacked CD4$^+$ helper cells due to a single recessive mutation was the first step in identifying ThPOK as the master regulator of CD4 lineage choice [29–30] (Figure 5.2a). Subsequent backcrossing of HD mice to an MHC class I-deficient background revealed that lack of CD4 T cells in HD mice was not due to a block in development of MHC class II-restricted thymocytes but rather to their redirection to the CD8 lineage. This finding led to the conclusion that lineage choice and positive selection were two mechanistically distinct processes [30]. Finally, in 2005 the mutation was mapped to a point mutation in the gene encoding the transcription factor ThPOK (T helper-inducing POZ Krueppel factor), which resulted in a

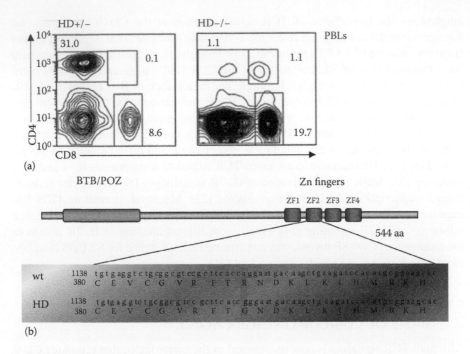

(a)

(b)

**FIGURE 5.2** ThPOK is a transcriptional regulator of CD4 T cell development. (a) FACS analysis of CD4, and CD8a expression in peripheral blood cells from mice heterozygous or homozygous for the HD mutation. Note absence of CD4 lymphocytes in homozygous mutant mice. (b) Schematic of ThPOK protein domain structure in a mouse, showing BTB/POZ domain, responsible for dimerization and interaction with chromatin modifiers, and Zn finger domains responsible for DNA interaction. The helper-deficient (HD) mouse strain carries a point mutation in ZF2 leading to loss of function.

single amino acid substitution (R to G) in its DNA binding domain [31] (Figure 5.2b). These and later studies from another group established that ThPOK functions as a "master regulator" of lineage commitment, whose presence or absence is necessary and sufficient to drive development of immature thymocytes to the CD4 or CD8 lineages, respectively [32–33]. The structure and expression pattern of the ThPOK gene appear to have been conserved since before the divergence of bony fish from other vertebrate lineages (Figure 5.3).

## 5.8   STRONG TCR SIGNALING IS LINKED TO ThPOK INDUCTION

The expression pattern of the ThPOK gene is tightly regulated during T cell development. At early DN and preselection DP stages ThPOK mRNA is essentially undetectable. ThPOK mRNA is first detected in $CD4^+CD8^{lo}$ thymocytes. Importantly, ThPOK mRNA is much higher in class II- than class I-restricted $CD4^+8^{lo}$ cells, suggesting that the ThPOK locus in these cells is responding differentially to MHC class I- or II-restricted TCR signals. ThPOK mRNA levels subsequently increase further during development to the SP CD4 stage, while diminishing to background levels

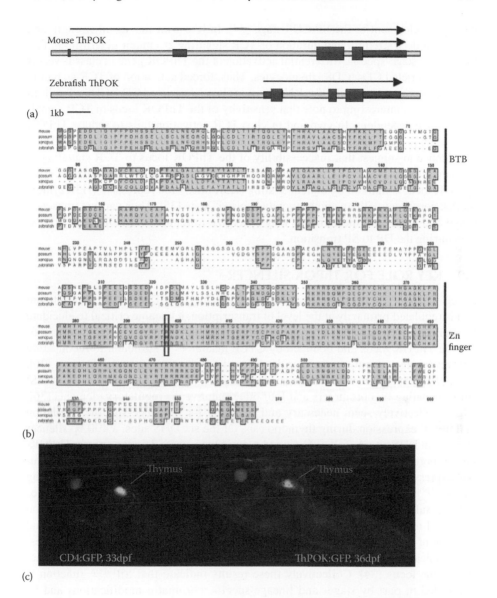

**FIGURE 5.3** ThPOK is evolutionarily conserved in vertebrates. (a) Schematic of ThPOK gene organization in a mouse and zebrafish, showing exon/intron organization and transcriptional orientation (brown and red boxes indicate noncoding and coding exons, respectively). The mouse ThPOK gene has two alternate start sites, as indicated. (b) Alignment of ThPOK proteins from different species, as indicated. Amino acid identity is indicated by gray shading. Note high homology within functionally important BTB and ZF domains. The residue altered in the HD mutant mouse is boxed in red. (c) Fluorescence microscopy analysis of reporter fish expressing zCD4:GFP or zThPOK-GFP reporter transgenes. Note bright reporter expression in thymus of both lines.

during development to the SP CD8 stage [34]. In line with the signal strength and kinetic models, strength and duration of TCR signaling are major determining factors for the stage-specific differential activation of the ThPOK gene in class I- versus class II-restricted CD4+CD8lo thymocytes. Thus, forced activation of TCR signaling induced expression of ThPOK mRNA even in class I-restricted CD4+CD8lo T cells in vivo [34]. It is interesting to note that sensitivity of the ThPOK locus to TCR signaling is stage dependent. In vivo and in vitro TCR activation fails to induce ThPOK expression in DP or in any pre-DP stage thymocytes, but only in CD4+8lo cells. There is evidence to suggest that inaccessibility of the ThPOK locus to TCR signaling in immature thymocytes may be epigenetically regulated. Thus, in DP thymocytes, the ThPOK locus is marked by both activating (H3K4me3) and repressive (H3K27me3) marks, consistent with an epigenetically "poised" state [35]. Thus various transcriptional and epigenetic mechanisms may synergize to regulate ThPOK gene expression during the DP to SP transition.

## 5.9    KEY ROLE FOR ThPOK SILENCER

Given that ThPOK functions as the master regulator of CD4 lineage commitment and is regulated primarily at the level of transcription, it became critical to elucidate the molecular mechanism of its transcriptional regulation. A 20kb region surrounding the ThPOK locus and bounded by CTCF insulators is sufficient to reconstitute normal lineage commitment when introduced as a transgene into HD mice [36]. Further investigation of this region identified six DNase hypersensitivity (DHS) sites, including two promoters and four other putative cis regulatory elements [36], which collectively seem necessary and sufficient for the stage-specific regulation of ThPOK expression during thymopoiesis. Of the six DHS sites, a 400 bp silencer element at DHS site A (Sil$^{ThPOK}$) gained prominence, as its deletion from a ThPOK-GFP reporter transgene abolished lineage-specific regulation, leading to promiscuous expression in both CD4 and CD8 T cells. Furthermore, knockout mice lacking the Sil$^{ThPOK}$ exhibit complete derepression of ThPOK in class I-restricted thymocytes, resulting in severe reduction in the number of CD8 T cells due to redirection of class I-restricted thymocytes to the CD4 lineage [36]. Interestingly, conditional deletion of the Sil$^{ThPOK}$ in mature CD8 T cells does not derepress ThPOK expression, indicating that CD8 lineage commitment entails permanent epigenetic silencing of the locus [34]. Collectively these results indicate that Sil$^{ThPOK}$ function is controlled in part by stage- and lineage-specific chromatin modifications and by sequential recruitment of DNA binding proteins. The precise transcription factors that control Sil$^{ThPOK}$ function remain incompletely characterized. One important factor is Runx3, which has previously been shown to be essential for CD8 differentiation [35,36]. The Sil$^{ThPOK}$ region has two putative Runx binding sites, and deletion of these motifs impairs silencer activity [36]. Interestingly, chromatin immunoprecipitation (ChIP) assays revealed that Runx complexes are bound to Sil$^{ThPOK}$ at all stages of thymic T cell development, indicating that Runx binding is not sufficient to account for stage-specific regulation of the Sil$^{ThPOK}$. Hence, the molecular basis by which differential TCR signals specifically modulate activity of the Sil$^{ThPOK}$ has yet to be established.

## 5.10 OPPOSING EFFECTS OF RUNX AND ThPOK ON CD4/CD8 LINEAGE SPECIFICATION

Several lines of evidence support that ThPOK functions as a master transcriptional regulator of CD4 lineage specification and as an antagonist of Runx3. ThPOK has been implicated in regulation of multiple genes involved in CD4 lineage specification, including CD4 and ThPOK, at least in some cases via direct binding to the target locus. Direct binding of ThPOK to CD4 and ThPOK silencers antagonizes their function and initiates positive-feedback loops that support CD4 surface expression and increased transcription of ThPOK itself. Induced expression of ThPOK in DN and CD4$^+$CD8$^{lo}$ thymocytes mediates derepression of CD4 expression, and in peripheral CD8 T cells causes activation of several CD4 lineage-restricted genes including GATA-3 [37–38]. Further, ectopic ThPOK expression leads to severe functional defects by peripheral CD8 T cells, due to downmodulation of many CD8 specific genes including CD8, Perforin, Granzyme B, and Eomes [39]. ThPOK has also been shown to bind to its own silencer and antagonize its function, indicative of a positive feedback regulatory loop [38]. Diminished ThPOK expression in mature CD4 T cells, on the other hand, results in downregulation of CD4 lineage-specific genes including CD4, upregulation of cytotoxic lineage genes such as Runx3 and Eomes, acquisition of high IFN-γ production capacity and transdifferentiation into CD8 T cells upon transfer into T cell deficient mice [38]. In vitro cotransfection assay showed that ThPOK antagonizes Runx-mediated repression of a reporter gene controlled by the CD4 silencer [37–39]. Furthermore, ThPOK represses Runx3 distal promoter activity during the differentiation of class II-restricted cells, although it is not clear whether the effect is direct [40–42]. Interestingly, the antagonistic effect of ThPOK on Runx-mediated silencing can be effectively ablated by the histone deacetylase inhibitor TsA, suggesting that antagonism depends on transcriptional repression by ThPOK of an unknown factor that acts in conjunction with Runx, a repressor-of-repressor model [43]. In summary, current data indicate that ThPOK and Runx3 antagonize each others' functions by supporting CD4 and CD8 lineage-specific gene expression programs, respectively. Regulatory loops initiated by ThPOK explain in part how the initial lineage specification signals that induce limited expression of ThPOK at the CD4$^+$CD8$^{lo}$ stage are amplified to drive full CD4 commitment.

## 5.11 ROLE OF ThPOK IN iNKT DEVELOPMENT AND FUNCTION

Recent reports indicate that iNKT cells, a distinct subset of T cells with unique antigen recognition and functional attributes, also depend on ThPOK for their differentiation. iNKT cells, like conventional T cells, develop from DP precursors in the thymus. However, while conventional CD4 T cells recognize a vast diversity of peptide antigens in the context of MHC class II, iNKT cells recognize a limited set of glycolipid antigens in the context of CD1d. iNKT cells express a semi-invariant TCR, consisting of Vα14Jα18 (mice) or Vα24Jα18 (humans), combined with Vβ8.2, Vβ7, or Vβ2 (mice), or Vβ11 (humans). An important hallmark of Vα14i NKT cells is their rapid response to cognate antigens, resulting in production of large amounts of Th1 and Th2 cytokines within minutes of antigen exposure in vivo, without prior priming. Because of this property, Vα14i NKT cells have been likened to innate immune cells or natural memory cell populations. iNKT cells play a powerful immunoregulatory role in a wide range of immune responses. Recently, a novel population of Vα14i NKT cells has been found that is distinguished by its exclusive production of Th17 cytokines, including IL-17 and IL-22. Such NKT17 cells originate as a distinct subset in thymus and comprise <10% of the CD1d tetramer+ Vα14i NKT cells in the liver and spleen, but a major proportion in the lymph nodes and skin. There is evidence to suggest that NKT17 cells act to promote collagen-induced arthritis, and that they contribute to airway hyperreactivity. A large proportion of mature iNKT thymocytes express CD4, suggesting the possibility that they are dependent on ThPOK for their normal maturation and function. Consistent with this notion, ThPOK expression can be detected in iNKT cells, although somewhat surprisingly in both DN and CD4+ subsets. In ThPOK-deficient mice, the SP CD4 iNKT cells are essentially all converted into a novel SP CD8 iNKT population. The most likely explanation for the appearance of CD8+ iNKT cells in ThPOK-deficient mice is that they arise through "redirection" of precursors that would normally adopt the CD4+ phenotype, although this remains to be directly demonstrated. In this context genetic evidence suggests that Gata3 is required for iNKT cells to express *ThPOK* [44]. In ThPOK-deficient mice it appears that CD8+ development is the default pathway for the subset of iNKT cells that would normally express CD4, while DN cells do not gain CD8 expression. Interestingly, ThPOK deficiency causes functional defects in both CD8 and DN Vα14*i* NKT cells, indicating that ThPOK is required for adoption of the iNKT program of gene expression by all iNKT cells. Whereas wt Vα14*i* NKT cells are highly responsive to stimulation with αGalCer [45], ThPOK deficiency results in hyporesponsiveness of both CD8 and DN Vα14*i* NKT cells [46]. This effect is not due to loss of CD4 surface expression, as Vα14*i* NKT cells from CD4-deficient mice show normal function [47], nor to gain of CD8 expression, as targeted deletion of the CD8α does not restore responsiveness of ThPOK-deficient Vα14*i* NKT [48]. RNA microarray analysis revealed that ThPOK-deficient iNKT cells increased expression of 253 genes, representing genes that are physiologically repressed, either directly or indirectly, by ThPOK in Vα14*i* NKT cells. Interestingly, *ThPOK*-deficient Vα14*i* NKT cells overexpressed the transcription factors RORγT, Ahr, Klf4, and Maf, which are all expressed by Th17 cells, as well as other markers of the NKT17 and Th17 subsets, including CD103, CD196, CD121a, Nrp-1, and IL-17RE. Even a twofold reduction in *ThPOK* gene dose is sufficient to partly alter the phenotype and function of Vα14*i* NKT cells toward the NKT17 program. Whereas ThPOK deficiency, therefore, supports NKT17 development, overexpression of ThPOK antagonizes NKT17 differentiation in part by inhibiting RORgT expression [49]. Hence, while ThPOK is not necessary for development of iNKT cells per se, the precise levels of ThPOK expression are critical for controlling iNKT subset specification and functional attributes.

## 5.12  ROLE OF ThPOK IN γδ T CELL DEVELOPMENT

An additional role for ThPOK has recently been revealed in development of γδ T cells, specifically in controlling development of the IFNγ-producing subset. Similar to CD4 commitment, it has been proposed that TCR signaling plays a critical role in the development of some γδ T cell subsets. Thus, thymocytes expressing a pre-arranged transgenic γδ TCR of defined specificity undergo maturation only in the presence of their specific intrathymic ligand, whereas in its absence they undergo alternative development to the αβ lineage [50–51]. TCR signal strength influences γδ development in at least two respects: (1) strong TCR signals seem to promote com-mitment to the γδ lineage at the expense of αβ commitment [52], and (2) within the γδ lineage, differentiation into specialized effector subsets seems to be controlled by differences in TCR signal strength, such that strong signals promote development of IFNγ-producing γδ cells [53]. ThPOK-GFP reporter mice have allowed detailed analysis of ThPOK expression at the single-cell level in the DN compartment. Surprisingly, GFP expression was detected in a significant fraction of γδ thymocytes, and in a developmentally regulated manner. Thus, only 10% to 15% of immature γδ TCR+ thymocytes express low GFP reporter levels, whereas most mature γδ thy-mocytes express high GFP levels, suggesting that the GFPlo immature cells are the major precursors of mature γδ thymocytes and that ThPOK might be an important regulator of γδ thymocyte maturation [54]. Indeed, ThPOK-deficient mice show a substantial reduction (50%–70%) in the absolute number of mature γδ thymocytes, demonstrating a key role for ThPOK in γδ thymocyte commitment/maturation and in proliferation/survival of mature γδ thymocytes. Mature γδ thymocytes in adult mice can be further subdivided into two major subsets based on surface marker and cytokine expression pattern. One subset predominantly expresses the Vγ1.1 TCR and the NK1.1 surface marker, and induces IFNγ upon stimulation (so-called NKT γδ cells) [55–56]; while the other preferentially expresses the Vγ2 TCR and the CCR6 surface marker, and preferentially secretes IL-17 upon stimulation [57–58]. The upstream signals and transcriptional pathways that promote alternate development to these two functionally distinct γδ subsets are poorly understood, although it has been reported that antigen-experienced γδ cells develop preferentially into the IFNγ-producing subset, suggesting that they depend on a TCR signal [50]. Comparison of NK1.1+ and CCR6+γδ thymocytes from ThPOK-GFP reporter mice shows that the former expresses higher levels of GFP, suggesting an important role for ThPOK in their development. Indeed, analysis of adult ThPOK-deficient mice showed a severe (four- to fivefold) reduction in absolute numbers of NK1.1+γδ thymocytes, but only a mild (<twofold) reduction of CCR6+ cells. In mice expressing a ThPOK-GFP reporter, GFP+ immature (DN3 and DN4) γδ cells already exhibit 40% Vγ 1.1 usage, suggesting that ThPOK expression marks these cells for maturation to the CD24⁻ stage. It is interesting that mice lacking the helix-loop-helix (HLH) tran-scriptional regulator Id3 exhibit a massive selective increase in the proportion and absolute number of Vγ1.1+ cells, similar to mice constitutively expressing ThPOK, suggesting a mechanistic link [59–60].

## 5.13 ROLE OF ThPOK IN REGULATORY T CELL DEVELOPMENT AND FUNCTION

During T cell development in the thymus, most immature T cells that express TCRs with high affinity for self-antigens are deleted during negative selection, but a few are able to escape by upmodulating the FoxP3 transcription factor and developing to the regulatory T (Treg) cell lineage. Treg cells comprise between 5% and 10% of peripheral CD4$^+$ T cells and play a critical role in the maintenance of peripheral tolerance by suppressing immune responses to self-antigens. FoxP3 is required for the development, maintenance, and function of Treg cells. Recent studies show that ThPOK can significantly influence Treg cell development and function. Thus, ThPOK transgenic mice show enhanced development of Treg cells, which express higher levels of CD25, and higher levels of IL-2 upon stimulation. As the development of CD4$^+$ FoxP3$^+$ Treg cells is dependent on IL-2, the ThPOK transgene may promote more efficient Treg development through increased IL-2 secretion. Treg cells from ThPOK transgenic mice produced less TGF-$\beta$ mRNA but more protein upon activation, resulting from more efficient translation of the TGF-$\beta$ transcript in these cells [61]. Indeed, Treg cells from ThPOK transgenic mice show higher suppressor activity when compared with Treg cells from WT controls. Finally, it was recently shown that ThPOK is responsible for physiological maintenance of Treg function. Upon migration to the intestinal epithelium, Tregs lose Foxp3 and convert to CD4IELs in a microbiota-dependent manner, an effect attributed to the loss of ThPOK [62].

## 5.14   ThPOK AND LYMPHOMAGENESIS

The POK family of transcription factors, to which ThPOK belongs, includes several factors that have been implicated in diverse hematological malignancies, including PLZF in AML [63], Bcl6 in B-cell lymphoma [64–65], and LRF/Pokemon in T-cell lymphoma and lung cancer [66–69]. Consequently, the Kappes group has explored whether ThPOK might exhibit similar oncogenic properties. Indeed, they recently reported that constitutive T cell-specific expression of ThPOK in transgenic mice (ThPOK$^{const}$ mice) leads to aggressive and highly penetrant thymic lymphoma, implicating ThPOK as a potent oncogene [70]. Phenotypically, ThPOK$^{const}$ lymphoma cells are immature in origin, resembling human T-cell acute lymphoblastic leukemia (T-ALL). Interestingly, 40% of immature human T-ALL lines express elevated ThPOK levels. The fact that ThPOK is normally silenced in immature DN and DP thymocyte stages in humans, as in mice, suggests that aberrant induction of ThPOK at early stages of thymopoiesis could be an important driver of human T-ALL initiation/development. Gene expression microarray analysis data revealed a close resemblance between ThPOK$^{const}$ tumors and thymic lymphomas from mice carrying a dominant negative mutation of Ikaros, suggesting that ThPOK gain-of-function and Ikaros loss-of-function mutants may in part affect the same intracellular pathways.

Genetic approaches precisely delineated the developmental requirements for ThPOK-mediated transformation, demonstrating that only immature thymocytes between the DN3 stage and DP stage are insensitive to ThPOK-mediated transformation. Lymphomagenesis is blocked in ThPOK$^{const}$ RAG−/− mice, but restored by a TCR transgene, indicating that TCR signaling is necessary for lymphomagenesis. It remains to be clarified whether this reflects a direct requirement for TCR signaling or for promoting development beyond the DN3 blockade caused by RAG deficiency. Strikingly, injection of a single dose of anti-CD3 antibody can induce lymphomagenesis in ThPOK$^{const}$ RAG–/– mice after 4 months. The long (4 months) time lag between antibody stimulation and lymphoma development indicates that a tumor progenitor population generated by antibody stimulation can survive and/ or propagate itself during this time, presumably allowing for accumulation of secondary mutations, like activating Notch1 mutations, that result in fully developed lymphomas. The tumor progenitor population appears to exhibit a DN4 phenotype, since 25,000 DN4 cells from 4-week-old ThPOK$^{const}$ mice can transfer aggressive lymphoma to adoptive hosts. Transferred cells partly maintain the DN4 phenotype, but also diverge to multiple, more mature subsets, suggesting that DN4 progenitor cells are capable of both self-renewal and differentiation. The underlying mechanism by which ThPOK promotes oncogenesis remains to be resolved, but may involve promoting immature stem cell-like properties.

## REFERENCES

1. Rothenberg, E. V., and Taghon, T. (2005). Molecular genetics of T cell development. *Annual Review of Immunology, 23*, 601–649.
2. Barthlott, T., Kohler, H., Pircher, H., and Eichmann, K. (1997). Differentiation of CD4(high)CD8(low) coreceptor-skewed thymocytes into mature CD8 single-positive cells independent of MHC class I recognition. *European Journal of Immunology, 27*(8), 2024–2032.
3. Chan, S., Correia-Neves, M., Dierich, A., Benoist, C., and Mathis, D. (1998). Visualization of CD4/CD8 T cell commitment. *The Journal of Experimental Medicine, 188*(12), 2321–2333.
4. Guidos, C. J., Danska, J. S., Fathman, C. G., and Weissman, I. L. (1990). T cell receptor-mediated negative selection of autoreactive T lymphocyte precursors occurs after commitment to the CD4 or CD8 lineages. *The Journal of Experimental Medicine, 172*(3), 835–845.
5. Lucas, B., and Germain, R. N. (1996). Unexpectedly complex regulation of CD4/CD8 coreceptor expression supports a revised model for CD4+ CD8+ thymocyte differentiation. *Immunity, 5*(5), 461–477.
6. Lundberg, K., Heath, W., Köntgen, F., Carbone, F. R., and Shortman, K. (1995). Intermediate steps in positive selection: Differentiation of CD4+ 8int TCRint thymocytes into CD4-8+ TCRhi thymocytes. *The Journal of Experimental Medicine, 181*(5), 1643–1651.
7. Suzuki, H., Punt, J. A., Granger, L. G., and Singer, A. (1995). Asymmetric signaling requirements for thymocyte commitment to the CD4+ versus CD8+ T cell lineages: A new perspective on thymic commitment and selection. *Immunity, 2*(4), 413–425.
8. Kydd, R., Lundberg, K., Vremec, D., Harris, A. W., and Shortman, K. (1995). Intermediate steps in thymic positive selection. Generation of CD4-8+ T cells in culture from CD4+ 8+, CD4int8+, and CD4+ 8int thymocytes with up-regulated levels of TCR-CD3. *The Journal of Immunology, 155*(8), 3806–3814.
9. Brugnera, E., Bhandoola, A., Cibotti, R., Yu, Q., Guinter, T. I., Yamashita, Y., … Singer, A. (2000). Coreceptor reversal in the thymus: Signaled CD4+ 8+ thymocytes initially terminate CD8 transcription even when differentiating into CD8+ T cells. *Immunity, 13*(1), 59–71.
10. Davis, C. B., Killeen, N., Crooks, M. C., Raulet, D., and Littman, D. R. (1993). Evidence for a stochastic mechanism in the differentiation of mature subsets of T lymphocytes. *Cell, 73*(2), 237–247.
11. Ravichandran, K. S., and Burakoff, S. J. (1994). Evidence for differential intracellular signaling via CD4 and CD8 molecules. *The Journal of Experimental Medicine, 179*(2), 727–732.

12. Veillette, A., Bookman, M. A., Horak, E. M., and Bolen, J. B. (1988). The CD4 and CD8 T cell surface antigens are associated with the internal membrane tyrosine-protein kinase p56 lck. *Cell*, *55*(2), 301-308.

13. Itano, A., Salmon, P., Kioussis, D., Tolaini, M., Corbella, P., and Robey, E. (1996). The cytoplasmic domain of CD4 promotes the development of CD4 lineage T cells. *The Journal of Experimental Medicine*, *183*(3), 731–741.

14. Hernández-Hoyos, G., Sohn, S. J., Rothenberg, E. V., and Alberola-Ila, J. (2000). Lck activity controls CD4/CD8 T cell lineage commitment. *Immunity*, *12*(3), 313–322.

15. Legname, G., Seddon, B., Lovatt, M., Tomlinson, P., Sarner, N., Tolaini, M., ... Zamoyska, R. (2000). Inducible expression of a p56 Lck transgene reveals a central role for Lck in the differentiation of CD4 SP thymocytes. *Immunity*, *12*(5), 537–546.

16. Negishi, I., Motoyama, N., Nakayama, K. I., Senju, S., Hatakeyama, S., Zhang, Q., ... Loh, D. Y. (1995). Essential role for ZAP-70 in both positive and negative selection of thymocytes. *Nature*, *376*(6539), 435–438.

17. Liu, X., and Bosselut, R. (2004). Duration of TCR signaling controls CD4-CD8 lineage differentiation in vivo. *Nature Immunology*, *5*(3), 280–288.

18. Kisielow, P., and Miazek, A. (1995). Positive selection of T cells: Rescue from pro-grammed cell death and differentiation require continual engagement of the T cell receptor. *The Journal of Experimental Medicine*, *181*(6), 1975–1984.

19. Liu, X., Adams, A., Wildt, K. F., Aronow, B., Feigenbaum, L., and Bosselut, R. (2003). Restricting Zap70 expression to CD4+ CD8+ thymocytes reveals a T cell receptor-dependent proofreading mechanism controlling the completion of positive selection. *The Journal of Experimental Medicine*, *197*(3), 363–373.

20. Saini, M., Sinclair, C., Marshall, D., Tolaini, M., Sakaguchi, S., and Seddon, B. (2010). Regulation of Zap70 expression during thymocyte development enables temporal sepa-ration of CD4 and CD8 repertoire selection at different signaling thresholds. *Science Signaling*, *3*(114), ra23–ra23.

21. Park, J. H., Adoro, S., Lucas, P. J., Sarafova, S. D., Alag, A. S., Doan, L. L., ... Feigenbaum, L. (2007). "Coreceptor tuning": Cytokine signals transcriptionally tai-lor CD8 coreceptor expression to the self-specificity of the TCR. *Nature Immunology*, *8*(10), 1049–1059.

22. Park, J. H., Adoro, S., Guinter, T., Erman, B., Alag, A. S., Catalfamo, M., ... Kubo, M. (2010). Signaling by intrathymic cytokines, not T cell antigen receptors, specifies CD8 lineage choice and promotes the differentiation of cytotoxic-lineage T cells. *Nature Immunology*, *11*(3), 257–264.

23. Setoguchi, R., Tachibana, M., Naoe, Y., Muroi, S., Akiyama, K., Tezuka, C., ... Taniuchi, I. (2008). Repression of the transcription factor Th-POK by Runx complexes in cytotoxic T cell development. *Science*, *319*(5864), 822–825.

24. Yu, Q., Park, J. H., Doan, L. L., Erman, B., Feigenbaum, L., and Singer, A. (2006). Cytokine signal transduction is suppressed in preselection double-positive thymocytes and restored by positive selection. *The Journal of Experimental Medicine*, *203*(1), 165–175.

25. McCaughtry, T. M., Etzensperger, R., Alag, A., Tai, X., Kurtulus, S., Park, J. H., ... Singer, A. (2012). Conditional deletion of cytokine receptor chains reveals that IL-7 and IL-15 specify CD8 cytotoxic lineage fate in the thymus. *The Journal of Experimental Medicine*, *209*(12), 2263–2276.

26. Bommhardt, U., Basson, M. A., Krummrei, U., and Zamoyska, R. (1999). Activation of the extracellular signal-related kinase/mitogen-activated protein kinase pathway discriminates CD4 versus CD8 lineage commitment in the thymus. *The Journal of Immunology*, *163*(2), 715–722.

27. Sharp, L. L., Schwarz, D. A., Bott, C. M., Marshall, C. J., and Hedrick, S. M. (1997). The influence of the MAPK pathway on T cell lineage commitment. *Immunity*, 7(5), 609–618.
28. Hedrick, S. M. (2002). T cell development: Bottoms-up. *Immunity*, 16(5), 619–622.
29. Dave, V. P., Allman, D., Keefe, R., Hardy, R. R., and Kappes, D. J. (1998). HD mice: A novel mouse mutant with a specific defect in the generation of CD4+ T cells. *Proceedings of the National Academy of Sciences*, 95(14), 8187–8192.
30. Keefe, R., Dave, V., Allman, D., Wiest, D., and Kappes, D. J. (1999). Regulation of lineage commitment distinct from positive selection. *Science*, 286(5442), 1149–1153.
31. He, X., He, X., Dave, V. P., Zhang, Y., Hua, X., Nicolas, E., ... Kappes, D. J. (2005). The zinc finger transcription factor Th-POK regulates CD4 versus CD8 T-cell lineage commitment. *Nature*, 433(7028), 826–833.
32. Kappes, D. J. (2010). Expanding roles for ThPOK in thymic development. *Immunological Reviews*, 238(1), 182–194.
33. Kappes, D. J., He, X., and He, X. (2006). Role of the transcription factor Th-POK in CD4: CD8 lineage commitment. *Immunological Reviews*, 209(1), 237–252.
34. Sun, G., Liu, X., Mercado, P., Jenkinson, S. R., Kypriotou, M., Feigenbaum, L., ... Bosselut, R. (2005). The zinc finger protein cKrox directs CD4 lineage differentiation during intrathymic T cell positive selection. *Nature Immunology*, 6(4), 373–381.
35. Tanaka, H., Naito, T., Muroi, S., Sco, W., Chihara, R., Miyamoto, C., ... Taniuchi, I. (2013). Epigenetic Thpok silencing limits the time window to choose CD4+ helper-lineage fate in the thymus. *The EMBO Journal*, 32(8), 1183–1194.
36. He, X., Park, K., Wang, H., Hc, X., Zhang, Y., Hua, X., ... Kappes, D. J. (2008). CD4-CD8 lineage commitment is regulated by a silencer element at the ThPOK transcription-factor locus. *Immunity*, 28(3), 346–358.
37. Wildt, K. F., Sun, G., Grueter, B., Fischer, M., Zamisch, M., Ehlers, M., and Bosselut, R. (2007). The transcription factor Zbtb7b promotes CD4 expression by antagonizing Runx-mediated activation of the CD4 silencer. *The Journal of Immunology*, 179(7), 4405–4414.
38. Wang, L., Wildt, K. F., Zhu, J., Zhang, X., Feigenbaum, L., Tessarollo, L., ... Bosselut, R. (2008). Distinct functions for the transcription factors GATA-3 and ThPOK during intrathymic differentiation of CD4+ T cells. *Nature Immunology*, 9(10), 1122–1130.
39. Liu, X., Taylor, B. J., Sun, G., and Bosselut, R. (2005). Analyzing expression of perforin, Runx3, and Thpok genes during positive selection reveals activation of CD8-differentiation programs by MHC II-signaled thymocytes. *The Journal of Immunology*, 175(7), 4465–4474.
40. Cruz-Guilloty, F., Pipkin, M. E., Djuretic, I. M., Levanon, D., Lotem, J., Lichtenheld, M. G., ... Rao, A. (2009). Runx3 and T-box proteins cooperate to establish the transcriptional program of effector CTLs. *The Journal of Experimental Medicine*, 206(1), 51–59.
41. Jenkinson, S. R., Intlekofer, A. M., Sun, G., Feigenbaum, L., Reiner, S. L., and Bosselut, R. (2007). Expression of the transcription factor cKrox in peripheral CD8 T cells reveals substantial postthymic plasticity in CD4-CD8 lineage differentiation. *The Journal of Experimental Medicine*, 204(2), 267–272.
42. Egawa, T., and Littman, D. R. (2008). ThPOK acts late in specification of the helper T cell lineage and suppresses Runx-mediated commitment to the cytotoxic T cell lineage. *Nature Immunology*, 9(10), 1131–1139.
43. Rui, J., Liu, H., Zhu, X., Cui, Y., and Liu, X. (2012). Epigenetic silencing of CD8 genes by ThPOK-mediated deacetylation during CD4 T cell differentiation. *The Journal of Immunology*, 189(3), 1380–1390.

44. Wang, L., Carr, T., Xiong, Y., Wildt, K. F., Zhu, J., Feigenbaum, L., ... Bosselut, R. (2010). The sequential activity of Gata3 and Thpok is required for the differentiation of CD1d-restricted CD4+ NKT cells. *European Journal of Immunology*, *40*(9), 2385–2390.

45. Kronenberg, M., and Gapin, L. (2002). The unconventional lifestyle of NKT cells. *Nature Reviews Immunology*, *2*(8), 557–568.

46. Engel, I., Hammond, K., Sullivan, B. A., He, X., Taniuchi, I., Kappes, D., and Kronenberg, M. (2010). Co-receptor choice by Vα14i NKT cells is driven by Th-POK expression rather than avoidance of CD8-mediated negative selection. *The Journal of Experimental Medicine*, *207*(5), 1015–1029.

47. Bendelac, A., Killeen, N., Littman, D. R., and Schwartz, R. H. (1994). A subset of CD4+ thymocytes selected by MHC class I molecules. *Science*, *263*(5154), 1774–1778.

48. Engel, I., and Kronenberg, M. (2012). Making memory at birth: Understanding the differentiation of natural killer T cells. *Current Opinion in Immunology*, *24*(2), 184–190.

49. Engel, I., Zhao, M., Kappes, D., Taniuchi, I., and Kronenberg, M. (2012). The transcription factor Th-POK negatively regulates Th17 differentiation in Vα14i NKT cells. *Blood*, *120*(23), 4524–4532.

50. Pereira, P., Zijlstra, M., McMaster, J., Loring, J. M., Jaenisch, R., and Tonegawa, S. (1992). Blockade of transgenic gamma delta T cell development in beta 2-microglobulin deficient mice. *The EMBO Journal*, *11*(1), 25.

51. Haks, M. C., Lefebvre, J. M., Lauritsen, J. P. H., Carleton, M., Rhodes, M., Miyazaki, T., ... Wiest, D. L. (2005). Attenuation of γδTCR signaling efficiently diverts thymocytes to the αβ lineage. *Immunity*, *22*(5), 595–606.

52. Lauritsen, J. P. H., Haks, M. C., Lefebvre, J. M., Kappes, D. J., and Wiest, D. L. (2006). Recent insights into the signals that control αβ/γδ-lineage fate. *Immunological Reviews*, *209*(1), 176–190.

53. Jensen, K. D., Su, X., Shin, S., Li, L., Youssef, S., Yamasaki, S., ... Baumgarth, N. (2008). Thymic selection determines γδ T cell effector fate: Antigen-naive cells make interleukin-17 and antigen-experienced cells make interferon γ. *Immunity*, *29*(1), 90–100.

54. Park, K., He, X., Lee, H. O., Hua, X., Li, Y., Wiest, D., and Kappes, D. J. (2010). TCR-mediated ThPOK induction promotes development of mature (CD24⁻) γδ thymocytes. *The EMBO Journal*, *29*(14), 2329–2341.

55. Lees, R. K., Ferrero, I., and MacDonald, H. R. (2001). Tissue-specific segregation of TCRγδ+ NKT cells according to phenotype TCR repertoire and activation status: Parallels with TCR α β+ NKT cells. *European Journal of Immunology*, *31*(10), 2901–2909.

56. Brennan, P. J., Brigl, M., and Brenner, M. B. (2013). Invariant natural killer T cells: An innate activation scheme linked to diverse effector functions. *Nature Reviews Immunology*, *13*(2), 101–117.

57. Haas, J. D., González, F. H. M., Schmitz, S., Chennupati, V., Föhse, L., Kremmer, E., ... Prinz, I. (2009). CCR6 and NK1.1 distinguish between IL-17A and IFN-γ-producing γδ effector T cells. *European Journal of Immunology*, *39*(12), 3488–3497.

58. Do, J. S., Fink, P. J., Li, L., Spolski, R., Robinson, J., Leonard, W. J., ... Min, B. (2010). Cutting edge: Spontaneous development of IL-17-producing γδ T cells in the thymus occurs via a TGF-β1-dependent mechanism. *The Journal of Immunology*, *184*(4), 1675–1679.

59. Ueda-Hayakawa, I., Mahlios, J., and Zhuang, Y. (2009). Id3 restricts the developmental potential of γδ lineage during thymopoiesis. *The Journal of Immunology*, *182*(9), 5306–5316.

60. Lauritsen, J. P. H., Wong, G. W., Lee, S. Y., Lefebvre, J. M., Ciofani, M., Rhodes, M., ... Wiest, D. L. (2009). Marked induction of the helix-loop-helix protein Id3 promotes the γδ T cell fate and renders their functional maturation Notch independent. *Immunity, 31*(4), 565–575.

61. Twu, Y. C., and Teh, H. S. (2014). The ThPOK transcription factor differentially affects the development and function of self-specific CD8+ T cells and regulatory CD4+ T cells. *Immunology, 141*(3), 431–445.

62. Sujino, T., London, M., Hoytema van Konijnenburg, D. P., Rendon, T., Buch, T., Silva, H. M., ... Mucida, D. (2016). Tissue adaptation of regulatory and intraepithelial CD4+ T cells controls gut inflammation. *Science, 352*(6293), 1581–1586.

63. Melnick, A., Carlile, G. W., McConnell, M. J., Polinger, A., Hiebert, S. W., and Licht, J. D. (2000). AML-1/ETO fusion protein is a dominant negative inhibitor of transcriptional repression by the promyelocytic leukemia zinc finger protein. *Blood, 96*(12), 3939–3947.

64. Kramer, M. H. H., Hermans, J., Wijburg, E., Philippo, K., Geelen, E., Van Krieken, J. H. J. M., ... Kluin, P. M. (1998). Clinical relevance of BCL2, BCL6, and MYC rearrangements in diffuse large B-cell lymphoma. *Blood, 92*(9), 3152–3162.

65. Ci, W., Polo, J. M., and Melnick, A. (2008). B-cell lymphoma 6 and the molecular pathogenesis of diffuse large B-cell lymphoma. *Current Opinion in Hematology, 15*(4), 381.

66. Maeda, T., Hobbs, R. M., Merghoub, T., Guernah, I., Zelent, A., Cordon-Cardo, C., ... Pandolfi, P. P. (2005). Role of the proto-oncogene Pokemon in cellular transformation and ARF repression. *Nature, 433*(7023), 278–285.

67. Bohn, O., Maeda, T., Filatov, A., Lunardi, A., Pandolfi, P. P., and Teruya-Feldstein, J. (2013). Utility of LRF/Pokemon and NOTCH1 protein expression in the distinction of nodular lymphocyte-predominant Hodgkin lymphoma and classical Hodgkin lymphoma. *International Journal of Surgical Pathology, 22*(1), 6–11.

68. Apostolopoulou, K., Pateras, I. S., Evangelou, K., Tsantoulis, P. K., Liontos, M., Kittas, C., ... Gorgoulis, V. G. (2007). Gene amplification is a relatively frequent event leading to ZBTB7A (Pokemon) overexpression in non-small cell lung cancer. *The Journal of Pathology, 213*(3), 294–302.

69. Liu, X. S., Liu, Z., Gerarduzzi, C., Choi, D. E., Ganapathy, S., Pandolfi, P. P., and Yuan, Z. M. (2016). Somatic human ZBTB7A zinc finger mutations promote cancer progression. *Oncogene, 35*(23), 3071–3078.

70. Lee, H. O., He, X., Mookerjee-Basu, J., Zhongping, D., Hua, X., Nicolas, E., ... Kappes, D. J. (2015). Disregulated expression of the transcription factor ThPOK during T-cell development leads to high incidence of T-cell lymphomas. *Proceedings of the National Academy of Sciences, 112*(25), 7773–7778.

56. Takaba, H., Morishita, Y., Tomofuji, Y., Danks, L., Nitta, T., Komatsu, N., Kodama, T., and Takayanagi, H. (2015). Fezf2 Orchestrates a Thymic Program of Self-Antigen Expression for Immune Tolerance. Cell 163, 975–987.

57. Passos, G. A., Mendes-da-Cruz, D. A., and Oliveira, E. H. (2015). The Thymic Orchestration Involving Aire, miRNAs, and Cell–Cell Interactions during the Induction of Central Tolerance. Front. Immunol. 6, 352.

58. Sansom, S. N., Shikama-Dorn, N., Zhanybekova, S., Nusspaumer, G., Macaulay, I. C., Deadman, M. E., Heger, A., Ponting, C. P., and Holländer, G. A. (2014). Population and single-cell genomics reveal the Aire dependency, relief from Polycomb silencing, and distribution of self-antigen expression in thymic epithelia. Genome Res. 24, 1918–1931.

59. Meredith, M., Zemmour, D., Mathis, D., and Benoist, C. (2015). Aire controls gene expression in the thymic epithelium with ordered stochasticity. Nat. Immunol. 16, 942–949.

60. Matsumoto, M., Nishikawa, Y., Nishijima, H., Morimoto, J., Matsumoto, M., and Mouri, Y. (2013). Which model better fits the role of aire in the establishment of self-tolerance: the transcription model or the maturation model? Front. Immunol. 4, 210.

61. Anderson, M. S., and Su, M. A. (2016). AIRE expands: new roles in immune tolerance and beyond. Nat. Rev. Immunol. 16, 247–258.

62. Abramson, J., and Goldfarb, Y. (2016). AIRE: From promiscuous molecular partnerships to promiscuous gene expression. Eur. J. Immunol. 46, 22–33.

63. Perniola, R. (2018). Twenty Years of AIRE. Front. Immunol. 9, 98.

64. Gardner, J. M., DeVoss, J. J., Friedman, R. S., Wong, D. J., Tan, Y. X., Zhou, X., Johannes, K. P., Su, M. A., Chang, H. Y., Krummel, M. F., et al. (2008). Deletional tolerance mediated by extrathymic Aire-expressing cells. Science 321, 843–847.

65. Yang, S., Fujikado, N., Kolodin, D., Benoist, C., and Mathis, D. (2015). Immune tolerance. Regulatory T cells generated early in life play a distinct role in maintaining self-tolerance. Science 348, 589–594.

# 6 TCR Signaling Circuits in αβ/γδ T Lineage Choice

*Shawn P. Fahl, Dietmar J. Kappes,*
*and David L. Wiest*

## CONTENTS

## ABSTRACT

Both of the major T lineages, αβ and γδ, arise from a common CD4⁻CD8⁻ double negative (DN) progenitor in the thymus. Although the molecular processes responsible for specification of these alternate cell fates remain incompletely understood, recent efforts have begun to shed light on the influence of the T cell antigen receptor (TCR) complex and the molecular effectors it employs. This chapter will summarize our current understanding of how distinct TCR signals drive an uncommitted progenitor to adopt the alternate αβ or γδ lineages, as well as how those TCR signals influence γδ effector fates. Particular attention is paid to the emerging role of the extracellular signal-regulated kinase (ERK)–early growth response (Egr)–inhibitor of DNA binding 3 (Id3) pathway, and how its influence on E protein function impacts fate specification. Outstanding questions and the advances in technology that can be used to address them will be highlighted.

T lymphocytes comprise two distinct lineages defined by the T cell antigen receptor (TCR) complex they express, αβ or γδ, and perform nonoverlapping roles in immune responses. αβ T cells are found primarily in secondary lymphoid organs, recognize peptide ligands in a major histocompatibility complex (MHC) restricted fashion, and respond to pathogen encounters by facilitating the production of antibodies or by lysing infected target cells. While γδ T cells comprise a much smaller percentage of lymphoid cells in secondary lymphoid tissues, they comprise a large proportion of the lymphocytes at epithelial surfaces that line the inside and outside of the body.[1,2] γδ T cells are not MHC restricted and instead recognize a wider variety of intact antigens including nonclassical MHC molecules, heat shock proteins, and lipids.[3,4] Although γδ T cell function remains less well understood than that of their αβ lineage counterparts, recent efforts have led to a much better understanding of their diverse roles in immune responses that appear to lie at the interface between innate and adaptive immunity.[5] γδ T cells have been found to play critical roles in the resistance to a number of bacterial and viral infections.[6–8]

## 6.1  TIMING OF LINEAGE SEPARATION

Both αβ and γδ lineage T cells arise from a common pool of CD4⁻CD8⁻ double negative (DN) progenitor thymocytes,[9,10] which can be further subdivided based on expression of CD44 and CD25. The developmental stage at which γδ lineage T cells diverge from αβ T cells has not been precisely defined, but recent evidence suggests it is complete upon arrival at the CD44⁻CD25⁺ (DN3) stage when β-selection occurs.[9–13] Accordingly, lineage separation occurs during the developmental transition between the CD44⁺CD25⁺ (DN2) stage and DN3 stage.[11] It is during this transition that rearrangement of the TCR loci begins (TCRγ, δ, and β).[14–16] Those precursors with productive TCRγ and δ rearrangements are eligible to become γδ T cells, which usually remain DN and exit the thymus to populate epithelial surfaces and, to a lesser extent, lymphoid organs.[17] Generation of γδ T cells in the mouse is more pronounced during fetal life and occurs in waves of cells expressing particular sets of Vγ and Vδ genes.[18–20] Precursors with productive TCRβ rearrangements express the TCRβ protein in association with pre-Tα and the CD3 signaling subunits (CD3γ,δ,ε, and TCR ζ) to form the pre-T cell receptor (pre-TCR) complex.[21,22] Pre-TCR assembly initiates a poorly understood ligand-independent signaling process that promotes adoption of the αβ fate and differentiation to the CD4⁺CD8⁺ (DP) stage, a differentiation step that γδ lineage thymocytes do not typically undergo.[23–28] Thus, while development of γδ and αβ lineage T cells clearly requires signals transduced by the γδTCR and pre-TCR complexes, respectively, the role of TCR signaling in orchestrating lineage separation remains controversial.

## 6.2  MARKERS OF LINEAGE COMMITMENT

One of the major challenges in gaining insight into the molecular processes orchestrating commitment of DN progenitors to the αβ or γδ lineage lies in distinguishing those progenitors that have made a fate decision from those that remain bipotential. The TCR complexes (i.e., pre-TCR for αβ lineage and γδTCR for γδ lineage) alone are no longer deemed reliable markers of the ultimate fate of developing DN thymocytes, because both pre-TCR and γδTCR signaling are able to support adoption of the αβ fate and development of progenitors to the CD4⁺CD8⁺ DP stage.[29] Accordingly, CD4 and CD8 expression must also be taken into consideration in assigning lineage, such that γδTCR-expressing cells that remain DN are assigned to the γδ lineage, whereas those developing to the DP stage in response to TCR signals from any receptor isotype are assigned to the αβ lineage.[28] Importantly, a progenitor that has adopted the αβ fate and differentiated to the DP stage in response to γδTCR signals downregulates the γδTCR by a combination of both transcriptional and posttranscriptional mechanisms.[30-32] Thus, assessing γδTCR expression on DP will not reveal whether αβ lineage commitment occurred in response to γδTCR signaling. We and others have utilized downregulation of CD24 (HSA) among γδTCR-expressing DN thymocytes as an additional marker of γδ commitment,[33-35] but this too has been questioned as many of the γδTCR⁺ cells exiting the thymus express CD24.[36-38] However, although some newly exported γδ cells (recent thymic emigrants, RTE) remain CD24⁺, they appear to make a minimal contribution to the long-lived peripheral γδ pool.[36-38] Recent efforts have identified CD73-induction as marking those γδTCR-expressing DN that have irreversibly committed to the γδ fate. Indeed, while CD73⁻ γδTCR-expressing DN remain bipotential and adopt the αβ fate upon separation from the selecting milieu, those expressing CD73 remain committed to the γδ fate.[32] The identification of CD73 as a marker of γδ lineage commitment should enable future efforts to focus on the molecular processes controlling lineage commitment.

## 6.3 ROLE OF THE TCR COMPLEX IN αβ/γδ LINEAGE COMMITMENT

The prevailing view is that αβ and γδ T cells arise from a common progenitor; however, the respective roles of the pre-TCR and γδ TCR complexes in specification of the αβ and γδ fates remain controversial.[9,11,13] Central to this controversy is whether the TCR complexes act to specify fate (instruction) or merely rescue the viability of cells whose fate was decided without input from the TCR (stochastic).[39,40] As originally proposed, neither of these models is consistent with the asymmetric depletion of in-frame TCR gene rearrangements in αβ and γδ lineage cells or with the lineage infidelity observed in TCR transgenic (Tg) and gene-targeted mice.[13,14,41-50] Specifically, in-frame TCRγ rearrangements are depleted from αβ lineage cells, but in-frame TCRβ rearrangements are not depleted from γδ T cells. Moreover, when pre-TCR expression is prevented by ablation of the pre-Tα subunit, the γδ TCR complex is able to support adoption of the αβ fate.[29] Thus, when instruction and stochastic models are linked to the TCR isotype, these models are inconsistent with the aforementioned evidence.

While pre-TCR signaling and γδTCR signaling are most often responsible for development of αβ and γδ lineage cells, respectively, the situations where lineage and TCR isotype are uncoupled provide clues to the nature of the differences in TCR signaling involved in these alternate fate choices and led to the proposal of the TCR signal strength model for αβ/γδ lineage commitment (Figure 6.1).[51] The TCR signal strength model posits that weak signals promote commitment to the αβ lineage, whereas comparatively strong signals promote commitment to the γδ lineage, irrespective of the TCR complex from which they originate. In support of this model, a single γδTCR (KN6) Tg was found capable of promoting either γδ or αβ lineage commitment depending on the nature of the TCR signal transduced. Indeed, adoption of the γδ fate was induced by strong TCR signals resulting from γδTCR-ligand engagement, whereas weakening those TCR signals, by removing either ligand (H-2T10ᵈ) or

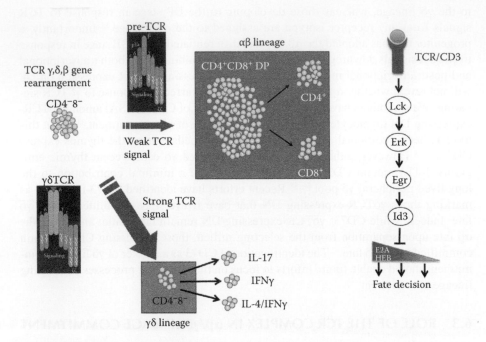

**FIGURE 6.1**   Schematic of αβ/γδ lineage commitment and specification of effector fate. The signal strength model proposes that the TCR signal strength dictates lineage choice, irrespective of the TCR isotype from which the signal originates. Strong and weak signals lead to adoption of the γδ and αβ fates, respectively. The differences in signal strength are manifested by changes in activation of the ERK–Egr–Id3 axis, which results in proportional suppression of E protein activity. Nevertheless, many critical unanswered questions remain: (1) How do different γδTCR complexes (Vγ subsets) lead to adoption of distinct fates? (2) What is the relationship between γδ lineage commitment and specification of effector fate? (3) How are TCR signals of differing strength able to promote distinct developmental outcomes through graded suppression of E protein activity and cooperation with other DNA-binding proteins?

the Src family kinase, p56[lck], diverted progenitors to the αβ fate.[34] Moreover, single cell progenitor analysis revealed that the differences in TCR signal strength that are orchestrating fate separation are doing so in an instructional manner.[32] Similar findings have been observed using a different γδTCR Tg model.[52] The TCR signal strength model is now widely accepted as providing the best explanation for the role of TCR signaling in orchestrating separation of the αβ and γδ fates.

## 6.4   BASIS FOR MORE ROBUST SIGNALING BY THE γδTCR

The requirement for TCR signals of greater intensity/longevity in promoting γδ lineage commitment engenders the question of how γδTCRs are able to transduce such signals. One possibility is that this capacity is simply intrinsic to the TCR. Consistent with this notion, the γδTCR is expressed at much higher levels than the pre-TCR complex.[53] Further, the γδTCR complex signals more robustly than the αβTCR upon equivalent stimulation by antibodies, perhaps because the γδTCR complex lacks CD3δ, instead possessing two CD3γε dimers.[54] Finally, TCRγδ pairs have been reported to signal in a ligand-independent fashion in vitro, as was previously shown for the pre-TCR;[24,55,56] however, the vast majority of the cells generated from ectopic expression were αβ lineage DP thymoctyes. Consistent with this perspective, when the ligand-independent signaling of the KN6 γδTCR Tg was tested in vivo, it also promoted commitment to the αβ lineage and development to the DP stage, suggesting that ligand-independent signaling by γδTCR complexes may be similar in intensity/duration to pre-TCR signals.[34]

At least some γδTCR complexes appear to depend on ligands to transduce the stronger or more prolonged signals required for γδ lineage commitment; however, addressing this issue has been challenging, as the number of thymic γδTCR ligands identified remains small. Nevertheless, there is clear evidence that such ligands play an important role in regulating development of some γδ cells. Dendritic epidermal γδ T cells (DETC) are a skin resident subset with a nearly invariant Vγ3Vδ1 γδTCR and their development appears to be ligand-dependent, as evidenced by the highly restricted TCR sequence and the inability of DETC γδTCR to exhibit ligand-independent signaling in vitro.[55,57,58] The Skint1 protein has been proposed as a putative ligand for the DETC γδTCR, since its ablation impairs the development of skin resident DETC; however, the function of Skint1 as a bona fide ligand has not been formally demonstrated.[59] The other ligands implicated in γδ T cell development are the highly homologous nonclassical MHC class 1 molecules H-2T10 and H-2T22 (T10/22). The expression of these ligands on the cell surface is dependent upon β2-microglobulin (β2M), and β2M deficiency has been shown to impair the development of thymocytes bearing T10/22-reactive γδTCR Tg (KN6 and G8), strongly suggesting that the development of these T-10/22 reactive γδ progenitors is dependent upon ligands.[33,60] More recently, the role of ligands in development of polyclonal T-10/22 reactive γδ cells was monitored using tetramer binding, which showed that tetramer-binding cells developed and exited the thymus even in the absence of their presumptive ligand.[55] However, effects on the γδTCR repertoire were not assessed, leaving open the possibility that even in this context, ligand binding may play a significant role in shaping the γδTCR repertoire. Resolving the discrepant effects of ligand loss on development of Tg and polyclonal T10/22-binding γδ progenitors will require both specific elimination of ligands, as β2M-deficiency eliminates all β2M-dependent molecules and not just T10/22, and deep-sequencing of the resultant γδTCR repertoires to determine if they are significantly altered by the absence of ligands. Finally, there is also indirect evidence in favor of the involvement of ligands in influencing γδ development. Indeed, the identification of CD73 induction as a marker of γδ lineage commitment is significant because its expression is induced by ligands on ~25% of γδTCR expressing thymocytes and more than 90% of γδ T cells in the periphery, suggesting that a large fraction of γδ progenitors encounter ligands in the thymus.[32] Moreover, as an illustration of the power of deep sequencing to provide insights into the potential role of ligand engagement in regulating γδTCR signaling and γδ T cell development, two recent studies have revealed that certain innate γδ T cell subsets exhibit extremely restricted TCR V region usage and even CDR3 sequences, which strongly suggest TCR-ligand mediated selection.[61,62]

*Question.* The question of how general ligand involvement is in controlling γδTCR signaling and development will not be resolved until more γδTCR ligands are identified and ablated. One approach that could be employed is to utilize soluble γδTCR complexes as probes to expression clone ligands or to assess effects on development of competitively inhibiting access by endogenous γδTCR during development.

## 6.5   ROLE OF TCR SIGNALING IN SPECIFICATION OF EFFECTOR FATE

One of the key differences between αβ and γδ T cells is the acquisition of effector fate during development. Conventional αβ T cells exit the thymus in a naïve state and acquire effector function in the periphery; however, many γδ T cells acquire their effector fate during development in the thymus. γδ T cell effector fates can be divided minimally into three subsets: IL-17 producers, IFNγ producers, and innate-like γδ T cells. Innate-like γδ T cells are defined either by their expression of the transcription factor PLZF and simultaneous production of IFNγ and IL-4,[63] or by their ability to rapidly produce either cytokine (IL-17 or IFNγ) in response to cytokine stimulation alone without TCR coengagement.[64,65] The effector fates of γδ T cell subsets can be delineated by surface expression of CD27, in that CD27+ subsets are enriched for IFNγ producers, while IL17-producing γδ T cells are restricted to the CD27– subset.[66] IFNγ/IL-4 coproducing innate-like γδ T cells coexpress CD27 and the NK cell-marker NK1.1. In addition to the phenotypic markers, effector fates are often linked to the use of particular Vγ genes, with Vγ2+ γδ T cells enriched for IL-17-producers, Vγ1+ γδ T cells enriched for IFNγ-producers, and IFNγ/IL-4 producing innate-like γδ T cells having an even more restricted TCR repertoire, expressing Vγ1.1 Vδ6.3/6.4.[67] The link between Vγ usage and effector fate raises the question of how TCR signaling influences the specification of effector fate, an issue that remains quite controversial.

As with the role of TCR signaling in γδ lineage commitment, two views have emerged: one suggesting TCR-dependence and the other TCR-independent predetermination. The predetermination model proposes that the γδ effector fate is not influenced by TCR signaling and may occur prior to or concurrent with γδ TCR expression. Consistent with this view, the capacity to generate IL-17-producing γδ T cells is largely dependent upon the context of the fetal thymus.[68] Moreover, gene expression profiling of immature γδ progenitors revealed that immature thymic subsets defined by Vγ usage were already enriched for fate-specifying transcription factors linked to their effector fate.[69] For example, Vγ2 expressing progenitors, which are enriched for those that will ultimately adopt the IL-17-producing effector fate, expressed elevated levels of the RORγt transcription factor required for IL-17 production.[69] IL-17 production has been linked to additional transcription factors including Sox13, Sox4, Tcf1 and Lef1, which are required for the generation of the Vγ2 subset of IL-17 producing γδ T cells.[70] These transcription factors are not required for Vγ4+ IL-17 producers, which instead depend on a partially distinct set of transcription factors. These data suggest that the Vγ2 and Vγ4 subsets of IL-17 producers arise from distinct developmental programs. Thus, these studies establish a connection between expression signatures, Vγ usage, and effector fate. Because the expression analysis was focused on immature CD24[high] γδ T cells, the authors advanced these findings as supporting a predetermination model, reasoning that immature CD24[high] γδ T cells have not yet been influenced by TCR signaling. However, because about 25% of immature CD24[high] γδ T cells express CD73, a TCR-ligand inducible molecule, it is likely that at least some of these immature cells have experienced γδ TCR signaling.[32]

An alternative to the predetermination model, the TCR-dependent model, has also been advanced, which proposes that a gradient of TCR signaling from weakest to strongest is involved in the generation of IL-17, IFNγ, and innate γδ T cells, respectively.[63] This model has been tested by manipulating TCR signaling, either via ligand engagement or mutation of signaling molecules, and assessing the impact on effector fate. These approaches have resulted in clear evidence supporting a role of TCR signaling in effector fate determination. Indeed, attenuation of surface T10/T22 through β2M deficiency diverted the effector fate of T10/T22-reactive γδ T cells from production of IFNγ to production of IL-17.[55] These results suggest that the relatively weak TCR signals generated in the absence of cognate ligand supported adoption of the IL-17-producing effector fate, while more intense, ligand-induced TCR signals supported adoption of the IFNγ-producing fate. Similar observations were made upon elimination of Skint1, which is necessary for the agonist selection of Vγ3Vδ1+ DETC γδ T cells.[59] Skint1-deficiency diverted Vγ3Vδ1+ progenitors from the IFNγ-producing effector fate to production of IL-17.[71] Thus, these findings clearly indicate that the nature of TCR signals transduced by developing progenitors can influence their effector fate, with more intense signaling promoting IFNγ production and weaker signaling promoting IL-17 production. It remains unclear whether the weak signals that facilitate adoption of the IL-17-producing fate are truly ligand-independent or involve interactions with ligands of lower affinity.

The most intense TCR signaling has been implicated in adoption of the innate-like γδ effector subtype, defined either by PLZF expression or by responsiveness to cytokine alone. The PLZF expressing innate γδ T cells express the natural killer marker NK1.1, utilize a restricted Vγ1.1+ Vδ6.3/6.4+ γδ TCR, and produce both IL-4 and IFNγ.[72,73] Two lines of evidence suggest that most intense TCR signaling supports adoption of this PLZF+ innate fate. First, KN6 γδTCR Tg progenitors adopt the PLZF+ innate fate upon engagement of high-affinity ligand (T10/T22b) but adopt the IFN-producing effector fate in response to lower affinity ligand (T10/22d).[35,74] Moreover, high-affinity antibody stimulation was reported to induce PLZF expression in multiple Vγ subsets, which clearly supports a role of intense TCR signaling in adoption of the innate fate, and is inconsistent with the notion of predetermination.[73] Development of cytokine-responsive, IL-17 and IFNγ-producing innate γδ T cells is also dependent upon strong TCR signals.[64] Indeed, the generation of these cells is impaired by attenuating the capacity of the γδTCR to transduce signals, either by mutating a key signaling molecule (i.e., Zap70 for IL-17 producers) or mutating a cofactor necessary for agonist selection (i.e., Skint1 for DETC).[64] Thus, while cellular context may influence the effector fate potential of γδ T cell progenitors, it is clear that the nature of TCR signaling during development in the thymus is able to influence the ultimate effector fate that they adopt.

## 6.6 NATURE OF THE SIGNALS THAT CONTROL LINEAGE AND EFFECTOR FATE SPECIFICATION

Although the TCR signal strength/duration model is widely accepted as providing the best explanation for how the TCR complex influences αβ/γδ lineage commitment, our understanding of the key events downstream of the TCR remains rudimentary. Evidence from gene-targeting approaches suggests that the signaling cascades that specify the αβ and γδ fates are genetically separable. For example, ablation of the adaptor protein LAT causes a severe blockade of both the αβ and γδ lineages;[75] however, mice expressing LAT molecules selectively defective in PLCγ recruitment exhibit a preferential block in development of αβ lineage cells.[76–78] Gene-targeting experiments have also identified numerous molecular effectors that are selectively required for either the αβ or γδ lineage development;[79] however, this information has not yet been unified into a comprehensive understanding of the signaling pathways that control fate. Nevertheless, substantial evidence suggests that differences in activation of the ERK–Egr–Id3 signaling axis are a key element of the differences in TCR signal strength/duration that influence αβ/γδ fate choice.[34,35,80]

### 6.6.1 ERK SIGNALING

Extracellular signal regulated kinase (ERK) is the terminal enzyme in a cascade of three protein kinases: MAP kinase kinase kinase (MAP3K; Raf), MAP kinase kinase (MAP2K; MEK), and MAP kinase (MAPK; ERK), which sequentially activate their downstream targets by phosphorylation at specific amino acid residues. Following activation, ERK is capable of phosphorylating numerous substrates in both the cytosol and nucleus.[81,82] TCR signaling leads to ERK/MAP kinase (MAPK) activation and is required for normal thymocyte development.[83] We, and others, have shown that ERK is more highly phosphorylated in developing γδ than in αβ lineage cells.[34,52] The greater ERK activation associated with adoption of the γδ fate is functionally important as γδ commitment is impaired by ERK1/2-deficiency, which redirects γδTCR-expressing progenitors to the αβ fate, as evidenced by development to the DP stage.[84] Differences in ERK activation have been implicated in dictating alternative fate choices for decades, but there has been relatively little progress in unraveling the molecular mechanism by which those differences in ERK signaling do so. A long-standing question has been whether the key differences in ERK activation reflect differences in intensity, duration, or both. This issue has long been studied in reductionist models such as PC12 cells, where epidermal growth factor (EGF) stimulation induces transient ERK activation and proliferation, while nerve growth factor (NGF) stimulation produces sustained ERK activation and differentiation into sympathetic neurons.[85–88] In an attempt to explain how differences in ERK activation might produce distinct biological outcomes, the Blenis lab advanced the immediate early gene (IEG) sensor hypothesis,[89] which posits that prolonged ERK activity produces altered biological outcomes by regulating the stability of IEG protein products (Figure 6.2).[90] The model proposes that transient ERK activation results in induction of IEG transcripts, but ERK activity decays prior to expression of IEG protein products, rendering them unstable and resulting in their rapid degradation.

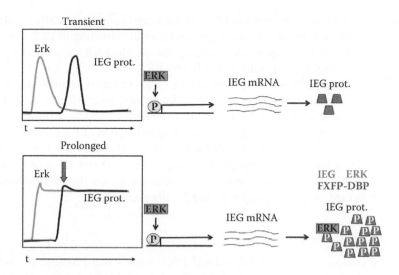

**FIGURE 6.2** Immediate early sensor model describing how prolonged ERK activation produces alternative fates. Transient activation of ERK induces a transcriptional response, but ERK activity decays prior to synthesis of immediate early gene (IEG) protein products derived from those transcripts. Consequently those nascent IEG protein products are unstable and rapidly degraded, diminishing their ability to influence cell behavior. Prolonged ERK activation induces a transcriptional response and persists until IEG proteins are produced. This enables active ERK to physically interact with IEG proteins via their DEF motifs (FXFP), phosphorylate them, and increase their stability. Their increased stability results in their accumulation and an enhanced ability to influence cell behavior. Proteins containing DEF motifs can be found in essentially all gene ontology classes, indicating that this mechanism could broadly alter cell behavior.

Conversely, prolonged ERK activation not only induces IEG transcripts but also persists until IEG protein products are expressed, resulting in ERK–IEG interactions, which stabilize the IEG proteins and cause them to accumulate. Accordingly, cells "perceive" a sustained ERK signal as one that results in the accumulation of IEG protein products. Nevertheless, this attractive hypothesis had not been tested in vivo. A direct test of this hypothesis was made possible by the observation that the ability of ERK to stabilize IEG protein products was found to depend on the FXFP (or DEF) motif on the IEG protein and the DEF binding pocket (DBP) of ERK.[91,92] Indeed, by mutating the DBP of ERK2, we demonstrated that the prolonged ERK signals associated with γδ lineage commitment, but not the transient ERK signals promoting αβ lineage commitment, were dependent upon the ability of ERK to dock with IEG.[84] Interestingly, the D domain of ERK, through which it interacts with most of its known targets, was dispensable for γδ lineage commitment.[84,93] One of the targets of prolonged ERK signals is the transcription factor Egr1, which accumulates and transactivates a set of gene targets not induced following transient ERK signaling.[84] While altering secondary transcriptional responses is likely to be one important way that increasing the stability of IEG proteins impacts development, there are likely to be many other cellular processes impacted as well, since proteins from virtually all cellular gene ontology classes possess DEF domains and so are potentially regulated by ERK. Nevertheless, because there is specificity in the targeting of DEF containing IEG by ERK, it will be important to identify the specific ERK substrates targeted at each developmental branch point controlled by ERK.

*Question.* The basis by which differences in ERK activation regulate αβ/γδ lineage commitment and other aspects of lymphoid development and function has been a longstanding question. Because ERK-substrate interactions appear to display tissue specificity, addressing this question will require identification of ERK substrates in cells at each developmental process under study. This will require the generation of mice expressing analog-sensitive ERK loci in which either D or DBP domain function has been disrupted. Subsequently, mass spectroscopy on the ERK substrates must be performed, as was recently reported for fibroblasts.[93]

### 6.6.2   Egr Proteins

One of the critical targets of the prolonged ERK signals that promote γδ lineage commitment is the transcription factor Egr1. The induction of Egr1 during γδ lineage commitment in response to strong signals results in part from a modest increase in transcript levels, but the majority of its induction occurs posttranscriptionally as a result of an ERK-mediated increase in protein stability.[84] Egr1 is a member of a family of zinc-finger transcription factors that we, and others, have shown to be critical for normal thymocyte development.[94–96] Importantly, Egr proteins are induced in proportion to the strength of a number of mitogenic signals, including those transduced by the TCR.[97,98] We determined that the marked induction of Egr expression not only correlates with the stronger and more prolonged TCR signals that direct adoption of the γδ fate, but it also plays an important role in this process, since elevating Egr levels through enforced expression promotes adoption of the γδ fate at the expense of the αβ lineage.[34,35] The superinduction of Egr1 by the prolonged ERK signals induced during γδ lineage commitment enables Egr1 to more potently transactivate a broader array of targets than is observed in response to the transient ERK activation induced during αβ lineage commitment.[84] Induction of this collection of Egr targets is one of the ways that prolonged ERK signals facilitate the adoption of the γδ fate. Although our understanding of how Egr targets collectively act to promote γδ lineage commitment is incomplete at present, it is clear that Id3 plays a critical role.

### 6.6.3   Id and E Proteins

Id3 is a helix-loop-helix (HLH) factor that is induced by TCR signaling through the action of Egr transcription factors.[99–101] Id3 is induced in proportion to signal strength, is more highly expressed in progenitors adopting the γδ fate, and is a critical component of the ERK–Egr–Id3 axis that regulates αβ/γδ lineage commitment.[34,35] Indeed, Id3 deficiency abrogates the ability of strong/prolonged TCR signals, or their mimicry by ectopically expressed Egr1, to promote development of γδ lineage cells.[34,35] Id3 deficiency also perturbs the development of γδ T cells, but does so in a Vγ-dependent manner. Indeed, Id3 deficiency impairs the development of both the Vγ2+ and Vγ3+ subsets,[35] while simultaneously leading to a massive expansion of innate type Vγ1.1Vδ6.3 γδ cells.[35,72,102,103] Interpretations of these findings differ; however, the most likely explanation is that Vγ2 and Vγ3 subsets are selected by strong TCR signals that depend on Id3 function to promote γδ lineage commitment. Conversely, the Vγ1.1Vδ6.3 γδ subset that expands in Id3-deficient mice is likely to do so because those cells are normally deleted by excessively intense signals, but Id3-deficiency enables them to escape deletion and expand. This perspective is supported by our analysis using the KN6 γδTCR transgenic (Tg) thymocytes generated by positive selection in the presence of moderate affinity T-10$^d$ ligand, since development of these cells is blocked by Id3 deficiency.[35] In contrast, exposure of KN6 γδTCR Tg thymocytes to exceedingly strong TCR signals, induced by high-affinity T-10$^b$ ligand, normally cause their deletion; however, Id3 deficiency enables them to escape deletion and expand, as has been observed for the Vγ1.1Vδ6.3 γδ subset.[33,35] While this interpretation has not yet been tested among the endogenous Vγ subsets listed earlier, these findings suggest a dichotomy of Id3 function, with Id3 promoting commitment and development of γδ cells by the stronger signals that typically accompany this process, while inducing the deletion of cells when signal strength exceeds a particular threshold. To validate this hypothesis, it will be important to replicate the aforementioned tests in endogenous Vγ subsets and investigate the molecular bases for the differing developmental outcomes.

These findings raise the question of how differential induction of Id3 influences αβ/γδ lineage choice. This is likely to occur through graded suppression of E proteins (Figure 6.3). E proteins are basic helix-loop-helix (bHLH) transcription factors that bind E-box motifs (CANNTG) either as homodimers or heterodimers with other bHLH proteins.[104] The two E protein family members expressed in T lineage progenitors are E2A and HEB, and they play critical roles in regulating T cell development, including enforcing a developmental checkpoint at the DN3 stage.[101] In fact, their activity must be repressed in order for T cell development to proceed beyond the DN3 stage.[105] E protein binding to DNA is blocked by pairing with Id3. Thus, Id3 induction by TCR signals reduces E protein activity in proportion to TCR signal strength.[35] The weak TCR signals promoting adoption of the αβ fate are predicted to result in a mild reduction in E protein activity, while the very strong signals that promote γδ lineage commitment should nearly extinguish E protein activity, perhaps mimicking the effect of E protein gene ablation. In support, both E2A and HEB deficiency impair the development of αβ T cells; however, E2A and HEB deficiency produce relatively mild alterations in γδ T cell development, and this effect is even more mild in E2A/HEB-double deficient mice.[106–109] Consequently, the lack of impairment of γδ development by E protein deficiency is consistent with the hypothesis that the strong TCR signals that promote γδ lineage commitment do so by profound, Id3-mediated repression of E protein function. Moreover, recent work from the Zhuang lab demonstrated using elegant genetic approaches that the expansion of the Vγ1.1Vδ6.3 γδ subset in Id3-deficient mice was caused by the preservation of E protein function, since expansion of these cells was blocked by E2A/HEB deficiency.[110] Altogether, these data suggest a model whereby graded reductions in E protein activity mediated by differences in TCR signal strength play an important role in αβ/γδ lineage commitment (Figure 6.3).

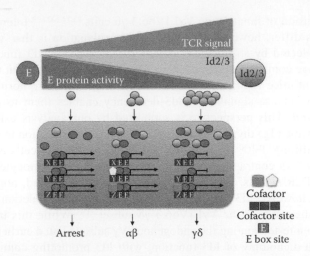

**FIGURE 6.3** Effect of TCR signals of differing strength on E protein DNA binding. The signal strength hypothesis proposes that the fate of thymic progenitors is determined by the strength of the TCR signal, which is manifested by graded suppression of E protein function. E protein function is suppressed by the E protein antagonist, Id3, which is induced in proportion to TCR signal strength. Prior to TCR signaling, E protein activity is maximal as is occupancy of genomic E protein binding sites. This enforces the developmental arrest at the DN3 stage. TCR signals of differing strength alter the activity of E proteins, with weak signals mildly repressing E protein binding to DNA and enabling adoption of the αβ fate, while strong signals profoundly repress E protein function and enable adoption of the γδ fate. The repression of E protein function by TCR signals is mediated by Id3, which dimerizes with E proteins and prevents them from binding to DNA. The ability of differences in the occupancy of E protein binding sites to orchestrate these alternative fates undoubtedly depends on cooperation with distinct transcription factors whose binding sites cluster nearby. Identification of the E protein binding sites and the transcription factors will enable the creation of global genomic networks that will enable a molecular dissection of the processes governing these fate decisions.

*Question.* Although this evidence strongly suggests that graded reductions in E protein function play a critical role in regulating αβ/γδ lineage commitment, the way these reductions in E proteins do so remains unclear. To gain insight into how graded reductions in E protein function influence fate, it will be necessary to build comprehensive, E-protein-focused genome-wide networks by identifying the regulatory elements differentially occupied by E proteins during αβ/γδ lineage commitment where their activity is reduced in a graded fashion. These networks serve to link differences in TCR signaling to the repertoire of enhancers modulated by changes in E protein function, to transcription factors with which they cooperate, and ultimately to the resultant changes in gene expression. To do so, chromatin immunoprecipitation sequencing (ChIP-Seq) must be performed to identify enhancers by virtue of their epigenetic marks (H3K4me3, p300, etc.) and determine which of these are differentially occupied by E proteins using anti-E protein antibodies. By combining this approach with HiC, which enables identification of the target elements controlled by the aforementioned enhancers, one can deduce how changes in E protein occupancy alter those interactions and translate into nuclear repositioning.[111] Finally, bioinformatic tools (e.g., the HOMER algorithm) can be used to mine these data to identify transcription factors whose binding sites cluster near elements differentially occupied by E proteins, as these factors are likely to cooperate with E proteins to regulate fate choice. Together, these analyses will enable the construction of global, E-protein focused molecular networks that provide a comprehensive view of how E protein targets link up to cooperating transcription factors and the signaling networks that regulate their expression and function. Such analysis has already been quite informative in dissecting the role of E proteins in regulating early B cell development.[112–115] It will also likely enable understanding of the specification of effector fate, since it, like αβ/γδ lineage fate, is tied to differences in γδTCR signal strength/duration. The specification of IL-17-producing γδT cells is linked to weaker signals that preserve greater E protein activity, and expression of the requisite E protein target, RORγt.[55,71,99] IFNγ-producing effectors are thought to be specified by more intense ligand-induced signaling, thought to cause greater E protein repression.[55,71] Consistent with this notion, IFNγ-producing T cells are induced by ablation of the E protein E47.[116] Finally, choice of the innate αβ and γδ fate is controlled by modulating the balance of Id and E protein activity.[110]

## 6.7 CONCLUDING REMARKS

Since the TCR signal strength model has been advanced, much progress has been made in understanding how TCR signals influence αβ/γδ lineage commitment; however, many important questions remain to be addressed. It remains unclear how the stronger and more prolonged TCR signals that promote adoption of the γδ fate are generated. A question of particular importance is whether intrathymic ligands play a role in generating such signals and therefore in shaping the γδ TCR repertoire. This issue remains controversial and is unlikely to be resolved until bona fide selecting ligands are identified and the effect of their elimination of γδ development is assessed. Another important question that remains to be addressed is how particular Vγ regions are linked to effector fate and whether this is controlled by the cellular context during the developmental window when the Vγ region is rearranged. It also remains unclear how adoption of effector fate by γδ progenitors relates to their commitment to the γδ fate. Do these events occur simultaneously or in a sequential manner, and if sequential, in what order? Resolving these questions will require a much better understanding of the molecular processes controlling fate specification. Clearly the application of cutting edge genomic and genetic approaches will lead to insights into how these processes are controlled. This may ultimately enable manipulation of γδ T cell generation or function for therapeutic benefit, as γδ lineage cells are increasingly understood to play important roles in both productive and pathologic immune responses.

## ACKNOWLEDGMENTS

The chapter derives in part from work supported by National Institutes of Health (NIH) grants P01AI102853, core grant P30CA006927, and an appropriation from the Commonwealth of Pennsylvania. The Commonwealth of Pennsylvania specifically disclaims responsibility for any analyses, interpretations, or conclusions.

# REFERENCES

1. Hayday, A. C., [gamma][delta] cells: A right time and a right place for a conserved third way of protection, *Annu Rev Immunol* 18, 975–1026, 2000.
2. Carding, S. R., and Egan, P. J., Gammadelta T cells: Functional plasticity and heterogeneity, *Nat Rev Immunol* 2(5), 336–345, 2002.
3. Salerno, A., and Dieli, F., Role of gamma delta T lymphocytes in immune response in humans and mice, *Crit Rev Immunol* 18(4), 327–357, 1998.
4. Born, W. K., Kemal Aydintug, M., and O'Brien, R. L., Diversity of gammadelta T-cell antigens, *Cell Mol Immunol* 10(1), 13–20, 2013.
5. Vantourout, P., and Hayday, A., Six-of-the-best: Unique contributions of gammadelta T cells to immunology, *Nat Rev Immunol* 13(2), 88–100, 2013.
6. Nishimura, H., Yajima, T., Kagimoto, Y., Ohata, M., Watase, T., Kishihara, K., Goshima, F., Nishiyama, Y., and Yoshikai, Y., Intraepithelial gammadelta T cells may bridge a gap between innate immunity and acquired immunity to herpes simplex virus type 2, *J Virol* 78(9), 4927–4930, 2004.
7. Lafarge, X., Merville, P., Cazin, M. C., Berge, F., Potaux, L., Moreau, J. F., and Dechanet-Merville, J., Cytomegalovirus infection in transplant recipients resolves when circulating gammadelta T lymphocytes expand, suggesting a protective antiviral role, *J Infect Dis* 184(5), 533–541, 2001.
8. Sutton, C. E., Mielke, L. A., and Mills, K. H., IL-17-producing gammadelta T cells and innate lymphoid cells, *Eur J Immunol* 42(9), 2221–2231, 2012.
9. Shortman, K., Wu, L., Kelly, K. A., and Scollay, R., The beginning and the end of the development of TCR gamma delta cells in the thymus, *Curr Top Microbiol Immunol* 173, 71–80, 1991.
10. Petrie, H. T., Scollay, R., and Shortman, K., Commitment to the T cell receptor-alpha beta or -gamma delta lineages can occur just prior to the onset of CD4 and CD8 expression among immature thymocytes, *Eur J Immunol* 22(8), 2185–2188, 1992.
11. Ciofani, M., Knowles, G. C., Wiest, D. L., von Boehmer, H., and Zuniga-Pflucker, J. C., Stage-specific and differential notch dependency at the alphabeta and gammadelta T lineage bifurcation, *Immunity* 25(1), 105–116, 2006.
12. Dudley, E. C., Petrie, H. T., Shah, L. M., Owen, M. J., and Hayday, A. C., T cell receptor beta chain gene rearrangement and selection during thymocyte development in adult mice, *Immunity* 1(2), 83–93, 1994.
13. Dudley, E. C., Girardi, M., Owen, M. J., and Hayday, A. C., Alpha beta and gamma delta T cells can share a late common precursor, *Curr Biol* 5(6), 659–669, 1995.
14. Wilson, A., de Villartay, J. P., and MacDonald, H. R., T cell receptor delta gene rearrangement and T early alpha (TEA) expression in immature alpha beta lineage thymocytes: Implications for alpha beta/gamma delta lineage commitment, *Immunity* 4(1), 37–45, 1996.
15. Capone, M., Hockett, R. D., Jr., and Zlotnik, A., Kinetics of T cell receptor beta, gamma, and delta rearrangements during adult thymic development: T cell receptor rearrangements are present in CD44(+)CD25(+) Pro-T thymocytes, *Proc Nat Acad Sci USA* 95(21), 12522–12527, 1998.
16. Livak, F., Tourigny, M., Schatz, D. G., and Petrie, H. T., Characterization of TCR gene rearrangements during adult murine T cell development, *J Immunol* 162(5), 2575–2580, 1999.
17. Schmolka, N., Wencker, M., Hayday, A. C., and Silva-Santos, B., Epigenetic and transcriptional regulation of gammadelta T cell differentiation: Programming cells for responses in time and space, *Semin Immunol* 27(1), 19–25, 2015.

18. Havran, W. L., and Allison, J. P., Developmentally ordered appearance of thymocytes expressing different T-cell antigen receptors, *Nature* 335(6189), 443–445, 1988.
19. Ito, K., Bonneville, M., Takagaki, Y., Nakanishi, N., Kanagawa, O., Krecko, E. G., and Tonegawa, S., Different gamma delta T-cell receptors are expressed on thymocytes at different stages of development, *Proc Nat Acad Sci USA* 86(2), 631–635, 1989.
20. Carding, S. R., Kyes, S., Jenkinson, E. J., Kingston, R., Bottomly, K., Owen, J. J., and Hayday, A. C., Developmentally regulated fetal thymic and extrathymic T-cell receptor gamma delta gene expression, *Genes Dev* 4(8), 1304–1315, 1990.
21. Berger, M. A., Dave, V., Rhodes, M. R., Bosma, G. C., Bosma, M. J., Kappes, D. J., and Wiest, D. L., Subunit composition of pre-T cell receptor complexes expressed by primary thymocytes: CD3 delta is physically associated but not functionally required, *J Exper Med* 186(9), 1461–1467, 1997.
22. Groettrup, M., and von Boehmer, H., A role for a pre-T-cell receptor in T-cell development, *Immunol Today* 14(12), 610–614, 1993.
23. Irving, B. A., Alt, F. W., and Killeen, N., Thymocyte development in the absence of pre-T cell receptor extracellular immunoglobulin domains, *Science* 280(5365), 905–908, 1998.
24. Yamasaki, S., Ishikawa, E., Sakuma, M., Ogata, K., Sakata-Sogawa, K., Hiroshima, M., Wiest, D. L., Tokunaga, M., and Saito, T., Mechanistic basis of pre-T cell receptor-mediated autonomous signaling critical for thymocyte development, *Nat Immunol* 7(1), 67–75, 2006.
25. Hoffman, E. S., Passoni, L., Crompton, T., Leu, T. M., Schatz, D. G., Koff, A., Owen, M. J., and Hayday, A. C., Productive T-cell receptor beta-chain gene rearrangement: Coincident regulation of cell cycle and clonality during development in vivo, *Genes Dev* 10(8), 948–962, 1996.
26. Aifantis, I., Buer, J., von Boehmer, H., and Azogui, O., Essential role of the pre-T cell receptor in allelic exclusion of the T cell receptor beta locus, *Immunity* 7(5), 601–607, 1997. [published erratum appears in *Immunity* 7(6), following 895, 1997]
27. Kruisbeek, A. M., Haks, M. C., Carleton, M., Michie, A. M., Zuniga-Pflucker, J. C., and Wiest, D. L., Branching out to gain control: How the pre-TCR is linked to multiple functions, *Immunol Today* 21(12), 637–644, 2000.
28. Kreslavsky, T., Garbe, A. I., Krueger, A., and von Boehmer, H., T cell receptor-instructed alphabeta versus gammadelta lineage commitment revealed by single-cell analysis, *J Exp Med* 205(5), 1173–1186, 2008.
29. Passoni, L., Hoffman, E. S., Kim, S., Crompton, T., Pao, W., Dong, M. Q., Owen, M. J., and Hayday, A. C., Intrathymic delta selection events in gammadelta cell development, *Immunity* 7(1), 83–95, 1997.
30. Ishida, I., Verbeek, S., Bonneville, M., Itohara, S., Berns, A., and Tonegawa, S., T-cell receptor gamma delta and gamma transgenic mice suggest a role of a gamma gene silencer in the generation of alpha beta T cells, *Proc Nat Acad Sci USA* 87(8), 3067–3071, 1990.
31. Ferrero, I., Mancini, S. J., Grosjean, F., Wilson, A., Otten, L., and MacDonald, H. R., TCRgamma silencing during alphabeta T cell development depends upon pre-TCR-induced proliferation, *J Immunol* 177(9), 6038–6043, 2006.
32. Coffey, F., Lee, S. Y., Buus, T. B., Lauritsen, J. P., Wong, G. W., Joachims, M. L., Thompson, L. F., Zuniga-Pflucker, J. C., Kappes, D. J., and Wiest, D. L., The TCR ligand-inducible expression of CD73 marks gammadelta lineage commitment and a metastable intermediate in effector specification, *J Exp Med* 211(2), 329–343, 2014.
33. Pereira, P., Zijlstra, M., McMaster, J., Loring, J. M., Jaenisch, R., and Tonegawa, S., Blockade of transgenic gamma delta T cell development in beta 2-microglobulin deficient mice, *EMBO J* 11(1), 25–31, 1992.

34. Haks, M. C., Lefebvre, J. M., Lauritsen, J. P., Carleton, M., Rhodes, M., Miyazaki, T., Kappes, D. J., and Wiest, D. L., Attenuation of gammadeltaTCR signaling efficiently diverts thymocytes to the alphabeta lineage, *Immunity* 22(5), 595–606, 2005.
35. Lauritsen, J. P., Wong, G. W., Lee, S. Y., Lefebvre, J. M., Ciofani, M., Rhodes, M., Kappes, D. J., Zuniga-Pflucker, J. C., and Wiest, D. L., Marked induction of the helix-loop-helix protein Id3 promotes the gammadelta T cell fate and renders their functional maturation Notch independent, *Immunity* 31(4), 565–575, 2009.
36. Zorbas, M. and Scollay, R., Development of gamma delta T cells in the adult murine thymus, *Eur J Immunol* 23(7), 1655–1660, 1993.
37. Kelly, K. A., Pearse, M., Lefrancois, L., and Scollay, R., Emigration of selected subsets of gamma delta + T cells from the adult murine thymus, *Int Immunol* 5(4), 331–335, 1993.
38. Tough, D. F., and Sprent, J., Lifespan of gamma/delta T cells, *J Exp Med* 187(3), 357–365, 1998.
39. MacDonald, H. R., and Wilson, A., The role of the T-cell receptor (TCR) in alpha beta/gamma delta lineage commitment: Clues from intracellular TCR staining, *Immunol Rev* 165, 87–94, 1998.
40. Lauritsen, J. P., Haks, M. C., Lefebvre, J. M., Kappes, D. J., and Wiest, D. L., Recent insights into the signals that control alphabeta/gammadelta-lineage fate, *Immunol Rev* 209, 176–90, 2006.
41. Kang, J., Baker, J., and Raulet, D. H., Evidence that productive rearrangements of TCR gamma genes influence the commitment of progenitor cells to differentiate into alpha beta or gamma delta T cells, *Eur J Immunol* 25(9), 2706–2709, 1995.
42. Livak, F., Petrie, H. T., Crispe, I. N., and Schatz, D. G., In-frame TCR delta gene rearrangements play a critical role in the alpha beta/gamma delta T cell lineage decision, *Immunity* 2(6), 617–627, 1995.
43. Burtrum, D. B., Kim, S., Dudley, E. C., Hayday, A. C., and Petrie, H. T., TCR gene recombination and alpha beta-gamma delta lineage divergence: Productive TCR-beta rearrangement is neither exclusive nor preclusive of gamma delta cell development, *J Immunol* 157 (10), 4293–4296, 1996.
44. Mertsching, E., and Ceredig, R., T cell receptor-gamma, delta-expressing fetal mouse thymocytes are generated without T cell receptor V beta selection, *Eur J Immunol* 26(4), 804–810, 1996.
45. Mertsching, E., Wilson, A., MacDonald, H. R., and Ceredig, R., T cell receptor alpha gene rearrangement and transcription in adult thymic gamma delta cells, *Eur J Immunol* 27(2), 389–396, 1997.
46. Aifantis, I., Azogui, O., Feinberg, J., Saint-Ruf, C., Buer, J., and von Boehmer, H., On the role of the pre-T cell receptor in alphabeta versus gammadelta T lineage commitment, *Immunity* 9(5), 649–655, 1998.
47. Bonneville, M., Ishida, I., Mombaerts, P., Katsuki, M., Verbeek, S., Berns, A., and Tonegawa, S., Blockage of alpha beta T-cell development by TCR gamma delta transgenes, *Nature* 342(6252), 931–934, 1989.
48. Dent, A. L., Matis, L. A., Hooshmand, F., Widacki, S. M., Bluestone, J. A., and Hedrick, S. M., Self-reactive gamma delta T cells are eliminated in the thymus, *Nature* 343(6260), 714–719, 1990.
49. Bruno, L., Fehling, H. J., and von Boehmer, H., The alpha beta T cell receptor can replace the gamma delta receptor in the development of gamma delta lineage cells, *Immunity* 5(4), 343–352, 1996.
50. Terrence, K., Pavlovich, C. P., Matechak, E. O., and Fowlkes, B. J., Premature expression of T cell receptor (TCR)alphabeta suppresses TCRgammadelta gene rearrangement but permits development of gammadelta lineage T cells, *J Exper Med* 192(4), 537–548, 2000.

51. Hayes, S. M., Shores, E. W., and Love, P. E., An architectural perspective on signaling by the pre-, alphabeta and gammadelta T cell receptors, *Immunol Rev* 191, 28–37, 2003.

52. Hayes, S. M., Li, L., and Love, P. E., TCR signal strength influences alphabeta/gammadelta lineage fate, *Immunity* 22(5), 583–593, 2005.

53. Borst, J., Jacobs, H., and Brouns, G., Composition and function of T-cell receptor and B-cell receptor complexes on precursor lymphocytes, *Curr Opin Immunol* 8(2), 181–190, 1996.

54. Hayes, S. M., and Love, P. E., Distinct structure and signaling potential of the gamma delta TCR complex, *Immunity* 16(6), 827–838, 2002.

55. Jensen, K. D., Su, X., Shin, S., Li, L., Youssef, S., Yamasaki, S., Steinman, L., Saito, T., Locksley, R. M., Davis, M. M., Baumgarth, N., and Chien, Y. H., Thymic selection determines gammadelta T cell effector fate: Antigen-naive cells make interleukin-17 and antigen-experienced cells make interferon gamma, *Immunity* 29(1), 90–100, 2008.

56. Mahtani-Patching, J., Neves, J. F., Pang, D. J., Stoenchev, K. V., Aguirre-Blanco, A. M., Silva-Santos, B., and Pennington, D. J., PreTCR and TCRgammadelta signal initiation in thymocyte progenitors does not require domains implicated in receptor oligomerization, *Sci Signal* 4(182), ra47, 2011.

57. Havran, W. L., Chien, Y. H., and Allison, J. P., Recognition of self antigens by skin-derived T cells with invariant gamma delta antigen receptors, *Science* 252(5011), 1430–1432, 1991.

58. Xiong, N., Kang, C., and Raulet, D. H., Positive selection of dendritic epidermal gammadelta T cell precursors in the fetal thymus determines expression of skin-homing receptors, *Immunity* 21(1), 121–131, 2004.

59. Boyden, L. M., Lewis, J. M., Barbee, S. D., Bas, A., Girardi, M., Hayday, A. C., Tigelaar, R. E., and Lifton, R. P., Skint1, the prototype of a newly identified immunoglobulin superfamily gene cluster, positively selects epidermal gammadelta T cells, *Nat Genet* 40(5), 656–662, 2008.

60. Wells, F. B., Gahm, S. J., Hedrick, S. M., Bluestone, J. A., Dent, A., and Matis, L. A., Requirement for positive selection of gamma delta receptor-bearing T cells, *Science* 253(5022), 903–905, 1991.

61. Wei, Y. L., Han, A., Glanville, J., Fang, F., Zuniga, L. A., Lee, J. S., Cua, D. J., and Chien, Y. H., A highly focused antigen receptor repertoire characterizes gammadelta T cells that are poised to make IL-17 rapidly in naive animals, *Front Immunol* 6, 118, 2015.

62. Kashani, E., Fohse, L., Raha, S., Sandrock, I., Oberdorfer, L., Koenecke, C., Suerbaum, S., Weiss, S., and Prinz, I., A clonotypic Vgamma4Jgamma1/Vdelta5Ddelta2Jdelta1 innate gammadelta T-cell population restricted to the CCR6(+)CD27(-) subset, *Nat Commun* 6, 6477, 2015.

63. Fahl, S. P., Coffey, F., and Wiest, D. L., Origins of gammadelta T cell effector subsets: A riddle wrapped in an enigma, *J Immunol* 193(9), 4289–4294, 2014.

64. Wencker, M., Turchinovich, G., Di Marco Barros, R., Deban, L., Jandke, A., Cope, A., and Hayday, A. C., Innate-like T cells straddle innate and adaptive immunity by altering antigen-receptor responsiveness, *Nat Immunol* 15(1), 80–87, 2014.

65. Sutton, C. E., Lalor, S. J., Sweeney, C. M., Brereton, C. F., Lavelle, E. C., and Mills, K. H., Interleukin-1 and IL-23 induce innate IL-17 production from gammadelta T cells, amplifying Th17 responses and autoimmunity, *Immunity* 31(2), 331–341, 2009.

66. Ribot, J. C., deBarros, A., Pang, D. J., Neves, J. F., Peperzak, V., Roberts, S. J., Girardi, M., Borst, J., Hayday, A. C., Pennington, D. J., and Silva-Santos, B., CD27 is a thymic determinant of the balance between interferon-gamma- and interleukin 17-producing gammadelta T cell subsets, *Nat Immunol* 10(4), 427–436, 2009.

67. Bonneville, M., O'Brien, R. L., and Born, W. K., Gammadelta T cell effector functions: A blend of innate programming and acquired plasticity, *Nat Rev Immunol* 10(7), 467–478, 2010.

68. Haas, J. D., Ravens, S., Duber, S., Sandrock, I., Oberdorfer, L., Kashani, E., Chennupati, V., Fohse, L., Naumann, R., Weiss, S., Krueger, A., Forster, R., and Prinz, I., Development of interleukin-17-producing gammadelta T cells is restricted to a functional embryonic wave, *Immunity* 37(1), 48–59, 2012.

69. Narayan, K., Sylvia, K. E., Malhotra, N., Yin, C. C., Martens, G., Vallerskog, T., Kornfeld, H., Xiong, N., Cohen, N. R., Brenner, M. B., Berg, L. J., and Kang, J., Intrathymic programming of effector fates in three molecularly distinct gammadelta T cell subtypes, *Nat Immunol* 13(5), 511–518, 2012.

70. Malhotra, N., Narayan, K., Cho, O. H., Sylvia, K. E., Yin, C., Melichar, H., Rashighi, M., Lefebvre, V., Harris, J. E., Berg, L. J., and Kang, J., A network of high-mobility group box transcription factors programs innate interleukin-17 production, *Immunity* 38(4), 681–693, 2013.

71. Turchinovich, G., and Hayday, A. C., Skint-1 identifies a common molecular mechanism for the development of interferon-gamma-secreting versus interleukin-17-secreting gammadelta T cells, *Immunity* 35(1), 59–68, 2011.

72. Alonzo, E. S., Gottschalk, R. A., Das, J., Egawa, T., Hobbs, R. M., Pandolfi, P. P., Pereira, P., Nichols, K. E., Koretzky, G. A., Jordan, M. S., and Sant'Angelo, D. B., Development of promyelocytic zinc finger and ThPOK-expressing innate gammadelta T cells is controlled by strength of TCR signaling and Id3, *J Immunol* 184(3), 1268–1279.

73. Kreslavsky, T., Savage, A. K., Hobbs, R., Gounari, F., Bronson, R., Pereira, P., Pandolfi, P. P., Bendelac, A., and von Boehmer, H., TCR-inducible PLZF transcription factor required for innate phenotype of a subset of gammadelta T cells with restricted TCR diversity, *Proc Natl Acad Sci USA* 106(30), 12453–12458, 2009.

74. Park, K., He, X., Lee, H. O., Hua, X., Li, Y., Wiest, D., and Kappes, D. J., TCR-mediated ThPOK induction promotes development of mature (CD24-) gammadelta thymocytes, *EMBO J* 29(14), 2329–2341, 2010.

75. Zhang, W., Sommers, C. L., Burshtyn, D. N., Stebbins, C. C., DeJarnette, J. B., Trible, R. P., Grinberg, A., Tsay, H. C., Jacobs, H. M., Kessler, C. M., Long, E. O., Love, P. E., and Samelson, L. E., Essential role of LAT in T cell development, *Immunity* 10(3), 323–332, 1999.

76. Nunez-Cruz, S., Aguado, E., Richelme, S., Chetaille, B., Mura, A. M., Richelme, M., Pouyet, L., Jouvin-Marche, E., Xerri, L., Malissen, B., and Malissen, M., LAT regulates gammadelta T cell homeostasis and differentiation, *Nat Immunol* 4(10), 999–1008, 2003.

77. Sommers, C. L., Park, C. S., Lee, J., Feng, C., Fuller, C. L., Grinberg, A., Hildebrand, J. A., Lacana, E., Menon, R. K., Shores, E. W., Samelson, L. E., and Love, P. E., A LAT mutation that inhibits T cell development yet induces lymphoproliferation, *Science* 296(5575), 2040–2043, 2002.

78. Aguado, E., Richelme, S., Nunez-Cruz, S., Miazek, A., Mura, A. M., Richelme, M., Guo, X. J., Sainty, D., He, H. T., Malissen, B., and Malissen, M., Induction of T helper type 2 immunity by a point mutation in the LAT adaptor, *Science* 296(5575), 2036–2040, 2002.

79. Hayes, S. M., Laird, R. M., and Love, P. E., Beyond alphabeta/gammadelta lineage commitment: TCR signal strength regulates gammadelta T cell maturation and effector fate, *Semin Immunol* 22(4), 247–251, 2010.

80. Lee, S. Y., Stadanlick, J., Kappes, D. J., and Wiest, D. L., Towards a molecular understanding of the differential signals regulating alphabeta/gammadelta T lineage choice, *Semin Immunol* 22(4), 237–246, 2010.

81. Lin, L. L., Wartmann, M., Lin, A. Y., Knopf, J. L., Seth, A., and Davis, R. J., cPLA2 is phosphorylated and activated by MAP kinase, *Cell* 72(2), 269–278, 1993.

82. Roux, P. P., and Blenis, J., ERK and p38 MAPK-activated protein kinases: A family of protein kinases with diverse biological functions, *Microbiol Mol Biol Rev* 68(2), 320–344, 2004.

83. Fischer, A. M., Katayama, C. D., Pages, G., Pouyssegur, J., and Hedrick, S. M., The role of erk1 and erk2 in multiple stages of T cell development, *Immunity* 23(4), 431–443, 2005.

84. Lee, S. Y., Coffey, F., Fahl, S. P., Peri, S., Rhodes, M., Cai, K. Q., Carleton, M., Hedrick, S. M., Fehling, H. J., Zuniga-Pflucker, J. C., Kappes, D. J., and Wiest, D. L., Noncanonical mode of ERK action controls alternative alphabeta and gammadelta T cell lineage fates, *Immunity* 41(6), 934–946, 2014.

85. Heasley, L. E., and Johnson, G. L., The beta-PDGF receptor induces neuronal differentiation of PC12 cells, *Mol Biol Cell* 3(5), 545–553, 1992.

86. Nguyen, T. T., Scimeca, J. C., Filloux, C., Peraldi, P., Carpentier, J. L., and Van Obberghen, E., Co-regulation of the mitogen-activated protein kinase, extracellular signal-regulated kinase 1, and the 90-kDa ribosomal S6 kinase in PC12 cells. Distinct effects of the neurotrophic factor, nerve growth factor, and the mitogenic factor, epidermal growth factor, *J Biol Chem* 268(13), 9803–9810, 1993.

87. Traverse, S., Seedorf, K., Paterson, H., Marshall, C. J., Cohen, P., and Ullrich, A., EGF triggers neuronal differentiation of PC12 cells that overexpress the EGF receptor, *Curr Biol* 4(8), 694–701, 1994.

88. Schlessinger, J., and Bar-Sagi, D., Activation of Ras and other signaling pathways by receptor tyrosine kinases, *Cold Spring Harb Symp Quant Biol* 59, 173–179, 1994.

89. Murphy, L. O., and Blenis, J., MAPK signal specificity: The right place at the right time, *Trends Biochem Sci* 31(5), 268–275, 2006.

90. Murphy, L. O., Smith, S., Chen, R. H., Fingar, D. C., and Blenis, J., Molecular interpretation of ERK signal duration by immediate early gene products, *Nat Cell Biol* 4(8), 556–564, 2002.

91. Dimitri, C. A., Dowdle, W., MacKeigan, J. P., Blenis, J., and Murphy, L. O., Spatially separate docking sites on ERK2 regulate distinct signaling events in vivo, *Curr Biol* 15(14), 1319–1324, 2005.

92. Shin, S., Dimitri, C. A., Yoon, S. O., Dowdle, W., and Blenis, J., ERK2 but not ERK1 induces epithelial-to-mesenchymal transformation via DEF motif-dependent signaling events, *Mol Cell* 38(1), 114–127, 2010.

93. Carlson, S. M., Chouinard, C. R., Labadorf, A., Lam, C. J., Schmelzle, K., Fraenkel, E., and White, F. M., Large-scale discovery of ERK2 substrates identifies ERK-mediated transcriptional regulation by ETV3, *Sci Signal* 4(196), rs11, 2011.

94. Miyazaki, T., Two distinct steps during thymocyte maturation from CD4–CD8– to CD4+CD8+ distinguished in the early growth response (Egr)-1 transgenic mice with a recombinase-activating gene-deficient background, *J Exper Med* 186(6), 877–885, 1997.

95. Carleton, M., Haks, M. C., Smeele, S. A., Jones, A., Belkowski, S. M., Berger, M. A., Linsley, P., Kruisbeek, A. M., and Wiest, D. L., Early growth response transcription factors are required for development of CD4(–)CD8(–) thymocytes to the CD4(+)CD8(+) stage, *J Immunol* 168(4), 1649–1658, 2002.

96. Xi, H., and Kersh, G. J., Early growth response gene 3 regulates thymocyte proliferation during the transition from CD4–CD8– to CD4+CD8+, *J Immunol* 172(2), 964–971, 2004.

97. Shao, H., Kono, D. H., Chen, L. Y., Rubin, E. M., and Kaye, J., Induction of the early growth response (Egr) family of transcription factors during thymic selection, *J Exp Med* 185(4), 731–744, 1997.

98. Bain, G., Cravatt, C. B., Loomans, C., Alberola-Ila, J., Hedrick, S. M., and Murre, C., Regulation of the helix-loop-helix proteins, E2A and Id3, by the Ras-ERK MAPK cascade, *Nat Immunol* 2(2), 165–171, 2001.

99. Xi, H., Schwartz, R., Engel, I., Murre, C., and Kersh, G. J., Interplay between RORgammat, Egr3, and E proteins controls proliferation in response to pre-TCR signals, *Immunity* 24(6), 813–826, 2006.

100. Rivera, R. R., Johns, C. P., Quan, J., Johnson, R. S., and Murre, C., Thymocyte selection is regulated by the helix-loop-helix inhibitor protein, Id3, *Immunity* 12(1), 17–26, 2000.

101. Engel, I., Johns, C., Bain, G., Rivera, R. R., and Murre, C., Early thymocyte development is regulated by modulation of e2a protein activity, *J Exp Med* 194(6), 733–746, 2001.

102. Ueda-Hayakawa, I., Mahlios, J., and Zhuang, Y., Id3 restricts the developmental potential of gammadelta lineage during thymopoiesis, *J Immunol* 182(9), 5306–5316, 2009.

103. Verykokakis, M., Boos, M. D., Bendelac, A., Adams, E. J., Pereira, P., and Kee, B. L., Inhibitor of DNA binding 3 limits development of murine slam-associated adaptor protein-dependent "innate" gammadelta T cells, *PLoS One* 5(2), e9303.

104. Murre, C., Helix-loop-helix proteins and lymphocyte development, *Nat Immunol* 6(11), 1079–1086, 2005.

105. Engel, I., and Murre, C., E2A proteins enforce a proliferation checkpoint in developing thymocytes, *Embo J* 23(1), 202–211, 2004.

106. Bain, G., Romanow, W. J., Albers, K., Havran, W. L., and Murre, C., Positive and negative regulation of V(D)J recombination by the E2A proteins, *J Exp Med* 189(2), 289–300, 1999.

107. Barndt, R., Dai, M. F., and Zhuang, Y., A novel role for HEB downstream or parallel to the pre-TCR signaling pathway during alpha beta thymopoiesis, *J Immunol* 163(6), 3331–3343, 1999.

108. Wojciechowski, J., Lai, A., Kondo, M., and Zhuang, Y., E2A and HEB are required to block thymocyte proliferation prior to pre-TCR expression, *J Immunol* 178(9), 5717–5726, 2007.

109. Barndt, R. J., Dai, M., and Zhuang, Y., Functions of E2A-HEB heterodimers in T-cell development revealed by a dominant negative mutation of HEB, *Mol Cell Biol* 20(18), 6677–6685, 2000.

110. Zhang, B., Lin, Y. Y., Dai, M., and Zhuang, Y., Id3 and Id2 act as a dual safety mechanism in regulating the development and population size of innate-like gammadelta T cells, *J Immunol* 192(3), 1055–1063, 2014.

111. Dekker, J., Marti-Renom, M. A., and Mirny, L. A., Exploring the three-dimensional organization of genomes: Interpreting chromatin interaction data, *Nat Rev Genet* 14(6), 390–403, 2013.

112. Lin, Y. C., Benner, C., Mansson, R., Heinz, S., Miyazaki, K., Miyazaki, M., Chandra, V., Bossen, C., Glass, C. K., and Murre, C., Global changes in the nuclear positioning of genes and intra- and interdomain genomic interactions that orchestrate B cell fate, *Nat Immunol* 13(12), 1196–1204, 2012.

113. Lin, Y. C., Jhunjhunwala, S., Benner, C., Heinz, S., Welinder, E., Mansson, R., Sigvardsson, M., Hagman, J., Espinoza, C. A., Dutkowski, J., Ideker, T., Glass, C. K., and Murre, C., A global network of transcription factors, involving E2A, EBF1 and Foxo1, that orchestrates B cell fate, *Nat Immunol* 11(7), 635–643, 2010.

114. Lin, Y. C., and Murre, C., Nuclear location and the control of developmental progression, *Curr Opin Genet Dev* 23(2), 104–108, 2013.

115. Murre, C., The epigenetics of early lymphocyte development, *Cold Spring Harb Symp Quant Biol* 78, 43–49, 2013.

116. Jones, M. E., and Zhuang, Y., Acquisition of a functional T cell receptor during T lymphocyte development is enforced by HEB and E2A transcription factors, *Immunity* 27(6), 860–870, 2007.

# 7 Phenotypic and Functional Characterization of Regulatory T Cell Populations

*Louis-Marie Charbonnier and Talal A. Chatila*

## CONTENTS

### ABSTRACT

Regulatory T (Treg) cells that express forkhead box P3 (Foxp3) are pivotal to the maintenance of peripheral immunological tolerance [1–3]. They play a critical role in controlling autoimmunity and limiting tissue destruction and inflammation. Failure of Treg cell differentiation in humans due to loss of function mutations in *FOXP3* results in fatal autoimmune lymphoproliferative disease, called immune dysregulation, polyendocrinopathy, enteropathy, X-linked (IPEX) [1,2,4]. A similar disease (scurfy phenotype) is also seen in *Foxp3* mutant mice [5–8]. Consistent with this role of Foxp3 in mediating Treg cell phenotypic stability, it has been shown in mice that Foxp3 deficiency does not impair Treg cell development per se, but rather abrogates immunoregulatory function [6,9]. In this chapter, we define methods to study Treg cell generation, phenotype, suppressive capacities, and stability. We also describe how to expand or deplete Treg cell population as well as the different *Foxp3*-related mouse strains available to study Treg cell biology.

## 7.1   REGULATORY T CELL DEVELOPMENT

Treg cell population could be divided in two main subsets, based on their origin in the thymus or the periphery. Thymic derived Treg (tTreg) cells are induced to express Foxp3 upon interaction with agonistic peptide antigens in specialized niches in the thymic medulla. Affinity to self-antigen by TCR of single positive CD4 thymocytes plays a crucial role in Foxp3 induction and the acquisition of a Treg cell epigenetic signature [1,10]. Peripherally induced Treg (pTreg) cells are differentiated from naïve CD4+ T cells exposed to non-self antigens (e.g., allergens, food, or commensal microbiota) within specialized niches in peripheral tissues, most notably at environmental interfaces in the lung, gut, and skin.

### 7.1.1   tTREG CELL DIFFERENTIATION

Recognition of self-antigens presented by MHC molecules on thymic antigen-presenting cells (APCs) is crucial for determining the fate of developing αβ T cells. Recognition of self can elicit diametrically opposite outcomes depending on the affinity of TCR-peptide/MHC interactions. On one hand, low affinity interactions are essential for thymocyte survival and commitment to either the CD4+ or CD8+ T cell lineage (positive selection of thymocytes). On the other hand, high affinity for self-antigens leads to apoptosis of thymocytes (negative selection). In between, intermediate affinity can skew cells to alternative fates, such as Treg cell differentiation (Figure 7.1a). Induction of *Foxp3* expression by Treg cell progenitors is mediated by a subtle combination of TCR, costimulation, cytokine, and calcium signaling (Figure 7.1b). After losing CD8 expression due to TCR affinity to MHC class II, CD4 single positive T cells can turn into tTreg cell progenitors by expressing GITR and CD25. Then, Foxp3 expression leads to the generation of recently differentiated thymic Treg cells and fully mature tTreg cells expressing neuropillin 1 (Nrp1) (Figure 7.1c).

Flow Cytometry Staining for tTreg Progenitors and Cells
- Stain thymic cells (2–3 millions) in suspension for surface markers in 300 μL of PBS/0.5%FCS for 30 min on ice.

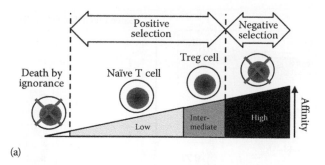

(a)

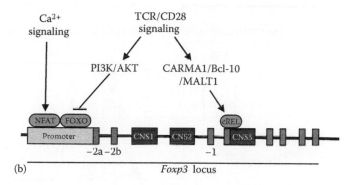

(b)

**FIGURE 7.1** Thymic Treg cell differentiation. (a) Schematic representation of thymic selection relative to T cell receptor (TCR) affinity to self-antigen/major histology complex (MHC). No affinity will induce death by ignorance. Low affinity will lead to the generation of naive T cells, whereas intermediate affinity will lead to the generation of Treg cells (positive selection). A high affinity will lead to apoptosis of autoreactive thymocytes (negative selection). (b) During tTreg cell development, the recognition of a self-antigen–MHC class II complex by the TCR triggers NF-κB family transcription factor cREL translocation to the nucleus and binding to the conserved noncoding sequence 3 (CNS3) region of the *Foxp3* locus, via the CARD-containing MAGUK protein 1 (CARMA1)/B cell lymphoma 10 (Bcl-10)/mucosa-associated lymphoid tissue lymphoma translocation protein 1 (MALT1) complex. The Forkhead box O (FOXO) transcription factors promote Foxp3 expression via binding to the promoter. A strong TCR/CD28 signaling will trigger the cytosolic translocation of FOXO1/3, via phosphoinositide 3-kinase (PI3K)/AKT signaling and inhibit Foxp3 induction. The nuclear factor of activated T cells (NFAT) pathway, which is activated downstream of $Ca^{2+}$ signaling, have a positive role in the induction of *Foxp3* through its binding to the Foxp3 promoter. *(Continued)*

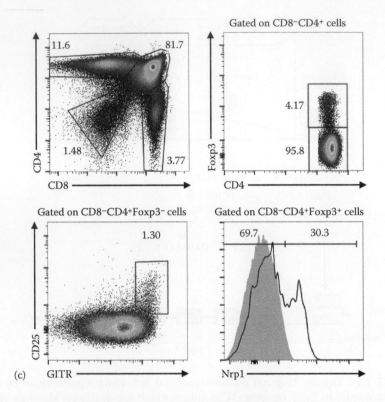

**FIGURE 7.1 (CONTINUED)**   Thymic Treg cell differentiation. (c) Upper left panel: CD4/CD8 expression on thymocytes reveals four distinct populations: CD4⁻CD8⁻ double nega-tive, CD4⁺CD8⁺ double positive, and CD8 and CD4 single positive T cells. Upper right panel: Among CD4 single positive cells (CD4⁺CD8⁻), tTreg cells represent the Foxp3⁺ cell population. Bottom left panel: Among the CD4⁺CD8⁻Foxp3⁻ cell population, tTreg cell progenitors represent the GITR⁺CD25⁺ population. Bottom right panel: Among the tTreg CD4⁺CD8⁻Foxp3⁺ cell population, mature tTreg cell progenitors represent the Nrp1⁺ population.

- Recommended cocktail and dilution:

| Antigen | Clone | Fluorochrome | Dilution |
| --- | --- | --- | --- |
| CD8 | 53-6.7 | FITC | 1/500 |
| CD4 | RM4-5 | BV605 | 1/500 |
| Nrp1 | 3DS304M | PE | 1/300 |
| GITR | DTA-1 | PE-Cy7 | 1/300 |
| CD25 | PC61 | APC | 1/500 |

- Wash with large volume of PBS/0.5%FCS, spin at 400 g 5 min and resuspend the pellet in 300 µL of eBioscience Fixation/Permeabilization buffer and incubate for 30 min on ice.
- Wash twice with 2 mL of eBioscience Permeabilization buffer and resuspend in 300 µL of eBioscience Permeabilization buffer with Foxp3 eF450 (FJK-16S clone, 1/500 dilution) for 30 min on ice.
- Wash twice with 2mL of eBioscience Permeabilization buffer and resuspend in 300 µL of PBS/0.5%FCS.
- Read on a flow cytometer.

Recommendations: Use single fluorochrome stained cells as control for compensation of spillover from one channel to another.

### 7.1.2 PERIPHERALLY INDUCED TREG CELL GENERATION

In order to investigate the potential of naïve T cells to convert into pTreg cells in periphery, one strategy consists in administrating chronically antigen in absence of adjuvant [11–13]. Antigen specific TCR transgenic DO11.10 (Jax Stock number: 003303) and OTII (Jax stock number: 004194) mice carry a MHC class II restricted rearranged T cell receptor transgene on BALB/c and C57BL/6 background, respectively, that react to ovalbumin peptide antigen (OVA$_{323-339}$). Sorted naive CD4$^+$CD62L$^{high}$CD44$^{lo}$ T cells are isolated from DO11.10 or OTII mice and adoptively transferred in BALB/c or C57BL/6 mice, respectively. A 1% OVA (grade II; Sigma-Aldrich) solution dissolved in drinking water for 5 consecutive days is administrated to the mice. The capacity of naïve T cells to convert in pTreg cells is analyzed by evaluating the frequency of Foxp3$^+$ cells among the adoptively transferred CD90.1$^+$CD4$^+$ T cells in the spleen, mesenteric lymph nodes, and small and large intestine.

### 7.1.3 IN VITRO INDUCED TREG CELL GENERATION

The capacity of CD4$^+$ T cells to convert into induced Treg (iTreg) cells can be evaluated in vitro [14]. Sorted mouse naive CD4$^+$CD62L$^{high}$CD44$^{lo}$ T cells (1 × 10$^6$/mL) are cultured with plate-bound anti-CD3/anti-CD28 (5 µg/mL each), recombinant TGF-β1 (dose response 0/0.5/1/5 ng/mL). After 4 days, the iTreg cell frequency is analyzed by flow cytometry for Foxp3 expression [15]. In addition to Foxp3 expression, iTreg cells also share with tTreg expression of markers such as CD25 and suppressive capacities that can be assessed in vitro or in vivo (see Section 7.3) [12,14]. The capacity of human naïve CD4$^+$CD45RA$^+$CCR7$^+$ T cells to become iTreg cells can also be evaluated using similar in vitro differentiation assay [16].

## 7.2 STUDYING REGULATORY T CELL POPULATIONS

Human and mouse Treg cells expressed common and specific canonical markers, including Foxp3, CD25, CTLA-4, Helios, or GITR [1].

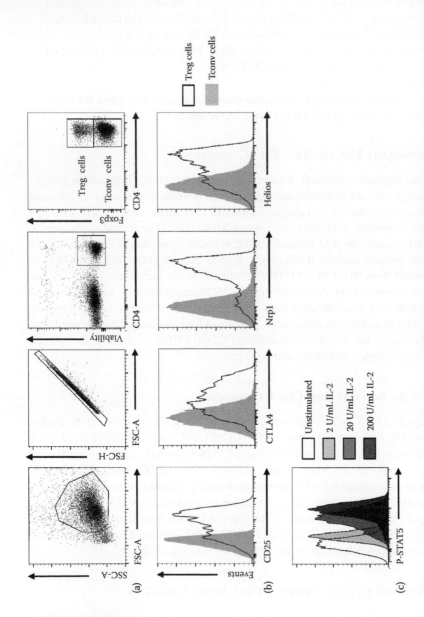

**FIGURE 7.2**  Treg cell phenotypic characterization. (a) Gating strategy including side scatter-height (SSC-H) versus forward scatter-height (FSC-H) for lymphocyte gate, FSC-H versus FSC-Area (FSC-A) to eliminate doublet events, CD4 versus viability to eliminate dead cells, and CD4 versus Foxp3 to discriminate conventional T cells (Tconv) and regulatory T (Treg) cells. (b) Overlays of Tconv and Treg cells for the different Treg cell canonical markers (CD25, CTLA-4, Nrp1, and Helios). (c) Phospho-STAT5 staining on Treg cells in response to different concentrations of IL-2 (0, 2, 20, 200 U/mL).

### 7.2.1 PHENOTYPIC CHARACTERIZATION OF TREG CELL POPULATIONS

Treg cells are characterized by the expression of canonical markers that are common regardless of their origin. In secondary lymphoid organs, Treg cells are characterized by high expression of Foxp3, CD25, GITR, Nrp1, Helios, and CTLA-4 as compared to conventional CD4 T cells (Figure 7.2a and b). Nrp1 and Helios expression have been described as markers to discriminate between pTreg (Nrp1⁻Helios⁻) and tTreg (Nrp1⁺Helios⁺) cell population. However, in the context of inflammation, pTreg cells can express Nrp1 [13]. For the staining per se, refer to Section 7.1.1.

| Flow Cytometry Staining | | | |
|---|---|---|---|
| **Surface Antigen** | **Clone** | **Fluorochrome** | **Dilution** |
| CD4 | RM4-5 | BV605 | 1/500 |
| Nrp1 | 3DS304M | PE | 1/300 |
| GITR | DTA-1 | PE-Cy7 | 1/300 |
| CD25 | PC61 | eF450 | 1/500 |
| **Intracellular Antigen** | **Clone** | **Fluorochrome** | **Dilution** |
| CTLA-4 | UC10-4B9 | FITC | 1/300 |
| Helios | 22F6 | APC | 1/300 |
| Foxp3 | FJK-16s | AF700 | 1/500 |

### 7.2.2 TRANSGENIC MICE

Several genetically modified mouse strains have been developed based on *Foxp3* locus during the last decade in order to investigate Treg cell biology using bacterial artificial chromosome (BAC) transgenes or germline gene knock in (KI) strategies. Some of these mouse strains allow the visualization and isolation of Treg cells by virtue of fluorescent protein expression under the control of *Foxp3* promoter, for example, *Foxp3*EGFP (Jax stock number: 006772) and *Foxp3*RFP mice (Jax stock number: 008374), whose Treg cells express the enhanced green fluorescent protein (EGFP) and red fluorescent protein (RFP), respectively [17,18]. Other strains have been developed to study Foxp3 deficiency like *Foxp3*K276X (Jax stock number: 021144), *Foxp3*ΔEGFP (Jax stock number: 020654), and *Foxp3*nullGFP mice [6,7,9]. The latter two strains also allow the tracking of Foxp3-deficient Treg cells due to mutant *Foxp3* allele-driven GFP expression. The introduction of a Cre recombinase activity under the control of *Foxp3* locus enables strategies that target or overexpress genes specifically in Treg cells. *Foxp3*YFPcre (Jax stock number: 016959) is a mouse strain that expresses a Cre recombinase and a yellow fluorescent protein from the endogenous *Foxp3* locus. Another mouse strain, *Foxp3*EGFPcre, carries a BAC transgene that drives expression of a Cre recombinase and EGFP under control of the *Foxp3* promoter (Jax stock number: 008694) [19,20]. The Rudensky group also generated an inducible *Foxp3*-driven Cre recombinase mutant mouse, *Foxp3*eGFPCreERT2 (Jax stock number: 016961) [21]. In these mice, the Cre recombinase is expressed only in

Foxp3+ cells and is fused with two modified estrogen receptors that selectively bind tamoxifen (ERT2). By administering tamoxifen to mice, the CreERT2 fusion protein expressed in Treg cells translocates from the cytosol to the nucleus and allows the recombination of marked (floxed) alleles.

Another set of mice allows the Treg cell-specific depletion by using diphtheria toxin. *Foxp3*DTR mice have been generated using both BAC and KI strategies by different laboratories, *Foxp3*DTR BAC from the Chatila laboratory, and *Foxp3*EGFP-DTR from the Rudensky laboratory (Jax stock number: 016958) [22]. The Rudensky laboratory also targeted the three different conserved noncoding sequences (CNS) within the *Foxp3* locus CNS1/2/3 knockout mice and highlighted the different roles of those three sequences in the Treg cell biology. CNS1 has been shown to be crucial for pTreg generation, CNS2 for maintenance of Foxp3 expression and Treg stability, and CNS3 for tTreg generation [23].

## 7.3 REGULATORY T CELL FUNCTIONS

### 7.3.1 EVALUATION OF STAT5 PHOSPHORYLATION CAPACITIES

Due to high expression of CD25, Treg cells are more sensitive to IL-2 than conventional T cells, and phosphorylation of STAT5 in response to low dose of IL-2 leads to proliferation of Treg cell population. Evaluation of the capacity of Treg cells to phosphorylate STAT5 in response to IL-2 is informative on their capacity to proliferate.

Spleen cells in suspension are stimulated for 30 min in complete medium at a concentration of 5 M/mL in the presence of recombinant mouse IL-2 (0, 2, 20, or 200 U/mL). After stimulation, cells are directly put on ice to stop stimulation and fixed by adding 1 volume of PBS/4% paraformaldehyde. After 20 min of incubation on ice, cells are washed twice with cold PBS and permeabilized by resuspension in ice-cold 90% methanol for 30 min. After two washes with ice-cold PBS, the cell pellets are resuspended with the antibody cocktail in ice-cold PBS and incubated on ice for 30 min. Then, the cells are washed two times in ice-cold PBS and can be read on a flow cytometer after being resuspended in PBS (Figure 7.2c).

| Flow Cytometry Staining | | | |
|---|---|---|---|
| Antigen | Clone | Fluorochrome | Dilution |
| Phospho-STAT5 | SRBCZX | APC | 1/100 |
| CD4 | RM4-5 | BV605 | 1/500 |
| Foxp3 | FJK-16s | eF450 | 1/500 |

### 7.3.2 EVALUATION OF IN VITRO SUPPRESSIVE CAPACITIES

In vitro suppression assay with Treg cells consists of mixing responder cells (CD4+ Tconv cells) with different amounts of Treg cells in the presence of anti-CD3/CD28 stimulation (Figure 7.3a and b). Tconv cells are stained for a proliferation dye following manufacturer instructions and $10^5$ cells per well of a round bottom, 96-well

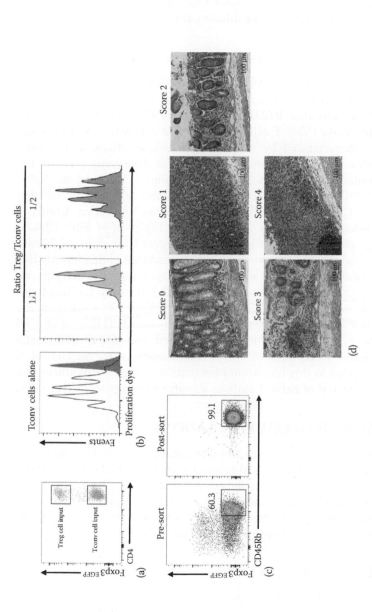

**FIGURE 7.3** Suppressive capacities of Treg cells. (a and b) In vitro suppression assay by Treg cells on Tconv cells proliferation. (a) Gating strategy to gate on Tconv cell input for proliferation analysis. (b) Dilution of the proliferation dye of T conv cells. Left panel: Tconv cells alone, unstimulated (dark gray) versus stimulated with anti-CD3/CD28 (black). Middle panel: 1/1 ratio Treg/Tconv cells. Right panel: 1/2 ratio Treg/Tconv cells. (c and d) In vivo suppression assay by Treg cells. (c) CD45Rb versus Foxp3$^{EGFP}$ staining among CD4$^+$ T cells prior and post sorting for CD45Rb$^{high}$CD4$^+$ T cells. (d) Examples of scores from 0 to 4 on colon section after H&E staining. Score 0: no inflammation; Score 1: mild, scattered infiltrates; Score 2: moderate infiltrates without loss of epithelium integrity; Score 3: moderate and diffuse or severe inflammation; Score 4: severe inflammation associated with loss of the epithelial barrier integrity.

plate are stimulated with previously plate-bound anti-CD3/anti-CD28 (2 µg/mL, 5 µg/mL, respectively). The addition of different amounts of Treg cells at a ratio of 1/1 to 1/4 Treg cell per Tconv cell is required to evaluate the capacity of Tregs cell to suppress Tconv cell proliferation. After 4 to 5 days, the proliferation of the Tconv cell population is evaluated by the dilution of the proliferation dye.

### 7.3.3 EVALUATION OF IN VIVO SUPPRESSIVE CAPACITIES

One of the well-accepted in vivo models for studying Treg cell suppressive functions exploits their capacity to control colitis induced by the transfer of naïve $CD45Rb^{high}$ $CD4^+$ T cells into RAG1-deficient mice [24]. In the absence of Treg cell cotransfer, naïve $CD4^+$ T cells induce severe colitis in RAG1-deficient mouse recipients, characterized by body weight loss, diarrhea, and colonic leucocyte infiltration and inflammation. Cotransfer of Treg cells prevents the development of inflammation and host morbidity (Figure 7.3c and d). Naïve ($CD4^+CD45Rb^{high}GFP^-$) and Treg ($CD4^+GFP^+$) cells are, respectively, isolated from the spleen of CD45.1 $Foxp3^{EGFP}$ and CD45.2 $Foxp3^{EGFP}$ mice. Colitis is induced in RAG1-deficient males by intraperitoneal injection of $5.10^5$ CD45.1 naïve $\pm 2.10^5$ Treg cells. Mice are weighed and monitored for signs of disease twice weekly. This protocol of in vivo lymphopenia-induced colitis has distinct advantages as compared to in vitro suppression assay. First, it is more physiologic than in vitro tissue culture-based assay. Second, it is useful for investigating the stability of Treg cells under inflammatory conditions by evaluating the percentage of $Foxp3^+$ cells coming from the Treg cell input (CD45.2) at the end of the experiment. Finally, as we previously reported, the role of pTreg cells derived from the naïve T cell input (CD45.1) is also essential for the control of the disease [25,26]. Due to this important observation, this protocol allows one to investigate the potential of naïve T cells as a source of competent pTreg cells.

## 7.4 MODULATING TREG CELL POPULATIONS

In the context of autoimmunity, allergy/asthma, and transplantation, efficient strategies have been described for educating and enhancing antigen-specific Treg cell population, leading to tolerance induction in the respective diseases. These approaches, pioneered in mouse models, have translated into promising human clinical trials [27,28]. On the other hand, depleting the Treg cell population represents an attractive strategy to restore antitumor adaptive immune responses [29–31].

### 7.4.1 IN VIVO EXPANSION OF TREG CELL POPULATION BY IL-2

IL-2 is a critical cytokine for immune regulation via its effects on proliferation of Treg cells. Treg cells are exquisitely sensitive to this cytokine as compared to other immune cell types due to their high expression of the IL-2 receptor alpha chain (IL-2Rα, known as CD25) [32]. Human studies using IL-2-based therapeutic regimens have been reported to successfully enhance Treg cell responses and induce favorable clinical responses in autoimmunity and graft versus host disease [33–35].

A method to specifically target IL-2 to Treg cells has been developed in mice by trapping IL-2 in immune complexes with a specific anti-IL2 monoclonal antibody [36]. IL-2/anti-IL-2 complexes are formed by incubating 1 µg recombinant mouse IL-2 and 9 µg of functional grade purified anti-mouse IL-2 (Clone: JES6–1A12) for 30 min at 37°C. The complexes are administered intraperitoneally for 3 consecutive days. On day 5, Treg cell population is expanded, representing approximately 40% of the CD4 compartment in secondary lymphoid organs, for example, spleen and peripheral lymph nodes. It has been shown that these expanded Treg cells are also activated and present a larger suppressive capacity than Treg cells isolated from unmanipulated mice [37,38].

### 7.4.2 Treg Cell Depletion

Acute in vivo ablation of Treg cells at the adult stage leads to the development of fatal autoimmunity, suggesting a permanent and sustained suppression by Treg cells of self-reactive T cells [22,39]. Transient depletion of Treg cells favors anti-tumor immunity [40]. In order to deplete Treg cell population, two strategies can be used. First, the use of anti-CD25 depleting monoclonal antibody (PC61) allows the depletion of the cells expressing high levels of CD25, Treg cells, and innate lymphoid cells. One single injection of 500 µg of the anti-IL-2Rα mono-clonal antibody PC61 intraperitoneally is performed and the peak of Treg cell depletion is observed in PC61-treated mice 6 days after, as compared to isotype control-treated mice [41]. The second strategy consists in using the $Foxp3^{DTR}$ mice. In this case, repetitive injection of 1 mg/kg of diphtheria toxin (DT) for 3 days allows the depletion of the Treg cell population with a peak of depletion of the Treg cell population observed at day 5 [42]. Ideally, this strategy requires two negative control groups—vehicle-treated $Foxp3^{DTR}$ mice and DT-treated WT mice—in order to exclude the effect of the injection of DT.

## 7.5 INVESTIGATION OF TREG CELL EPIGENETIC STATUS AND STABILITY

### 7.5.1 Treg Specific Demethylation Region

Treg methylation signature has been shown to be independent of Foxp3 expression in tTreg cells and affects several genes like *Foxp3*, *Ctla4*, *Il2ra*, *Eos*, *Helios*, and *Gitr* [15]. Studying the methylation status of CpG within those loci represents an attractive strategy to evaluate the capacity of cells to express those Treg cell markers. Genomic DNA from sorted cells is treated by bisulfite to convert unmethylated cytosines into uracil leaving methylated cytosines unchanged. Bisulfite conversion of DNA is performed using a DNA methylation kit according to the manufacturer instructions. Converted DNA is amplified by methylation-specific primer sequences for the different loci (see following table). The PCR product is purified and inserted into a TOPO TA cloning vector. The ligation product is used to transform competent bacteria (10-beta compe-tent *E. coli*) and clones are selected on kanamycin and/or ampicillin (50 mg/mL each). Plasmid DNA was extracted by the mini-prep spin columns method and clones that present a fragment at the same size of the PCR amplicon after EcoRI digestion are selected. Sanger sequencing is done with M13R primer. Blast analysis was done by comparing the M13R sequence and converted gene sequence.

| Gene Locus | Primer Forward (5'→3') | Primer Reverse (5'→3') |
| --- | --- | --- |
| *Foxp3* CNS2 | TTTTGGGTTTTTTTGGTATTTAAGA | TTAACCAAATTTTTCTACCATTAAC |
| *Ctla4* Exon 2 | TGGTGTTGGTTAGTAGTTATGGTGT | AAAATTCCACCTTACAAAAATACAATC |
| *Il2ra* Intron1a | TTTTAGAGTTAGAAGATAGAAGGTAT GGAA | TCCCAATACTTAACAAAACCACATAT |
| *Eos* Intron1b | TAAGAAATTGGGTGTGGTATATGTA | TTTCCCCTACTAAAACTCCTTAAAC |
| *Helios* Intron3a | AGGATGGTTTTTATTGAAGGTGAT | ATACACACCAAACAAACACTACACC |
| *Gitr* Exon5 | GAGGTGTAGTTGTTAGTTGAGGATGT | AACCCCTACTCTCACCAAAAATATAA |

## 7.5.2　Ex-Treg Quantification and Pathogenicity

Several studies reported the instability of Foxp3 expression in Treg cells in inflamed tissues, leading to the generation of pathogenic ex-Treg cells that produce inflammatory cytokines and contribute to disease [12,43–45]. By crossing mice carrying a loxP-flanked STOP sequence followed by the yellow fluorescent protein gene (YFP) inserted into the *Rosa26* locus (*Rosa26*[YFP]) (Jax stock number: 006148) with those carrying a *Foxp3*[EGFPCre] BAC transgene (Jax stock number: 008694), Treg and ex-Treg cell populations could be tracked [44]. Whereas Treg cells express both fluorescent proteins EGFP and YFP, ex-Treg cells became single positive by losing the expression of the EGFP driven by *Foxp3* promoter (Figure 7.4). Treg and ex-Treg cell populations can be analyzed for cytokine production and can be also sorted and tested for pathogenicity after adoptive transfer [43].

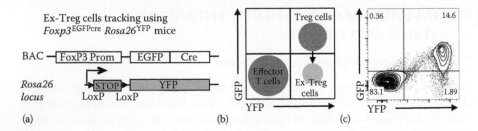

(a)　　　　　　　　　　　　　　　　(b)　　　　　　　　　　　　　　(c)

**FIGURE 7.4** Treg cell stability. (a) Schematic representation of the mouse line with Treg cell-specific expression of yellow fluorescent protein (YFP), allowing fate mapping of Treg cells. *Foxp3*[EGFPCre]*Rosa26*[YFP] mice were obtained by crossing mice harboring *Foxp3*[EGFPCre] BAC with those carrying a knock-in allele at the Rosa26 locus encoding the YFP under control of a floxed stop signal (*Rosa26*[YFP]). (b) Schematic representation of flow cytometry for CD4+ T cell subsets of *Foxp3*[EGFPCre]*Rosa26*[YFP] mice based on GFP and YFP expression. Effector T cells (Teff) never expressed Foxp3 and are GFP⁻YFP⁻. Treg cells express both GFP and YFP, whereas ex-Treg cells lost the expression of GFP but still express YFP. (c) Example of frequencies of Teff, Treg, and ex-Treg cells in the spleen of *Foxp3*[EGFPCre]*Rosa26*[YFP] mice in steady state.

## REFERENCES

1. Josefowicz, S. Z., L. F. Lu, and A. Y. Rudensky, Regulatory T cells: Mechanisms of differentiation and function. *Annu Rev Immunol*, 2012. **30**: 531–564.
2. Verbsky, J. W., and T. A. Chatila, Immune dysregulation, polyendocrinopathy, enteropathy, X-linked (IPEX) and IPEX-related disorders: An evolving web of heritable autoimmune diseases. *Curr Opin Pediatr*, 2013. **25**(6): 708–714.
3. Hori, S., T. Nomura, and S. Sakaguchi, Control of regulatory T cell development by the transcription factor Foxp3. *Science*, 2003. **299**(5609): 1057–1061.
4. Chatila, T. A. et al., JM2, encoding a fork head-related protein, is mutated in X-linked autoimmunity-allergic disregulation syndrome. *J Clin Invest*, 2000. **106**(12): R75–R81.
5. Fontenot, J. D., M. A. Gavin, and A. Y. Rudensky, Foxp3 programs the development and function of CD4+CD25+ regulatory T cells. *Nat Immunol*, 2003. **4**(4): 330–336.
6. Lin, W. et al., Regulatory T cell development in the absence of functional Foxp3. *Nat Immunol*, 2007. **8**(4): 359–368.
7. Lin, W. et al., Allergic dysregulation and hyperimmunoglobulinemia E in Foxp3 mutant mice. *J Allergy Clin Immunol*, 2005. **116**(5): 1106–1115.
8. Wildin, R. S. et al., X-linked neonatal diabetes mellitus, enteropathy and endocrinopathy syndrome is the human equivalent of mouse scurfy. *Nat Genet*, 2001. **27**(1): 18–20.
9. Gavin, M. A. et al., Foxp3-dependent programme of regulatory T-cell differentiation. *Nature*, 2007. **445**(7129): 771–775.
10. Morikawa, H., and S. Sakaguchi, Genetic and epigenetic basis of Treg cell development and function: From a FoxP3-centered view to an epigenome-defined view of natural Treg cells. *Immunol Rev*, 2014. **259**(1): 192–205.
11. Mucida, D. et al., Oral tolerance in the absence of naturally occurring Tregs. *J Clin Invest*, 2005. **115**(7): 1923–1933.
12. Noval Rivas, M. et al., Regulatory T cell reprogramming toward a Th2-cell-like lineage impairs oral tolerance and promotes food allergy. *Immunity*, 2015. **42**(3): 512–523.
13. Weiss, J. M. et al., Neuropilin 1 is expressed on thymus-derived natural regulatory T cells, but not mucosa-generated induced Foxp3+ T reg cells. *J Exp Med*, 2012. **209**(10): 1723–1742, S1.
14. Fantini, M. C. et al., In vitro generation of CD4+ CD25+ regulatory cells from murine naive T cells. *Nat Protoc*, 2007. **2**(7): 1789–1794.
15. Charbonnier, L. M. et al., Control of peripheral tolerance by regulatory T cell-intrinsic Notch signaling. *Nat Immunol*, 2015. **16**(11): 1162–1173.
16. Charbonnier, L. M. et al., Regulatory T-cell deficiency and immune dysregulation, polyendocrinopathy, enteropathy, X-linked-like disorder caused by loss-of-function mutations in LRBA. *J Allergy Clin Immunol*, 2015. **135**(1): 217–227.
17. Haribhai, D. et al., Regulatory T cells dynamically control the primary immune response to foreign antigen. *J Immunol*, 2007. **178**(5): 2961–2972.
18. Wan, Y. Y., and R. A. Flavell, Identifying Foxp3-expressing suppressor T cells with a bicistronic reporter. *Proc Natl Acad Sci USA*, 2005. **102**(14): 5126–5131.
19. Rubtsov, Y. P. et al., Regulatory T cell-derived interleukin-10 limits inflammation at environmental interfaces. *Immunity*, 2008. **28**(4): 546–558.
20. Zhou, X. et al., Selective miRNA disruption in T reg cells leads to uncontrolled autoimmunity. *J Exp Med*, 2008. **205**(9): 1983–1991.
21. Rubtsov, Y. P. et al., Stability of the regulatory T cell lineage in vivo. *Science*, 2010. **329**(5999): 1667–1671.
22. Kim, J. M., J. Rasmussen, and A. Y. Rudensky, Regulatory T cells prevent catastrophic autoimmunity throughout the lifespan of mice. *Nat Immunol*, 2007. **8**(2): 191–197.
23. Zheng, Y. et al., Role of conserved non-coding DNA elements in the Foxp3 gene in regulatory T-cell fate. *Nature*, 2010. **463**(7282): 808–812.

24. Singh, B. et al., Control of intestinal inflammation by regulatory T cells. *Immunol Rev*, 2001. **182**: 190–200.

25. Schmitt, E. G. et al., IL-10 produced by induced regulatory T cells (iTregs) controls colitis and pathogenic ex-iTregs during immunotherapy. *J Immunol*, 2012. **189**(12): 5638–5648.

26. Haribhai, D. et al., A requisite role for induced regulatory T cells in tolerance based on expanding antigen receptor diversity. *Immunity*, 2011. **35**(1): 109–122.

27. Gregori, S., L. Passerini, and M. G. Roncarolo, Clinical outlook for type-1 and FOXP3(+) T regulatory cell-based therapy. *Front Immunol*, 2015. **6**: 593.

28. Safinia, N. et al., Regulatory T cells: Serious contenders in the promise for immunological tolerance in transplantation. *Front Immunol*, 2015. **6**: 438.

29. Bulliard, Y. et al., OX40 engagement depletes intratumoral Tregs via activating FcgammaRs, leading to antitumor efficacy. *Immunol Cell Biol*, 2014. **92**(6): 475–480.

30. Simpson, T. R. et al., Fc-dependent depletion of tumor-infiltrating regulatory T cells co-defines the efficacy of anti-CTLA-4 therapy against melanoma. *J Exp Med*, 2013. **210**(9): 1695–1710.

31. Byrne, W. L. et al., Targeting regulatory T cells in cancer. *Cancer Res*, 2011. **71**(22): 6915–6920.

32. Cheng, G., A. Yu, and T. R. Malek, T-cell tolerance and the multi-functional role of IL-2R signaling in T-regulatory cells. *Immunol Rev*, 2011. **241**(1): 63–76.

33. Pham, M. N., M. G. von Herrath, and J. L. Vela, Antigen-specific regulatory t cells and low dose of IL-2 in treatment of type 1 diabetes. *Front Immunol*, 2015. **6**: 651.

34. Koreth, J. et al., Interleukin-2 and regulatory T cells in graft-versus-host disease. *N Engl J Med*, 2011. **365**(22): 2055–2066.

35. Rosenzwajg, M. et al., Low-dose interleukin-2 fosters a dose-dependent regulatory T cell tuned milieu in T1D patients. *J Autoimmun*, 2015. **58**: 48–58.

36. Boyman, O. et al., Selective stimulation of T cell subsets with antibody-cytokine immune complexes. *Science*, 2006. **311**(5769): 1924–1927.

37. Webster, K. E. et al., In vivo expansion of T reg cells with IL-2-mAb complexes: Induction of resistance to EAE and long-term acceptance of islet allografts without immunosuppression. *J Exp Med*, 2009. **206**(4): 751–760.

38. Charbonnier, L. M. et al., CTLA4-Ig restores rejection of MHC class-II mismatched allografts by disabling IL-2-expanded regulatory T cells. *Am J Transplant*, 2012. **12**(9): 2313–2321.

39. Kim, J. et al., Cutting edge: Depletion of Foxp3+ cells leads to induction of autoimmunity by specific ablation of regulatory T cells in genetically targeted mice. *J Immunol*, 2009. **183**(12): 7631–7634.

40. Bos, P. D. et al., Transient regulatory T cell ablation deters oncogene-driven breast cancer and enhances radiotherapy. *J Exp Med*, 2013. **210**(11): 2435–2466.

41. Vokaer, B. et al., Critical role of regulatory T cells in Th17-mediated minor antigen-disparate rejection. *J Immunol*, 2010. **185**(6): 3417–3425.

42. Arvey, A. et al., Inflammation-induced repression of chromatin bound by the transcription factor Foxp3 in regulatory T cells. *Nat Immunol*, 2014. **15**(6): 580–587.

43. Komatsu, N. et al., Pathogenic conversion of Foxp3+ T cells into TH17 cells in autoimmune arthritis. *Nat Med*, 2014. **20**(1): 62–68.

44. Zhou, X. et al., Instability of the transcription factor Foxp3 leads to the generation of pathogenic memory T cells in vivo. *Nat Immunol*, 2009. **10**(9): 1000–1007.

45. Bailey-Bucktrout, S. L. et al., Self-antigen-driven activation induces instability of regulatory T cells during an inflammatory autoimmune response. *Immunity*, 2013. **39**(5): 949–962.

# 8 Natural Killer T (NKT) Cells in Mice and Men

*Nicola M. Heller, Rosa Berga-Bolanos,*
*Lynette Naler, and Jyoti Misra Sen*

## CONTENTS

### ABSTRACT

Natural killer T (NKT) cells are innate natural killer (NK)-like cells that express a limited repertoire of T cell receptors (TCRs). Akin to other innate cells, NKT cells acquire functional characteristics during development in the thymus. Similar to mature effector T cells, murine NKT cells have been subdivided into effector subsets based on expression of transcription factors Gata3, RORγt, or Tbet, and by the cytokine expression. In this chapter we have attempted to place the available research on human and murine NKT cells side by side even though making direct connections remains challenging, as research in the two species has progressed independently.

## 8.1 INTRODUCTION

Natural killer T (NKT) cells are a group of immune cells that are poised to respond to infection and shape the innate and adaptive immune response by rapidly producing cytokines.[1] NKT cells develop in the thymus from CD4+CD8+ double-positive (DP) thymocytes and migrate to peripheral locations such as the liver, spleen, lung, and gut. NKT cells are so called because they share characteristics with T cells and natural killer (NK) cells. Phenotypically, NKT cells share expression of molecules with conventional T cells, such as T cell receptor (TCR), and CD4 or CD8 and NK cells, such as NK1.1 (CD161), CD16, and CD56. In contrast to the diverse TCR repertoire of conventional T cells, NKT cells present a very limited TCR-α chain and β-chain repertoire.[2] Unlike conventional T cells, NKT cells recognize glycolipid antigen in the context of MHC-like CD1d molecules expressed on antigen-presenting cells (APCs).[3] Like NK cells, and in contrast to conventional T cells, NKT cells acquire functional characteristics, such as cytokine production during maturation, and rapidly respond upon encountering antigen. Recent studies have revealed that, like conventional effector T cells, NKT cells differentiate to produce type 1, type 2, or type 17 cells that produce IFN-γ, IL-4, and IL-17, respectively. Thus, NKT cells represent one of the earliest and best-studied cells in the growing family of innate lymphocytes. In this chapter, we summarize the current understanding of the development and function of NKT cells in mice and humans.

## 8.2 NATURAL KILLER T (NKT) CELLS IN MICE

Type 1 NKT cells in mice express a semi-invariant αβ TCR such that an invariant Vα14-Jα18 TCRα chain pairs with a limited Vβ repertoire consisting of Vβ2, Vβ7, or Vβ8 chains to form a functional TCR.[4–6] NKT cells rapidly produce cytokines and potentiate subsequent signaling cascades, which profoundly influence innate and adaptive immune responses from other cell types.

The prototypic antigen for invariant NKT cells is α-galactosylceramide (αGalCer), a synthetic form of a marine sponge glycolipid,[7] but they have also been shown to recognize other microbial glycolipids and self-antigens such as isoglobotrihexosylceramide (iGb3).[8,9] Cells expressing semivariant TCR with Vα10 recognize a distinct glycolipid antigen indicating diversity in NKT cell antigen range.[10] NKT cells can be easily identified by flow cytometry with anti-TCRβ antibodies and CD1d tetramers loaded with the antigen αGalCer. Likewise, NKT cells can be purified from the thymus and other organs by positive selection of CD1d-tetramer-positive cells using magnetic-activated cell sorting (MACS) methods.[11]

NKT cell development was originally defined based on surface expression of CD24, NK1.1, and CD44, and shown to proceed in a linear pattern of stages 0, 1, 2, and 3 (Table 8.1[12]). In this classification, NK1.1− CD44+ CD24+ NKT cells were designated as the most immature NKT precursors and NK1.1+ CD44+ CD24− NKT cells were designated as mature NKT cells.[13–15] NKT cells express CD4, or neither CD4 nor CD8 and are called CD4− CD8− double negative (DN). By contrast human NKT cells have a subset of NKT cells that coexpress CD8,[16] further described in following sections.

**TABLE 8.1**

**Mouse NKT Cell Subsets**

| Developmental Stages | Surface Markers | Functional Stages | Transcription Factors | Other Surface Markers |
|---|---|---|---|---|
| Stage 0/1 | CD24$^{hi/lo}$ CD44$^{lo}$ NK1.1$^-$ | NKT0 | PLZF$^{lo}$ | CD4$^-$ |
| Stage 2$^b$ | CD24$^{lo}$ CD44$^{hi}$ NK1.1$^-$ | NKT2$^c$ | PLZF$^{hi}$T-bet$^-$GATA3$^+$ROR$\gamma$t$^-$ | CD4$^+$ IL-17RB$^{+/-}$ |
| | | NKT17 | PLZF$^{int}$ T-bet$^-$ GATA3$^+$ ROR$\gamma$t$^+$ | CD4$^-$ IL-17RB$^{+/-}$ |
| Stage 3 | CD24$^{lo}$ CD44$^{hi}$ NK1.1$^{+a}$ | NKT1$^d$ | PLZF$^{lo}$T-bet$^+$GATA3$^{lo}$ROR$\gamma$t$^-$ | CD4$^{+/-}$ IL-17RB$^-$ |

*Sources:* Benlagha, K., Kyin, T., Beavis, A., Teyton, L., and Bendelac, A., *Science* 296(5567), 553–555, 2002; Benlagha, K., Wei, D. G., Veiga, J., Teyton, L., and Bendelac, A., *J Exp Med* 202(4), 485–492, 2005; Pellicci, D. G., Hammond, K. J., Uldrich, A. P., Baxter, A. G., Smyth, M. J., and Godfrey, D. I., *J Exp Med* 195(7), 835–844, 2002; Watarai, H., Sekine-Kondo, E., Shigeura, T., Motomura, Y., Yasuda, T., Satoh, R. et al., *PLoS Biol* 10(2), e1001255, 2012; Terashima, A., Watarai, H., Inoue, S., Sekine, E., Nakagawa, R., Hase, K. et al., *J Exp Med* 205(12), 2727–2733, 2008; Lee, Y. J., Holzapfel, K. L., Zhu, J., Jameson, S. C., and Hogquist, K. A., *Nat Immunol* 14(11), 1146–1154, 2013.

[a] NK1.1 is not expressed by Balb/c mice.

[b] Stage 2 NKT cells comprise NKT2 and NKT17 cells.

[c] NKT2 cells are more abundant in Balb/c mice than C57BL/6 mice.

[d] NKT1 cells are more abundant in C57BL/6 mice than Balb/c mice.

More recently, NKT cells were functionally divided into T-helper (Th)1-like, Th2-like, and Th17-like cells based on the cytokines produced upon antigenic activation[17] (Table 8.1). In contrast to the linear model of NKT cell development (stages 0, 1, 2, and 3) mentioned earlier, this model of NKT cell development was based on the common progenitor NKT0 that differentiates into NKT1, 2, and 17 subsets. Thus, NKT cells were separated into three distinctive subsets by expression of a combination of transcription factors PLZF, T-bet, Gata3, and ROR$\gamma$t. Transcription factor PLZF has been proposed to be a master regulator for NKT cell lineage.[18,19] Expression of PLZF was low in NKT1 cells, intermediate in NKT17 cells, and highest in NKT2 cells. Expression of GATA-3 and IRF4 was high in NKT2 and NKT17 cells, although *Gata3* mRNA levels were found to be similar in all three NKT subsets.[17] Other NKT cell subsets have also been defined, including cells that express programmed cell death protein 1 (PD1), CXCR5 and that produce IL-21,[20,21] IL-10-producing NKT10 cells,[22] and Forkhead box P3 (Foxp3)-expressing regulatory NKT cells have also been described.[23] IL-17RB along with surface expression of CD4 and NK1.1 defined another NKT subset found in the thymus and the periphery.[24]

## 8.3   REGULATION OF DEVELOPMENT AND DIFFERENTIATION OF MOUSE NKT CELLS

NKT cells develop from DP cells in the thymic cortex.[25–27] DP thymocytes survive in the cortex for 3.8 to 4 days, during which time TCRα chain rearrangement takes place with proximal Vα segments being rearranged first followed by rearrangement of distal Vα segments. The rearrangement of Vα14-Jα18, a critical factor in the development of NKT cells, is a late event in the lifetime of precursor DP thymocytes. Therefore, mutant mice that have DP thymocytes with decreased lifetime do not rearrange Vα14-Jα18 gene segments and fail to develop into NKT cells.[28] Several transcription factors are required for the DP-to-NKT transition: HEB, RORγt, Runx1, TCF1, c-Myb, and Bcl11b. Deletion of the E protein HEB resulted in absence of NKT cells due to reduced lifetime of precursor DP thymocytes and a failure to generate Vα-Jα rearrangements.[29] Impaired expression of HEB, RORγt or Runx1 that regulate the half life of DP cells impacts NKT cell development.[26] Likewise, decreased lifetime of DP thymocytes in mice deficient for TCF1 expression fail to develop NKT cells.[30,31] c-Myb also controls the lifetime of DP cells and is required for the expression of CD1d and signaling molecules SLAMF1, SLAMF6, and SAP.[32] In addition, signaling through the adaptor molecule signaling lymphocytic activation molecule (SLAM)-associated protein (SAP) and Fyn tyrosine play a role in NKT cell development.[33–36] Selection of NKT cells requires DP–DP cell interactions that upregulate expression of early growth response proteins 1 (Egr1) and Egr2, key factors that activate the PLZF expression.[18,19,37] Transcription factor c-Myc is required to promote proliferation at the DP-to-NKT transition.[38] As NKT cells undergo blasting and cell division during development, their metabolic needs are also regulated by the PI3K/mTOR pathway.[39–41]

Several other transcription factors have been shown to be differentially expressed in NKT cell subsets and have been implicated in the generation of mature NKT cells. These include PLZF, T-bet, GATA-3, RORγt, Th-POK, TOX, c-Myb, Runx3, Bcl11b, c-Myc, E proteins, and Id proteins.[42–45] T-bet, Runx3, and Hobit (a Blimp-1 homolog) were shown to be highly expressed NKT1 cells,[46–48] and RORγt was shown to be required for the development of NKT17 cells.[49–51] GATA-3 was shown to be required for the generation of NKT2 and NKT17 cells[12,52] but was also expressed in NKT0 and NKT1 populations.[17] Th-POK has been related to CD4+ NKT cell generation and to NKT17 cells.[53,54] Deletion of the transcription factor TOX led to a profound reduction in thymic NKT cells but the mechanism remains unclear.[55] Bcl11b deletion represses expression of *Zbtb7b* (Th-POK) and *Runx3* resulting in increase in NK1.1+ NKT1 cells.[56,57] Another transcription factor, KLF-2, was shown to control CD62L and S1P₁ expression after positive selection controls NKT2 cells.[58] Interestingly, Id3 proteins that negatively regulate E proteins were highly expressed in NKT2 cells.[59] As mentioned earlier, there is a reduction in the number of NKT cells that develop in TCF1-deficient mice.[30,31] In addition, the frequency of NKT2 and NKT17 is greatly diminished with lesser effect on the development of NKT1 cells when TCF1 is conditionally deleted from NKT cell precursors.[60] Recently, β-catenin was shown to regulate differentiation of NKT2 and NKT17 but not NKT1 cells in the thymus.[60] LEF1 also affects postselection NKT cell development in a manner not redundant with TCF1.[61] Thus, a number of transcription factors are differentially expressed in NKT cell subsets, and some have been shown to regulate NKT cell development and differentiation into functionally diverse subsets. However, further analysis will help solidify understanding of the development of this very influential group of cells that have profound impact on the well-being of the organism.

## 8.4 FUNCTIONAL ROLE FOR MOUSE NKT CELLS

Despite being present in low abundance, the function of NKT cells dramatically impacts the control of tumorigenesis and disease. The diverse NKT cell populations express chemokine receptors and other homing receptors that regulate their distribution to different tissues.[62]

Liver derived-NKT1 cells show antitumor activity that may be attributed to potent IFN-γ cytokine production in several tumor models.[63,64] At a mechanistic level, activation of NKT cells has been postulated to be through presentation of self-lipids by CD1d-positive tumors[65] followed by tumor lysis by a perforin-dependent mechanism[66] or intracellular granzyme B expression.[67] NKT2 cells localize predominantly in mucosal surfaces such as in the lungs and the intestines.[24,68] Due to their IL-17RB expression, NKT2 cells can be activated by IL-25; upon activation, they produce IL-4, IL-9, IL-10, and IL-13. In the lungs, they contribute to Th2-type airway hyperreactivity in an IL-25-dependent manner.[17,24,69] NKT2 cells, together with NKT17 cells, have also been associated with the disruption of mucosal homeostasis in the intestines. More commonly found in the small intestine, they protect from NKT cell-mediated pathogenic responses in colitis.[70-72] NKT17 cells are also found in the intestinal tissues as well as in the lymph nodes, the lungs, and the skin where they produce IL-17A, IL-21, and IL-22.[50,73] They can be activated by IL-23 due to their IL-23R expression.[17,51] In the lungs, they may also be activated by a microbial infection, leading to airway hyperreactivity with neutrophilia.[74]

Finally, NKT cells also accumulate in adipose tissue where they decrease with increased adiposity and insulin resistance but are restored with weight loss.[75-77] The production of IL-4, IL-13, and IL-10 by adipose-tissue resident NKT cells promotes alternative activation of macrophages (to the M2 phenotype), suppresses inflammation, and protects against type 2 diabetes. As a result, mice lacking NKT cells and those with high-fat diet-induced obesity have more proinflammatory M1 macrophages at the steady state, and promote inflammation, insulin resistance, and hepatic steatosis,[78] demonstrating that NKT cells may have a pivotal role in adipose homeostasis.

## 8.5 NKT CELLS INTERACT WITH COMMENSAL MICROBES AT MUCOSAL SURFACES

NKT cells closely interact with commensal microbes at mucosal surfaces in the intestinal lumen and the lungs; these interactions control NKT cell development and function.[68,79] Interestingly, NKT cell competence is acquired in neonatal mice upon exposure to commensal microbes and is sustained throughout life.[71,80] In the absence of microbes, the numbers of NKT cells in the small and large intestine are significantly increased, and these NKT cell populations show impaired maturation and function.[72,80] Recent experimental observation showed that monocolonization of germ-free mice with *Bacteriodes fragilis*, an intestinal commensal rich in sphingolipids, supported generation of mature functional NKT cells comparable to mice housed under specific pathogen-free conditions.[81] These NKT cell and commensal microbial interactions are bidirectional,[82,83] such that the crosstalk provides critical immune defense against infection and contributes to inflammation and tissue damage at mucosal surfaces.[84,85] Future studies will detail the diversity of microbial lipids and other by-products that regulate the development and function of NKT cells.

## 8.6 NKT CELLS IN HUMANS

The discovery of NKT cells in humans[86,87] revealed a high degree of similarity to mouse NKT cells. TCR usage, ligand specificity and functionality are similar between the species. However, the nomenclature of human NKT cells differs from the mouse nomenclature and it is summarized in Table 8.2.

## 8.7 HUMAN NKT CELLS INTERACT WITH MULTIPLE CELL TYPES AND PLAY A ROLE IN HUMAN DISEASE

NKT cells serve to bridge the innate and adaptive immune systems. They accomplish this by different mechanisms, including secretion of soluble factors, such as cytokines, and by direct cell–cell interactions. Human NKT cells have been shown to interact with and affect multiple immune and nonimmune cell types, and influence the outcome of many human disease processes. The following section describes these cell–cell interactions with associated commentary on how they either contribute to disease pathogenesis or could be harnessed as therapy for disease resolution or improved outcomes. All the studies described next utilize human NKT cells. Therefore, they are mainly in vitro studies or studies that correlate the presence, frequency, or in vitro function of NKT cells with parameters of disease.

### 8.7.1 Dendritic Cells

Crosstalk between dendritic cells (DCs) and NKT cells occurs in a bidirectional fashion through direct interaction and cytokine production. DCs are considered to be the major APC that presents lipid antigens to type 1 NKT in vivo. In general, interaction of NKT cells with immature DCs promotes tolerance, while interaction with mature DCs promotes IFN-γ and IL-4 by NKT cells.[88] Additionally, human

**TABLE 8.2**
**Human NKT and MAIT Subsets**

| | Type I ("Invariant" or "NKT") Typically 0.1%–0.2% of T-Cells in Peripheral Blood* | Type II ("Non-Invariant" or "Non-NKT") | NKT-Like MAIT | NKT-Like γδ Cells |
|---|---|---|---|---|
| TCRα | Vα24 Jα18 | ? Variable repertoire | Vα7.2-Jα33 | Variable γδ repertoire |
| TCRβ | Vβ11 | ? Variable repertoire | Vβ6 and 8; Vβ2 and 13 | |
| MHC I-like | CD1d-restricted | CD1d-restricted | MR1 | MHC, CD1a, b, c, d |
| CD4/8 | CD4+CD8αβ- ("CD4+"); CD4-CD8αβ- ("double negative"); CD4-CD8α+ ("CD8α+") CD161+; CD4- CD8αβ+ CD8αβ | CD4+ and CD4-CD8- | CD8+ or DN; CD161+IL-18Rα | DN or CD8+ |
| Cytokines released | Th1, Th2, and Th17 | Th2 | Th1, Th2, and Th17 | Mainly Th1 (TNF-α, IFN-γ) and IL-17 |
| Cytotoxicity | Cytotoxic | | Cytotoxic potential | Cytotoxic |
| Antigens | αGalCer | Not αGalCer | Secreted microbial vitamin B metabolites; α-Mannosyl ceramide? | Sulfatide and small phosphorylated bacterial metabolites |
| Endogenous antigens | βGlcCer lysophosphatidylcholine lysosphingomyelin plasmalogen lysophosphatidyl-ethanolamine | Sulfatide lysophosphatidylcholine cardiolipin | | |

*Note:* ? unknown, not yet determined.

* Highly variable range: <0.1% and >2%. In human omental fat: ~10% of T-cells.

CD4$^+$CD8β$^-$ NKT cells that interact with αGalCer-loaded DCs express high levels of IL-4, IL-5, IL-10, IL-13, and TNF-α as compared to the CD4$^-$CD8β$^-$ (double negative) subset.[89] Furthermore, tolerogenic monocyte-derived DCs promoted IL-10 production and a tolerogenic NKT phenotype from Vα24+ NKT cells. This tolerogenic phenotype was characterized by anergy of the NKT cells upon restimulation with αGalCer-loaded monocyte-derived DCs and low cytotoxicity to target cells.[90] When the αGalCer-stimulated DCs were cultured with the CD4 subset, IL-12p70 secretion was higher than when cultured with the DN subset. DC maturation and cytokine production is stimulated by direct contact through CD28/CD80/86 or αGalCer-loaded CD40/CD40L crosstalk to NKT cells.[91] However, as opposed to mouse NKT cells, αGalCer treatment of tumor antigen loaded-human monocyte-derived DCs (i.e., "double-loaded" moDCs) did not reverse the suppression of antigen-specific cytotoxic T lymphocytes (CTL) responses[92] that were observed with free αGalCer treatment.[93,94] These studies underscore the need for more investigation into human NKT cell interaction with dendritic cells.

Detailed studies of the DC subsets in human blood revealed the mechanism by which NKT cells interact with plasmacytoid (p)DCs and myeloid (m)DCs to augment antiviral responses.[95] Human pDCs do not express CD1d, in contrast to the mDC subset,[96] and human NKT cells do not express Toll-like receptors (TLRs). Therefore, TLR9-mediated (CpG) activation of pDC conditioned NKT cells allowing interaction with mDCs to produce more IFN-γ and less IL-4. This "licensing" of the NKT cells to interact with mDCs required a soluble factor in the conditioned medium from the TLR9-activated pDCs, whereas NKT cell interaction with mDCs occurred in a CD1d-restricted manner.

Immature dendritic cells (iDCs) when infected with *Leishmania infantum* are also targets of NKT-mediated cell lysis.[97] αGalCer-expanded and purified NKT cells were cytotoxic to the infected iDCs, most likely through recognition of the endogenous glycosphingolipid, iGb3, presented in CD1d. The αGalCer-expanded NKT cells produced IFN-γ and very low amounts of IL-4 in response to the infected iDCs. Because *Leishmania* is able to evade NK recognition, this rapid NKT response in the early phase of infection is critical for immunity to this protozoan parasite, possibly due to initiation of a robust Th1 response following infection and NKT production of IFN-γ. Another study focused on the delivery of CD169-targeting liposomal αGalCer to human moDCs to activate NKT cell release of IFN-γ.[98] By targeting the CD169/sialoadhesin pathway for presentation of lipid antigens by moDCs to NKT cells, human NKT cells released much more IFN-γ. This targeting strategy would be useful for NKT cell-based therapies against sialylated pathogens that include group B *Streptococcus* and *Campylobacter jejuni*.

### 8.7.2  MONOCYTES, MACROPHAGES, AND MYELOID-DERIVED SUPPRESSOR CELLS

Human NKT cells cause monocyte lysis in vitro, dependent upon CD1d expression.[99] Crosstalk between monocytes and NKT cells is mediated by the 4-1BB/L interaction, which promotes survival of both monocytes and nonactivated NKT cells under homeostatic conditions.[100] This interaction is important for controlling the numbers of infiltrating monocytes at a site of inflammation, such as to the lungs during viral infection. Under these conditions, NKT cell activation overrides the 4-1BB/L survival signal, resulting in monocyte apoptosis. When cocultured in vitro, αGalCer-activated NKT cells elicited IL-12 and TNF-α production from human monocytes. In humans that are susceptible to infection or go on to develop asthma, it is unknown whether NKT cell control of monocyte numbers and lung injury[101] or their production of lung-protective IL-22[102] is diminished during flu infection.

Human studies examining the interaction of NKT cells with macrophages found that NKT cells directly interacted with *Brucella suis 1330*-infected macrophages to control infection.[103] When killing infected macrophages, NKT cells also required Fas-FasL interaction, granule exocytosis from the NKT cells, and ligation of CD1d and CD4. In addition, cytotoxicity was improved when the macrophages were loaded with αGalCer. Furthermore, Kim et al. demonstrated that NKT-macrophage interactions could contribute to pathogenesis in chronic obstructive pulmonary disease (COPD).[104] The authors found an increased number of IL-13+ CD68+ macrophages and Vα24-positive NKT cells in lung sections from patients with COPD as compared to healthy controls. Interestingly, they also showed that mouse NKT cells and lung macrophages interacted to elicit IL-13 secretion from the macrophages. IL-13 is thought to contribute to pathogenesis in COPD. However, the IL-13 secretion required CD1d expression on the macrophages and Jα14 expression on the NKT cells. Other mouse studies have also demonstrated direct interaction between CD169+ macrophages and NKT cells,[105] but more in-depth analyses of human NKT-macrophage interactions are warranted.

NKT cells have immunomodulatory activity on the function of myeloid-derived suppressor cells (MDSC). This was highlighted in an in vitro study of the behavior of MDSCs sorted from the blood of influenza A-infected adults.[106] MDSCs are characterized by their ability to suppress T cell proliferation by producing peroxynitrites and expressing two L-arginine scavenging enzymes, iNOS and ArgI.[107] However, activated human NKT cells had an opposite effect on the T cell suppressive activity of the MDSCs; T cell proliferation in a mixed lymphocyte reaction was enhanced by NKT cells. These findings highlight a strategy where the immune responses to influenza could be augmented by harnessing the activating potential of NKT cells to relieve MDSC suppression of viral-specific T cells. Furthermore, MDSCs diminish CTL responses to tumors.[108] When NKT cell agonists were injected in a mouse model, there was a reduction in the frequency of circulating MDSCs and tumor regression occurred.[108] In phase I trials of αGalCer-loaded APC injection in humans with metastatic disease, there was an increase in circulating NKT cells, serum IFN-γ,[109,110] and IL-12.[110] As a result, tumor necrosis and decreased tumor markers in serum were observed in patients with intrahepatic tumors.[109] Therefore, such studies suggest that NKT-directed therapies for cancer treatment could help and hold great potential.

### 8.7.3 NEUTROPHILS

Human NKT cells can interact with human neutrophils to alter neutrophil release of cytokines. Studies by De Santo et al.[111] demonstrated that NKT–neutrophil interaction was promoted by serum amyloid A 1 (SAA-1) and, interestingly, also lowered IL-10 production and enhanced IL-12 production from the neutrophils. Crosstalk occurs in both directions, as cytotoxicity and expression of IFN-γ and T-bet were diminished in αGalCer-stimulated human NKT cells in the presence of neutrophils.[112] Inhibition of human NKT cell function by neutrophils could play an important regulatory role in diseases where highly activated or abundant NKT cells aggravate pathogenesis, such as sepsis or asthma.[2,113] On the other hand, neutrophils or their soluble factors could counteract beneficial NKT functions in eliminating tumor cells[92] or bacteria.[114] More studies are required to uncover the extent of NKT–neutrophil crosstalk and its impact on the pathogenesis of some human diseases.

### 8.7.4    NK Cells

Human NKT cells can enhance the functional activity of human NK cells. When activated NKT cells were incubated with NKT-depleted human peripheral blood mononuclear cells (PBMC), the cytotoxic effector function of the NK cells was augmented.[115] The authors demonstrated a role for IFN-γ and possibly other soluble factors, such as TNFα or IL-2, in the enhanced lytic capacity of the NK cells. Cell–cell contact between NKT and NK cells moderately enhanced cytokine secretion and activation of the NK cells. In addition to CTLs, NK cells are one of the most effective tumor cell-lysing cell types and their cytotoxicity is particularly important when tumor cells lose MHC I expression and can no longer be recognized by CTLs.[116,117] Since cancer patients have reduced circulating NKT cell numbers,[118–121] therapeutic strategies aimed at expanding NKT cell numbers to promote NK cell tumor cell lysis are particularly important.

### 8.7.5    B Cells

NKT cells' benefit to B cell function in humans was first described by Galli et al.[122] and reviewed by Dellabona et al.[123] The influence of NKT cells on B cells depends upon the type of NKT that interacts with the B cell. Galli et al. and Zeng et al. demonstrated that human CD4+, CD8α+, and CD4−CD8α− (double negative) NKT cells led to secretion of IgM from autologous B cells in vitro.[122,124] The latter study also described IgA and IgG release from B cells cocultured with all three types of human NKT cells.[124] Interestingly, this enhanced Ig release required direct contact but not αGalCer, CD40 ligation, or Th2 cytokines. Although IgE production by the B cells was also described in this study,[124] other NKT-B cell studies did not detect IgE. This may be a result of the presence or absence of αGalCer.[125] In addition, Zeng et al. described differential actions of NKT cells on naïve versus memory B cells. For example, human CD4+ NKT cells caused differentiation of naïve B cells and plasma cell differentiation of class-switched memory B cells. Human NKT cells also influenced B cell function by suppressing the ability of B cells to cause proliferation of T cells. When cocultured with CD4+ NKT cells, B cells produced IL-4 and IL-10. However, in terms of B cell influence on NKT function, αGalCer-loaded human B cells are not as potent as DCs at inducing NKT cell secretion of Th1 and Th2 cytokines.

NKT cells from patients with disease can have effects on B cell function that enhance pathogenesis. For example, the frequency of NKT cells in the blood of individuals with systemic lupus erythematosus (SLE) is lower than in healthy subjects.[126–130] However, the NKT cells from affected individuals are strong inducers of IgG and anti-dsDNA IgG autoantibodies from autologous B cells, as compared to NKT cells from healthy individuals.[130] There is also a reduction in the frequency and functionality of human NKT cells in the peripheral blood in common variable immunodeficiency (CVID),[131–133] a disease characterized by hypogammaglobulinemia. In contrast to lupus NKT cells, the reduction in numbers or help provided by NKT cells in CVID could play a role in the deficiency of B cell antibody production due to diminished direct or indirect interactions between the NKT cells and B cells. Taken together, the aforementioned studies highlight the need to understand NKT-B cell biology in more detail, so this interaction can either be augmented or diminished to influence the contribution of B cells to the disease process.

## 8.7.6  T CELLS

NKT cells are capable of secreting large amounts of Th1-, Th2-, and Th17-polarizing cytokines. As a result, NKT-potentiated T helper cell polarization can either augment or suppress the adaptive immune response and T cell polarization that promotes immunity or pathogenesis. These Th-polarizing cytokines produced by NKT cells influence the outcome of naïve T-cell differentiation. Additionally, the interaction of different human NKT cell subsets with DC can influence the polarization of any naïve T-cells the DC encounters thereafter. For example, double-negative NKT cell interaction with αGalCer-DC results in IL-5- and IL-13-secreting Th2 cells, whereas CD4[+] NKT cell interaction with αGalCer-DC produces IFN-γ-producing Th1 cells.[89] Blocking antibody experiments further demonstrated that this Th1-polarizing effect was dependent on CD40L or IL-12.

Using expanded NKT cells purified from PBMCs, the outcome of interactions between human NKT cells and human T regulatory (Treg) cells was determined in a series of in vitro assays.[134] For instance, Foxp3+ Tregs suppressed NKT cell proliferation in a manner dependent upon the nature of the activating NKT ligand. Generally, Treg suppression of NKT proliferation is dependent upon both cell contact and IL-10 secretion from the Treg cells. However, strong NKT agonists, like αGalCer, diminished the capacity of the Tregs to suppress proliferation of NKT cells. In addition, another study showed that cytokine secretion from NKT cells was suppressed by Tregs.[135] The Tregs reduced secretion of IFN-γ, IL-4, IL-13, and IL-10 from all subsets of human NKT cells, and this required cell–cell contact. Furthermore, contact with NKT cells enhanced Foxp3 expression and the suppressive ability of the Treg cells. This study suggests that the role of Tregs in dampening the function of NKT should be considered when designing vaccines or therapeutics based on augmenting NKT function. In contrast, another study of human NKT–Treg interactions described no effect on the suppressive function of Tregs but rather an enhancement of Treg proliferation.[136] This was due to abundant IL-2 secretion from the activated CD4[+] but not CD4[−] NKT cells. The NKT-mediated increase in Treg numbers is thought to be helpful in suppressing autoimmunity in mouse models. In humans, reduced NKT numbers and/or defects in cytokine secretion have been linked to most autoimmune diseases, such as multiple sclerosis (MS), T1D, SLE, Sjogren's syndrome, rheumatoid arthritis (RA), and sarcoidosis.[137–142] These correlative studies suggest a link between NKT cells and maintenance of immune tolerance. For example, in patients with MS, there is a reduction in the number of NKT cells,[140] which increases on treatment with type I IFN-β, thus preventing relapse.[143] In RA, there is a reduction in the number of circulating NKT cells, which are unresponsive to antigenic stimulation in the presence of αGalCe.[126]

In terms of NKT–Treg interactions in disease, CD4+ NKT cells from allergic asthmatic individuals were cytotoxic for Tregs, with cytotoxicity increasing according to the classification of disease severity.[144] Moreover, NKT cells from allergic asthmatics were more efficient at killing Tregs than NKT cells from healthy controls. In addition, expression of granzyme B and perforin was higher in the NKT cells from allergic asthmatics compared to the healthy controls. Interestingly, following a course of oral corticosteroid treatment of allergic asthmatics, the percentage of dead Tregs diminished suggesting that treatment impacted NKT cell-dependent killing of Treg cells. Furthermore, autologous Tregs from the allergic asthmatics were more susceptible to cytotoxicity by allergic asthmatic NKT cells than healthy control Tregs. The studies described in this section clearly demonstrate the positive and negative influence of NKT cells on Treg numbers and function. Therefore, the impact of NKT cells on Tregs must be thoughtfully considered when designing approaches based on increasing or decreasing NKT cell numbers or function in a therapeutic setting.

### 8.7.7  NONIMMUNE CELLS: ADIPOCYTES, KERATINOCYTES, AND EPITHELIAL CELLS

Adipose tissue is composed of a variety of immune cells and fat-storing adipocytes that regulate lipid availability in response to energy demands. Accumulating evidence suggests that metabolic syndromes are associated with a proinflammatory state within adipose tissue, yet the regulation of the immune response in adipose tissue is still being uncovered. NKT cells are abundant in human adipose tissue in individuals of healthy weight, while obese humans have diminished NKT cells in omental fat[75–77,145] that are restored with weight loss.[76] NKT cells bind different lipid antigens presented in CD1d molecules, and recent work has revealed that they interact with adipocytes loaded with αGalCer.[46] This interaction resulted in enhanced IL-2 secretion from the NKT cells dependent upon direct cell contacts and CD1d expression on the adipocytes. Although this was an in vitro cell culture model in mouse cells, the authors demonstrated NKT cell numbers and CD1d expression in adipose tissue of obese mice were reduced, much like in humans. Furthermore, in human visceral fat, expression of mRNA for Vα24 and CD1d was negatively correlated with body mass index.[46] These data hint that a reduction in NKT cell–adipocyte interactions may play a role in immune dysregulation and adipose inflammation that contributes to obesity in humans. Once NKT–adipocyte interactions are fully understood, they could be exploited for controlling weight gain and obesity in the future.

Additionally, a study of skin biopsies from patients with allergic contact dermatitis demonstrated that NKT cells target keratinocytes for lysis.[146] αGalCer-pulsed primary keratinocytes or human keratinocytes, from the HaCat cell line, were lysed by activated NKT-cell clone F6 in vitro. The recognition of keratinocytes for lysis required CD1d expression, as CD1d knockdown in the target cells significantly decreased keratinocyte lysis. Furthermore, NKT cells containing the cytotoxic granule proteins perforin, granzyme B, and granzyme K were increased in biopsies from the lesional skin of allergic contact dermatitis patients. In contrast, there were no NKT cells detected in normal skin biopsies, leading the authors to conclude that the interaction of NKT cells with keratinocytes may play a role in the pathogenesis of allergic contact dermatitis. Another study found that patients with SLE and discoid lupus erythematosus (DLE) had increased numbers of NKT cells expressing IFN-γ and CCR4 in lesional skin, while unaffected skin from the same patients lacked NKT cells.[147] Interestingly, the blood of these patients had a lower percentage of NKT cells than healthy controls. This observation highlights a potential mechanism implicating recruitment of NKT cells from the blood to the target site of inflammation in the skin and further suggests that NKT cells may be involved in the skin pathologies associated with SLE.

NKT cells have also been implicated in the pathogenesis of a number of lung diseases. Akbari et al. found that in patients with asthma, 60% of pulmonary $CD4^+CD3^+$ cells were NKT cells and not MHC II-restricted $CD4^+$ T cells.[148] COPD patients have increased numbers of NKT cells in their lungs and blood.[104,149–151] When NKT cells were cocultured with human airway epithelial cells that had been exposed to cigarette smoke extract (CSE) for 24 h, production of IFN-γ by the NKT cells increased significantly.[152] Expression of ST3GAL5 and UGCG, enzymes involved in synthesis of glycolipid activators on NKT cells, was increased in the human airway epithelial cells exposed to CSE; this was mitigated by addition of the antioxidant N-acetyl cysteine (NAC). Whether direct interaction with airway epithelial cells was required for NKT cell elaboration of IFN-γ was not investigated. Since airway epithelial cells are the critical barrier and sentinel cells of the lung to inhaled toxins, irritants, allergens, and microbes in many lung diseases, further studies of NKT cell–airway epithelial cell crosstalk and interactions are warranted.

A further example of overactivated NKT cell-mediated cytotoxicity on epithelial cell barriers is from studies of patients with ulcerative colitis.[153] In one study, noninvariant NKT cells were purified from the patient's lamina propria mononuclear cells. The investigators found that these particular NKT cells did not express Vα24 nor respond to αGalCer, but they did recognize the sulfatide family of self-glycolipids. When these NKT cells were cocultured with the HT-29 human intestinal epithelial cell line in the presence of lyso-sulfatide, the percentage of dead epithelial cells doubled. Interestingly, the ulcerative colitis NKT cells secreted large amounts of IL-13 and upregulated IL-13Rα2 on their surface. However, the role of the secreted IL-13 and IL-13Rα2 on the pathogenic NKT cells in the disease process and epithelial cell cytotoxicity was not investigated.

Human cholangiocytes, the epithelial cells of the bile duct, are active participants in immune responses within that organ. They express CD1d, which is increased in patients with primary biliary cirrhosis[154] or infected with hepatitis C.[155] Coincubation of the human cholangiocyte cell line, H69, with two different human NKT cell clones, JC2.7 and J3N.5, stimulated IL-4 secretion after 24 hours, and IFN-γ and IL-13 secretion at 72 hours.[156] Similar results were also observed with primary cholangiocytes from patients with primary sclerosing cholangitis and nonalcoholic steatohepatitis. Not all diseases cause an increase in CD1d expression, as CD1d expression on cholangiocytes was decreased in three types of liver disease as compared to healthy controls.

## 8.8   MODULATING HUMAN NKT CELLS FOR THERAPEUTIC PURPOSE IN HUMAN DISEASE

NKT cells have been implicated as pathogenic, protective, or defective in many different diseases. Alterations in their numbers or function have been correlated with the presence, absence, or severity of disease. For example, the frequency of total circulating NKT cells is decreased in MS,[142,157] SLE,[129,142] RA,[124,126,158–160] and cancer.[161] In some cases, there are conflicting reports about correlations of increased or decreased NKT cells in the peripheral blood of patients with disease, as in asthma.[124,162,163] Inappropriate activation of NKT cells has been linked to psoriasis[164,165] and atherosclerosis.[166–168] Furthermore, in patients with SLE, the NKT cells are also functionally defective.[126,129] Therefore, therapeutic strategies targeted at either increasing or decreasing the number and function of NKT cells have been proposed and tested. αGalCer treatment stimulates human NKT cell activation[109,169] and this strategy has been tested in mouse models of experimental autoimmune encephalomyelitis (EAE), SLE, RA, and type 1 diabetes. αGalCer has been tested for augmenting human type 1 NKT cell responses to tumors in phase I and phase II clinical trials, with promising results so far.[109,110,169–171] However, mouse models of some autoimmune diseases have shown that αGalCer treatment may exacerbate the disease,[172–175] so analogues of αGalCer have been developed. These analogues have modifications to their structure that cause a shift in the profile of type 1 and type 2 cytokines secreted when they are bound by NKT cells.[176] The other approaches to augmenting NKT cell numbers include expanding NKT cells or subsets of NKT cells in vitro for reinfusion. This has been tested in a phase I–II trial for lung cancer.[171] Another strategy to increase NKT function is to enhance the presentation of self-ligands to NKT cells, either by activating the biosynthetic pathways that produce self-glycolipids or by modulating DCs to upregulate CD1d and CD40 expression.

In diseases where downregulation of NKT cell numbers or function is required, NKT cell depletion strategies have been tried. These include antibodies to CD1d and to NKT markers. For example, neutralizing anti-CD1d and antihuman V24J18 was protective in peanut- and *Aspergillus*-induced experimental eosinophilic esophagitis.[177] A second depletion strategy uses a humanized monoclonal NKT cell-depleting antibody, NKTT120, which binds human invariant T-cell receptor and FcRI and III. Promising studies in macaques showed that NKTT120 specifically and effectively depleted NKT cells but not T, B, and NK cells.[178] NKT cells were depleted within 24 hours with a single dose of NKTT120, and the duration until reappearance of NKT cells was dependent on the dose of NKTT120 used. There were no toxic effects, suggesting this antibody may be a strong candidate for clinical trials. The studies described in this section are a tantalizing glimpse into the potential for NKT cells as cellular immunotherapy for a wide range of human diseases.

## 8.9 UNANSWERED QUESTIONS AND FUTURE CHALLENGES FOR HUMAN HEALTH

As is clear from Table 8.2, careful characterization and a deeper understanding of human NKT cell subsets, their ligands, and distribution in different tissues is required. It is critical to understand the development, maturation, and migration of NKT cells in the human body under conditions of health and disease in order to understand how to manipulate their generation and function for therapeutic use. To this end, two animal models have been developed: a human CD1d knock-in mouse[179,180] (a humanized mouse model) and a Cxcr6[gfp/+] animal. Human CD1d knock-in mice possess a population of type I NKT cells that appear to mimic human type I NKT cells. Through breeding the knock-in mice to mice with a genetic background susceptible to a particular disease, such as autoimmunity, or to transgenic mice that express a particular disease antigen, human diseases have been effectively modeled. The latter strategy has been used to model human hepatitis B infection through introduction of the hepatitis B virus (HBV) transgene into the CD1d knock-in mice. A decrease in the frequency of NKT cells was observed in these mice,[181] as is found in human HBV disease.[182] The human CD1d knock-in mice have been further improved recently by addition of the human Va24 transgene.[183] Using an immune deficient mouse engrafted with human liver, thymus, and CD34+ hematopoietic stem cells, Lockridge et al. generated a humanized mouse with a competent iNKT cell system.[184] These mice with human immune systems will be useful tools to understanding human iNKT cell development and teasing out the role of these cells in disease. Because human NKT cells express high levels of CXCR6, the Cxcr6[gfp/+] animals are useful proxies for the behavior of human NKT cells in vivo. The migration and recirculation of GFP-positive NKT cells can be tracked in vivo in the circulation and in different tissues at homeostasis[185] and during disease.[186] In summary, more in-depth studies of the biology and development of human NKT cell subsets are crucial to understanding the role of these cells in human disease and to harnessing their potential as a therapy for many different diseases.

In conclusion we have summarized the available information on murine and human NKT cells. Overall a better understanding of the similarities and differences between mouse and human NKT cells will facilitate mechanistic studies and promote development of therapeutic strategies for use in human diseases.

## ACKNOWLEDGMENTS

This research was supported by the Intramural Research Program (IRP) in the National Institute on Aging (NIA) of the National Institutes of Health (NIH) to JMS and by NIH grant R01 HL124477 to NMH.

## REFERENCES

1. Kinjo, Y., Kitano, N., and Kronenberg, M., The role of invariant natural killer T cells in microbial immunity, *J Infect Chemother* 19(4), 560–570, 2013.

2. Bendelac, A., Savage, P. B., and Teyton, L., The biology of NKT cells, *Annu Rev Immunol* 25, 297–336, 2007.

3. Rossjohn, J., Pellicci, D. G., Patel, O., Gapin, L., and Godfrey, D. I., Recognition of CD1d-restricted antigens by natural killer T cells, *Nat Rev Immunol* 12(12), 845–857, 2012.

4. Benlagha, K., Weiss, A., Beavis, A., Teyton, L., and Bendelac, A., In vivo identification of glycolipid antigen-specific T cells using fluorescent CD1d tetramers, *J Exp Med* 191(11), 1895–1903, 2000.

5. Lantz, O., and Bendelac, A., An invariant T cell receptor alpha chain is used by a unique subset of major histocompatibility complex class I-specific CD4+ and CD4-8- T cells in mice and humans, *J Exp Med* 180(3), 1097–1106, 1994.

6. Matsuda, J. L., Naidenko, O. V., Gapin, L., Nakayama, T., Taniguchi, M., Wang, C. R., Koezuka, Y., and Kronenberg, M., Tracking the response of natural killer T cells to a glycolipid antigen using CD1d tetramers, *J Exp Med* 192(5), 741–754, 2000.

7. Kawano, T., Cui, J., Koezuka, Y., Toura, I., Kaneko, Y., Motoki, K., Ueno, H., Nakagawa, R., Sato, H., Kondo, E., Koseki, H., and Taniguchi, M., CD1d-restricted and TCR-mediated activation of valpha14 NKT cells by glycosylceramides, *Science* 278(5343), 1626–1629, 1997.

8. Joyce, S., Woods, A. S., Yewdell, J. W., Bennink, J. R., De Silva, A. D., Boesteanu, A., Balk, S. P., Cotter, R. J., and Brutkiewicz, R. R., Natural ligand of mouse CD1d1: Cellular glycosylphosphatidylinositol, *Science* 279(5356), 1541–1544, 1998.

9. Zhou, D., Mattner, J., Cantu, C., 3rd, Schrantz, N., Yin, N., Gao, Y., Sagiv, Y., Hudspeth, K., Wu, Y. P., Yamashita, T., Teneberg, S., Wang, D., Proia, R. L., Levery, S. B., Savage, P. B., Teyton, L., and Bendelac, A., Lysosomal glycosphingolipid recognition by NKT cells, *Science* 306(5702), 1786–1789, 2004.

10. Uldrich, A. P., Patel, O., Cameron, G., Pellicci, D. G., Day, E. B., Sullivan, L. C., Kyparissoudis, K., Kjer-Nielsen, L., Vivian, J. P., Cao, B., Brooks, A. G., Williams, S. J., Illarionov, P., Besra, G. S., Turner, S. J., Porcelli, S. A., McCluskey, J., Smyth, M. J., Rossjohn, J., and Godfrey, D. I., A semi-invariant Valpha10+ T cell antigen receptor defines a population of natural killer T cells with distinct glycolipid antigen-recognition properties, *Nat Immunol* 12(7), 616–623, 2011.

11. Exley, M. A., Balk, S. P., and Wilson, S. B., Isolation and functional use of human NK T cells, *Curr Protoc Immunol* Chapter 14, Unit 14.11, 2003.

12. Lee, Y. J., Holzapfel, K. L., Zhu, J., Jameson, S. C., and Hogquist, K. A., Steady-state production of IL-4 modulates immunity in mouse strains and is determined by lineage diversity of iNKT cells, *Nat Immunol* 14(11), 1146–1154, 2013.

13. Benlagha, K., Kyin, T., Beavis, A., Teyton, L., and Bendelac, A., A thymic precursor to the NK T cell lineage, *Science* 296(5567), 553–555, 2002.

14. Benlagha, K., Wei, D. G., Veiga, J., Teyton, L., and Bendelac, A., Characterization of the early stages of thymic NKT cell development, *J Exp Med* 202(4), 485–492, 2005.

15. Pellicci, D. G., Hammond, K. J., Uldrich, A. P., Baxter, A. G., Smyth, M. J., and Godfrey, D. I., A natural killer T (NKT) cell developmental pathway involving a thymus-dependent NK1.1(-)CD4(+) CD1d-dependent precursor stage, *J Exp Med* 195(7), 835–844, 2002.

16. Takahashi, T., Chiba, S., Nieda, M., Azuma, T., Ishihara, S., Shibata, Y., Juji, T., and Hirai, H., Cutting edge: Analysis of human V alpha 24+CD8+ NK T cells activated by alpha-galactosylceramide-pulsed monocyte-derived dendritic cells, *J Immunol* 168(7), 3140–3144, 2002.

17. Watarai, H., Sekine-Kondo, E., Shigeura, T., Motomura, Y., Yasuda, T., Satoh, R., Yoshida, H., Kubo, M., Kawamoto, H., Koseki, H., and Taniguchi, M., Development and function of invariant natural killer T cells producing T(h)2- and T(h)17-cytokines, *PLoS Biol* 10(2), e1001255, 2012.

18. Kovalovsky, D., Uche, O. U., Eladad, S., Hobbs, R. M., Yi, W., Alonzo, E., Chua, K., Eidson, M., Kim, H. J., Im, J. S., Pandolfi, P. P., and Sant'Angelo, D. B., The BTB-zinc finger transcriptional regulator PLZF controls the development of invariant natural killer T cell effector functions, *Nat Immunol* 9(9), 1055–1064, 2008.

19. Savage, A. K., Constantinides, M. G., Han, J., Picard, D., Martin, E., Li, B., Lantz, O., and Bendelac, A., The transcription factor PLZF directs the effector program of the NKT cell lineage, *Immunity* 29(3), 391–403, 2008.

20. Chang, P. P., Barral, P., Fitch, J., Pratama, A., Ma, C. S., Kallies, A., Hogan, J. J., Cerundolo, V., Tangye, S. G., Bittman, R., Nutt, S. L., Brink, R., Godfrey, D. I., Batista, F. D., and Vinuesa, C. G., Identification of Bcl-6-dependent follicular helper NKT cells that provide cognate help for B cell responses, *Nat Immunol* 13(1), 35–43, 2012.

21. King, I. L., Fortier, A., Tighe, M., Dibble, J., Watts, G. F., Veerapen, N., Haberman, A. M., Besra, G. S., Mohrs, M., Brenner, M. B., and Leadbetter, E. A., Invariant natural killer T cells direct B cell responses to cognate lipid antigen in an IL-21-dependent manner, *Nat Immunol* 13(1), 44–50, 2012.

22. Sag, D., Krause, P., Hedrick, C. C., Kronenberg, M., and Wingender, G., IL-10-producing NKT10 cells are a distinct regulatory invariant NKT cell subset, *J Clin Invest* 124(9), 3725–3740, 2014.

23. Monteiro, M., Almeida, C. F., Caridade, M., Ribot, J. C., Duarte, J., Agua-Doce, A., Wollenberg, I., Silva-Santos, B., and Graca, L., Identification of regulatory Foxp3+ invariant NKT cells induced by TGF-beta, *J Immunol* 185(4), 2157–2163, 2010.

24. Terashima, A., Watarai, H., Inoue, S., Sekine, E., Nakagawa, R., Hase, K., Iwamura, C., Nakajima, H., Nakayama, T., and Taniguchi, M., A novel subset of mouse NKT cells bearing the IL-17 receptor B responds to IL-25 and contributes to airway hyperreactivity, *J Exp Med* 205(12), 2727–2733, 2008.

25. Bezbradica, J. S., Hill, T., Stanic, A. K., Van Kaer, L., and Joyce, S., Commitment toward the natural T (iNKT) cell lineage occurs at the CD4+8+ stage of thymic ontogeny, *Proc Natl Acad Sci USA* 102(14), 5114–5119, 2005.

26. Egawa, T., Eberl, G., Taniuchi, I., Benlagha, K., Geissmann, F., Hennighausen, L., Bendelac, A., and Littman, D. R., Genetic evidence supporting selection of the Valpha14i NKT cell lineage from double-positive thymocyte precursors, *Immunity* 22(6), 705–716, 2005.

27. Gapin, L., Matsuda, J. L., Surh, C. D., and Kronenberg, M., NKT cells derive from double-positive thymocytes that are positively selected by CD1d, *Nat Immunol* 2(10), 971–978, 2001.

28. Guo, J., Hawwari, A., Li, H., Sun, Z., Mahanta, S. K., Littman, D. R., Krangel, M. S., and He, Y. W., Regulation of the TCRalpha repertoire by the survival window of CD4(+)CD8(+) thymocytes, *Nat Immunol* 3(5), 469–476, 2002.

29. D'Cruz, L. M., Knell, J., Fujimoto, J. K., and Goldrath, A. W., An essential role for the transcription factor HEB in thymocyte survival, Tcra rearrangement and the development of natural killer T cells, *Nat Immunol* 11(3), 240–249, 2010.

30. Sharma, A., Berga-Bolanos, R., and Sen, J. M., T cell factor-1 controls the lifetime of CD4+ CD8+ thymocytes in vivo and distal T cell receptor alpha-chain rearrangement required for NKT cell development, *PLoS One* 9(12), e115803, 2014.

31. Sharma, A., Chen, Q., Nguyen, T., Yu, Q., and Sen, J. M., T cell factor-1 and beta-catenin control the development of memory-like CD8 thymocytes, *J Immunol* 188(8), 3859–3868, 2012.

32. Hu, T. S., Simmons, A., Yuan, J., Bender, T. P., and Alberola-Ila, J., The transcription factor c-Myb primes CD4(+)CD8(+) immature thymocytes for selection into the iNKT lineage, *Nat Immunol* 11(5), 435–441, 2010.

33. Chan, B., Lanyi, A., Song, H. K., Griesbach, J., Simarro-Grande, M., Poy, F., Howie, D., Sumegi, J., Terhorst, C., and Eck, M. J., SAP couples Fyn to SLAM immune receptors, *Nat Cell Biol* 5(2), 155–160, 2003.

34. Gadue, P., Morton, N., and Stein, P. L., The Src family tyrosine kinase Fyn regulates natural killer T cell development, *J Exp Med* 190(8), 1189–1196, 1999.

35. Griewank, K., Borowski, C., Rietdijk, S., Wang, N., Julien, A., Wei, D. G., Mamchak, A. A., Terhorst, C., and Bendelac, A., Homotypic interactions mediated by Slamf1 and Slamf6 receptors control NKT cell lineage development, *Immunity* 27(5), 751–762, 2007.

36. Nichols, K. E., Hom, J., Gong, S. Y., Ganguly, A., Ma, C. S., Cannons, J. L., Tangye, S. G., Schwartzberg, P. L., Koretzky, G. A., and Stein, P. L., Regulation of NKT cell development by SAP, the protein defective in XLP, *Nat Med* 11(3), 340–345, 2005.

37. Seiler, M. P., Mathew, R., Liszewski, M. K., Spooner, C., Barr, K., Meng, F., Singh, H., and Bendelac, A., Elevated and sustained expression of the transcription factors Egr1 and Egr2 controls NKT lineage differentiation in response to TCR signaling, *Nat Immunol* 13(3), 264–271, 2012.

38. Dose, M., Sleckman, B. P., Han, J., Bredemeyer, A. L., Bendelac, A., and Gounari, F., Intrathymic proliferation wave essential for Valpha14+ natural killer T cell development depends on c-Myc, *Proc Natl Acad Sci USA* 106(21), 8641–8646, 2009.

39. Prevot, N., Pyaram, K., Bischoff, E., Sen, J. M., Powell, J. D., and Chang, C. H., Mammalian target of rapamycin complex 2 regulates invariant NKT cell development and function independent of promyelocytic leukemia zinc-finger, *J Immunol* 194(1), 223–230, 2015.

40. Wei, J., Yang, K., and Chi, H., Cutting edge: Discrete functions of mTOR signaling in invariant NKT cell development and NKT17 fate decision, *J Immunol* 193(9), 4297–4301, 2014.

41. Zhang, L., Tschumi, B. O., Corgnac, S., Ruegg, M. A., Hall, M. N., Mach, J. P., Romero, P., and Donda, A., Mammalian target of rapamycin complex 1 orchestrates invariant NKT cell differentiation and effector function, *J Immunol* 193(4), 1759–1765, 2014.

42. D'Cruz, L. M., Yang, C. Y., and Goldrath, A. W., Transcriptional regulation of NKT cell development and homeostasis, *Curr Opin Immunol* 22(2), 199–205, 2010.

43. Das, R., Sant'Angelo, D. B., and Nichols, K. E., Transcriptional control of invariant NKT cell development, *Immunol Rev* 238(1), 195–215, 2010.

44. Hu, T., Wang, H., Simmons, A., Bajana, S., Zhao, Y., Kovats, S., Sun, X. H., and Alberola-Ila, J., Increased level of E protein activity during invariant NKT development promotes differentiation of invariant NKT2 and invariant NKT17 subsets, *J Immunol* 191(10), 5065–5073, 2013.

45. Michel, M. L., Mendes-da-Cruz, D., Keller, A. C., Lochner, M., Schneider, E., Dy, M., Eberl, G., and Leite-de-Moraes, M. C., Critical role of ROR-gammat in a new thymic pathway leading to IL-17-producing invariant NKT cell differentiation, *Proc Natl Acad Sci USA* 105(50), 19845–19850, 2008.

46. Huh, J. Y., Kim, J. I., Park, Y. J., Hwang, I. J., Lee, Y. S., Sohn, J. H., Lee, S. K., Alfadda, A. A., Kim, S. S., Choi, S. H., Lee, D. S., Park, S. H., Seong, R. H., Choi, C. S., and Kim, J. B., A novel function of adipocytes in lipid antigen presentation to iNKT cells, *Mol Cell Biol* 33(2), 328–339, 2013.

47. Townsend, M. J., Weinmann, A. S., Matsuda, J. L., Salomon, R., Farnham, P. J., Biron, C. A., Gapin, L., and Glimcher, L. H., T-bet regulates the terminal maturation and homeostasis of NK and Valpha14i NKT cells, *Immunity* 20(4), 477–494, 2004.

48. van Gisbergen, K. P., Kragten, N. A., Hertoghs, K. M., Wensveen, F. M., Jonjic, S., Hamann, J., Nolte, M. A., and van Lier, R. A., Mouse Hobit is a homolog of the transcriptional repressor Blimp-1 that regulates NKT cell effector differentiation, *Nat Immunol* 13(9), 864–871, 2012.

49. Coquet, J. M., Chakravarti, S., Kyparissoudis, K., McNab, F. W., Pitt, L. A., McKenzie, B. S., Berzins, S. P., Smyth, M. J., and Godfrey, D. I., Diverse cytokine production by NKT cell subsets and identification of an IL-17-producing CD4-NK1.1- NKT cell population, *Proc Natl Acad Sci USA* 105(32), 11287–11292, 2008.

50. Doisne, J. M., Bartholin, L., Yan, K. P., Garcia, C. N., Duarte, N., Le Luduec, J. B., Vincent, D., Cyprian, F., Horvat, B., Martel, S., Rimokh, R., Losson, R., Benlagha, K., and Marie, J. C., iNKT cell development is orchestrated by different branches of TGF-beta signaling, *J Exp Med* 206(6), 1365–1378, 2009.

51. Rachitskaya, A. V., Hansen, A. M., Horai, R., Li, Z., Villasmil, R., Luger, D., Nussenblatt, R. B., and Caspi, R. R., Cutting edge: NKT cells constitutively express IL-23 receptor and RORgammat and rapidly produce IL-17 upon receptor ligation in an IL-6-independent fashion, *J Immunol* 180(8), 5167–5171, 2008.

52. Kim, P. J., Pai, S. Y., Brigl, M., Besra, G. S., Gumperz, J., and Ho, I. C., GATA-3 regulates the development and function of invariant NKT cells, *J Immunol* 177(10), 6650–6659, 2006.

53. Enders, A., Stankovic, S., Teh, C., Uldrich, A. P., Yabas, M., Juelich, T., Altin, J. A., Frankenreiter, S., Bergmann, H., Roots, C. M., Kyparissoudis, K., Goodnow, C. C., and Godfrey, D. I., ZBTB7B (Th-POK) regulates the development of IL-17-producing CD1d-restricted mouse NKT cells, *J Immunol* 189(11), 5240–5249, 2012.

54. Wang, L., Carr, T., Xiong, Y., Wildt, K. F., Zhu, J., Feigenbaum, L., Bendelac, A., and Bosselut, R., The sequential activity of Gata3 and Thpok is required for the differentiation of CD1d-restricted CD4+ NKT cells, *Eur J Immunol* 40(9), 2385–2390, 2010.

55. Aliahmad, P., and Kaye, J., Development of all CD4 T lineages requires nuclear factor TOX, *J Exp Med* 205(1), 245–256, 2008.

56. Albu, D. I., VanValkenburgh, J., Morin, N., Califano, D., Jenkins, N. A., Copeland, N. G., Liu, P., and Avram, D., Transcription factor Bcl11b controls selection of invariant natural killer T-cells by regulating glycolipid presentation in double-positive thymocytes, *Proc Natl Acad Sci USA* 108(15), 6211–6216, 2011.

57. Kastner, P., Chan, S., Vogel, W. K., Zhang, L. J., Topark-Ngarm, A., Golonzhka, O., Jost, B., Le Gras, S., Gross, M. K., and Leid, M., Bcl11b represses a mature T-cell gene expression program in immature CD4(+)CD8(+) thymocytes, *Eur J Immunol* 40(8), 2143–2154, 2010.

58. Weinreich, M. A., Takada, K., Skon, C., Reiner, S. L., Jameson, S. C., and Hogquist, K. A., KLF2 transcription-factor deficiency in T cells results in unrestrained cytokine production and upregulation of bystander chemokine receptors, *Immunity* 31(1), 122–130, 2009.

59. D'Cruz, L. M., Stradner, M. H., Yang, C. Y., and Goldrath, A. W., E and Id proteins influence invariant NKT cell sublineage differentiation and proliferation, *J Immunol* 192(5), 2227–2236, 2014.

60. Berga-Bolanos, R., Zhu, W. S., Steinke, F. C., Xue, H. H., and Sen, J. M., Cell-autonomous requirement for TCF1 and LEF1 in the development of natural killer T cells, *Mol Immunol* 68(2), 2015.

61. Carr, T., Krishnamoorthy, V., Yu, S., Xue, H. H., Kee, B. L., and Verykokakis, M., The transcription factor lymphoid enhancer factor 1 controls invariant natural killer T cell expansion and Th2-type effector differentiation, *JEM* 212(5), 793–807, 2015.

62. Johnston, B., Kim, C. H., Soler, D., Emoto, M., and Butcher, E. C., Differential chemokine responses and homing patterns of murine TCR alpha beta NKT cell subsets, *J Immunol* 171(6), 2960–2969, 2003.

63. Crowe, N. Y., Coquet, J. M., Berzins, S. P., Kyparissoudis, K., Keating, R., Pellicci, D. G., Hayakawa, Y., Godfrey, D. I., and Smyth, M. J., Differential antitumor immunity mediated by NKT cell subsets in vivo, *J Exp Med* 202(9), 1279–1288, 2005.

64. Smyth, M. J., Thia, K. Y., Street, S. E., MacGregor, D., Godfrey, D. I., and Trapani, J. A., Perforin-mediated cytotoxicity is critical for surveillance of spontaneous lymphoma, *J Exp Med* 192(5), 755–760, 2000.

65. Wu, D. Y., Segal, N. H., Sidobre, S., Kronenberg, M., and Chapman, P. B., Cross-presentation of disialoganglioside GD3 to natural killer T cells, *J Exp Med* 198(1), 173–181, 2003.

66. Kawano, T., Nakayama, T., Kamada, N., Kaneko, Y., Harada, M., Ogura, N., Akutsu, Y., Motohashi, S., Iizasa, T., Endo, H., Fujisawa, T., Shinkai, H., and Taniguchi, M., Antitumor cytotoxicity mediated by ligand-activated human V alpha24 NKT cells, *Cancer Res* 59(20), 5102–5105, 1999.

67. Coquet, J. M., Kyparissoudis, K., Pellicci, D. G., Besra, G., Berzins, S. P., Smyth, M. J., and Godfrey, D. I., IL-21 is produced by NKT cells and modulates NKT cell activation and cytokine production, *J Immunol* 178(5), 2827–2834, 2007.

68. Zeissig, S., and Blumberg, R. S., Commensal microbiota and NKT cells in the control of inflammatory diseases at mucosal surfaces, *Curr Opin Immunol* 25(6), 690–696, 2013.

69. Stock, P., Lombardi, V., Kohlrautz, V., and Akbari, O., Induction of airway hyperreactivity by IL-25 is dependent on a subset of invariant NKT cells expressing IL-17RB, *J Immunol* 182(8), 5116–5122, 2009.

70. Fuss, I. J. and Strober, W., The role of IL-13 and NK T cells in experimental and human ulcerative colitis, *Mucosal Immunol* 1 Suppl 1, S31–S33, 2008.

71. Olszak, T., An, D., Zeissig, S., Vera, M. P., Richter, J., Franke, A., Glickman, J. N., Siebert, R., Baron, R. M., Kasper, D. L., and Blumberg, R. S., Microbial exposure during early life has persistent effects on natural killer T cell function, *Science* 336(6080), 489–493, 2012.

72. Wingender, G., Stepniak, D., Krebs, P., Lin, L., McBride, S., Wei, B., Braun, J., Mazmanian, S. K., and Kronenberg, M., Intestinal microbes affect phenotypes and functions of invariant natural killer T cells in mice, *Gastroenterology* 143(2), 418–428, 2012.

73. Michel, M. L., Keller, A. C., Paget, C., Fujio, M., Trottein, F., Savage, P. B., Wong, C. H., Schneider, E., Dy, M., and Leite-de-Moraes, M. C., Identification of an IL-17-producing NK1.1(neg) iNKT cell population involved in airway neutrophilia, *J Exp Med* 204(5), 995–1001, 2007.

74. Pichavant, M., Goya, S., Meyer, E. H., Johnston, R. A., Kim, H. Y., Matangkasombut, P., Zhu, M., Iwakura, Y., Savage, P. B., DeKruyff, R. H., Shore, S. A., and Umetsu, D. T., Ozone exposure in a mouse model induces airway hyperreactivity that requires the presence of natural killer T cells and IL-17, *J Exp Med* 205(2), 385–393, 2008.

75. Ji, Y., Sun, S., Xu, A., Bhargava, P., Yang, L., Lam, K. S., Gao, B., Lee, C. H., Kersten, S., and Qi, L., Activation of natural killer T cells promotes M2 Macrophage polarization in adipose tissue and improves systemic glucose tolerance via interleukin-4 (IL-4)/STAT6 protein signaling axis in obesity, *J Biol Chem* 287(17), 13561–13571, 2012.

76. Lynch, L., Nowak, M., Varghese, B., Clark, J., Hogan, A. E., Toxavidis, V., Balk, S. P., O'Shea, D., O'Farrelly, C., and Exley, M. A., Adipose tissue invariant NKT cells protect against diet-induced obesity and metabolic disorder through regulatory cytokine production, *Immunity* 37(3), 574–587, 2012.

77. Lynch, L., O'Shea, D., Winter, D. C., Geoghegan, J., Doherty, D. G., and O'Farrelly, C., Invariant NKT cells and CD1d(+) cells amass in human omentum and are depleted in patients with cancer and obesity, *Eur J Immunol* 39(7), 1893–1901, 2009.

78. Wu, L., Parekh, V. V., Gabriel, C. L., Bracy, D. P., Marks-Shulman, P. A., Tamboli, R. A., Kim, S., Mendez-Fernandez, Y. V., Besra, G. S., Lomenick, J. P., Williams, B., Wasserman, D. H., and Van Kaer, L., Activation of invariant natural killer T cells by lipid excess promotes tissue inflammation, insulin resistance, and hepatic steatosis in obese mice, *Proc Natl Acad Sci USA* 109(19), E1143–E1152, 2012.

79. Dowds, C. M., Blumberg, R. S., and Zeissig, S., Control of intestinal homeostasis through crosstalk between natural killer T cells and the intestinal microbiota, *Clin Immunol*, 2015.

80. Wei, B., Wingender, G., Fujiwara, D., Chen, D. Y., McPherson, M., Brewer, S., Borneman, J., Kronenberg, M., and Braun, J., Commensal microbiota and CD8+ T cells shape the formation of invariant NKT cells, *J Immunol* 184(3), 1218–1226, 2010.

81. An, D., Oh, S. F., Olszak, T., Neves, J. F., Avci, F. Y., Erturk-Hasdemir, D., Lu, X., Zeissig, S., Blumberg, R. S., and Kasper, D. L., Sphingolipids from a symbiotic microbe regulate homeostasis of host intestinal natural killer T cells, *Cell* 156(1–2), 123–133, 2014.

82. Tremaroli, V., and Backhed, F., Functional interactions between the gut microbiota and host metabolism, *Nature* 489(7415), 242–249, 2012.

83. Hooper, L. V., Littman, D. R., and Macpherson, A. J., Interactions between the microbiota and the immune system, *Science* 336(6086), 1268–1273, 2012.

84. Liao, C. M., Zimmer, M. I., Shanmuganad, S., Yu, H. T., Cardell, S. L., and Wang, C. R., Dysregulation of CD1d-restricted type ii natural killer T cells leads to spontaneous development of colitis in mice, *Gastroenterology* 142(2), 326–334.e1–e2, 2012.

85. Liao, C. M., Zimmer, M. I., and Wang, C. R., The functions of type I and type II natural killer T cells in inflammatory bowel diseases, *Inflamm Bowel Dis* 19(6), 1330–1338, 2013.

86. Porcelli, S., Yockey, C. E., Brenner, M. B., and Balk, S. P., Analysis of T-Cell antigen receptor (Tcr) expression by human peripheral-blood Cd4-8-alpha/beta T-cells demonstrates preferential use of several V-beta genes and an invariant Tcr alpha-chain, *J Exp Med* 178(1), 1–16, 1993.

87. Dellabona, P., Padovan, E., Casorati, G., Brockhaus, M., and Lanzavecchia, A., An invariant V alpha 24-J alpha Q/V beta 11 T cell receptor is expressed in all individuals by clonally expanded CD4-8-T cells, *J Exp Med* 180(3), 1171–1176, 1994.

88. Kadowaki, N., Antonenko, S., Ho, S., Rissoan, M. C., Soumelis, V., Porcelli, S. A., Lanier, L. L., and Liu, Y. J., Distinct cytokine profiles of neonatal natural killer T cells after expansion with subsets of dendritic cells, *J Exp Med* 193(10), 1221–1226, 2001.

89. Liu, T. Y., Uemura, Y., Suzuki, M., Narita, Y., Hirata, S., Ohyama, H., Ishihara, O., and Matsushita, S., Distinct subsets of human invariant NKT cells differentially regulate T helper responses via dendritic cells, *Eur J Immunol* 38(4), 1012–1023, 2008.

90. Yamaura, A., Hotta, C., Nakazawa, M., Van Kaer, L., and Minami, M., Human invariant Valpha24+ natural killer T cells acquire regulatory functions by interacting with IL-10-treated dendritic cells, *Blood* 111(8), 4254–4263, 2008.

91. Vincent, M. S., Leslie, D. S., Gumperz, J. E., Xiong, X., Grant, E. P., and Brenner, M. B., CD1-dependent dendritic cell instruction, *Nat Immunol* 3(12), 1163–1168, 2002.

92. Moreno, M., Molling, J. W., von Mensdorff-Pouilly, S., Verheijen, R. H., Hooijberg, E., Kramer, D., Reurs, A. W., van den Eertwegh, A. J., von Blomberg, B. M., Scheper, R. J., and Bontkes, H. J., IFN-gamma-producing human invariant NKT cells promote tumor-associated antigen-specific cytotoxic T cell responses, *J Immunol* 181(4), 2446–2454, 2008.

93. Ho, L. P., Urban, B. C., Jones, L., Ogg, G. S., and McMichael, A. J., CD4(–) CD8alphaalpha subset of CD1d-restricted NKT cells controls T cell expansion, *J Immunol* 172(12), 7350–7358, 2004.

94. Osada, T., Morse, M. A., Lyerly, H. K., and Clay, T. M., Ex vivo expanded human CD4+ regulatory NKT cells suppress expansion of tumor antigen-specific CTLs, *Int Immunol* 17(9), 1143–1155, 2005.

95. Montoya, C. J., Jie, H. B., Al-Harthi, L., Mulder, C., Patino, P. J., Rugeles, M. T., Krieg, A. M., Landay, A. L., and Wilson, S. B., Activation of plasmacytoid dendritic cells with TLR9 agonists initiates invariant NKT cell-mediated cross-talk with myeloid dendritic cells, *J Immunol* 177(2), 1028–1039, 2006.

96. Marschner, A., Rothenfusser, S., Hornung, V., Prell, D., Krug, A., Kerkmann, M., Wellisch, D., Poeck, H., Greinacher, A., Giese, T., Endres, S., and Hartmann, G., CpG ODN enhance antigen-specific NKT cell activation via plasmacytoid dendritic cells, *Eur J Immunol* 35(8), 2347–2357, 2005.

97. Campos-Martin, Y., Colmenares, M., Gozalbo-Lopez, B., Lopez-Nunez, M., Savage, P. B., and Martinez-Naves, E., Immature human dendritic cells infected with Leishmania infantum are resistant to NK-mediated cytolysis but are efficiently recognized by NKT cells, *J Immunol* 176(10), 6172–6179, 2006.

98. Kawasaki, N., Vela, J. L., Nycholat, C. M., Rademacher, C., Khurana, A., van Rooijen, N., Crocker, P. R., Kronenberg, M., and Paulson, J. C., Targeted delivery of lipid antigen to macrophages via the CD169/sialoadhesin endocytic pathway induces robust invariant natural killer T cell activation, *Proc Natl Acad Sci USA* 110(19), 7826–7831, 2013.

99. Kok, W. L., Denney, L., Benam, K., Cole, S., Clelland, C., McMichael, A. J., and Ho, L. P., Pivotal advance: Invariant NKT cells reduce accumulation of inflammatory monocytes in the lungs and decrease immune-pathology during severe influenza A virus infection, *J Leukoc Biol* 91(3), 357–368, 2012.

100. Cole, S. L., Benam, K. H., McMichael, A. J., and Ho, L. P., Involvement of the 4-1BB/4-1BBL pathway in control of monocyte numbers by invariant NKT cells, *J Immunol* 192(8), 3898–3907, 2014.

101. Aldridge, J. R., Jr., Moseley, C. E., Boltz, D. A., Negovetich, N. J., Reynolds, C., Franks, J., Brown, S. A., Doherty, P. C., Webster, R. G., and Thomas, P. G., TNF/iNOS-producing dendritic cells are the necessary evil of lethal influenza virus infection, *Proc Natl Acad Sci USA* 106(13), 5306–5311, 2009.

102. Paget, C., Ivanov, S., Fontaine, J., Renneson, J., Blanc, F., Pichavant, M., Dumoutier, L., Ryffel, B., Renauld, J. C., Gosset, P., Gosset, P., Si-Tahar, M., Faveeuw, C., and Trottein, F., Interleukin-22 is produced by invariant natural killer T lymphocytes during influenza A virus infection: Potential role in protection against lung epithelial damages, *J Biol Chem* 287(12), 8816–8829, 2012.

103. Bessoles, S., Dudal, S., Besra, G. S., Sanchez, F., and Lafont, V., Human CD4+ invariant NKT cells are involved in antibacterial immunity against Brucella suis through CD1d-dependent but CD4-independent mechanisms, *Eur J Immunol* 39(4), 1025–1035, 2009.

104. Kim, E. Y., Battaile, J. T., Patel, A. C., You, Y., Agapov, E., Grayson, M. H., Benoit, L. A., Byers, D. E., Alevy, Y., Tucker, J., Swanson, S., Tidwell, R., Tyner, J. W., Morton, J. D., Castro, M., Polineni, D., Patterson, G. A., Schwendener, R. A., Allard, J. D., Peltz, G., and Holtzman, M. J., Persistent activation of an innate immune response translates respiratory viral infection into chronic lung disease, *Nat Med* 14(6), 633–640, 2008.

105. Barral, P., Polzella, P., Bruckbauer, A., van Rooijen, N., Besra, G. S., Cerundolo, V., and Batista, F. D., CD169(+) macrophages present lipid antigens to mediate early activation of iNKT cells in lymph nodes, *Nat Immunol* 11(4), 303–312, 2010.

106. De Santo, C., Salio, M., Masri, S. H., Lee, L. Y., Dong, T., Speak, A. O., Porubsky, S., Booth, S., Veerapen, N., Besra, G. S., Grone, H. J., Platt, F. M., Zambon, M., and Cerundolo, V., Invariant NKT cells reduce the immunosuppressive activity of influenza A virus-induced myeloid-derived suppressor cells in mice and humans, *J Clin Invest* 118(12), 4036–4048, 2008.

107. Bronte, V., Serafini, P., De Santo, C., Marigo, I., Tosello, V., Mazzoni, A., Segal, D. M., Staib, C., Lowel, M., Sutter, G., Colombo, M. P., and Zanovello, P., IL-4-induced arginase 1 suppresses alloreactive T cells in tumor-bearing mice, *J Immunol* 170(1), 270–278, 2003.

108. Mussai, F., De Santo, C., and Cerundolo, V., Interaction between invariant NKT cells and myeloid-derived suppressor cells in cancer patients: Evidence and therapeutic opportunities, *J Immunother* 35(6), 449–459, 2012.

109. Nieda, M., Okai, M., Tazbirkova, A., Lin, H., Yamaura, A., Ide, K., Abraham, R., Juji, T., Macfarlane, D. J., and Nicol, A. J., Therapeutic activation of Valpha24+Vbeta11+ NKT cells in human subjects results in highly coordinated secondary activation of acquired and innate immunity, *Blood* 103(2), 383–389, 2004.

110. Chang, D. H., Osman, K., Connolly, J., Kukreja, A., Krasovsky, J., Pack, M., Hutchinson, A., Geller, M., Liu, N., Annable, R., Shay, J., Kirchhoff, K., Nishi, N., Ando, Y., Hayashi, K., Hassoun, H., Steinman, R. M., and Dhodapkar, M. V., Sustained expansion of NKT cells and antigen-specific T cells after injection of alpha-galactosyl-ceramide loaded mature dendritic cells in cancer patients, *J Exp Med* 201(9), 1503–1517, 2005.

111. De Santo, C., Arscott, R., Booth, S., Karydis, I., Jones, M., Asher, R., Salio, M., Middleton, M., and Cerundolo, V., Invariant NKT cells modulate the suppressive activity of IL-10-secreting neutrophils differentiated with serum amyloid A, *Nat Immunol* 11(11), 1039–1046, 2010.

112. Wingender, G., Hiss, M., Engel, I., Peukert, K., Ley, K., Haller, H., Kronenberg, M., and von Vietinghoff, S., Neutrophilic granulocytes modulate invariant NKT cell function in mice and humans, *J Immunol* 188(7), 3000–3008, 2012.

113. Godfrey, D. I. and Kronenberg, M., Going both ways: Immune regulation via CD1d-dependent NKT cells, *J Clin Invest* 114(10), 1379–1388, 2004.

114. Tupin, E., Kinjo, Y., and Kronenberg, M., The unique role of natural killer T cells in the response to microorganisms, *Nat Rev Microbiol* 5(6), 405–417, 2007.

115. Moreno, M., Molling, J. W., von Mensdorff-Pouilly, S., Verheijen, R. H., von Blomberg, B. M., van den Eertwegh, A. J., Scheper, R. J., and Bontkes, H. J., In vitro expanded human invariant natural killer T-cells promote functional activity of natural killer cells, *Clin Immunol* 129(1), 145–154, 2008.

116. Moretta, L., Bottino, C., Pende, D., Castriconi, R., Mingari, M. C., and Moretta, A., Surface NK receptors and their ligands on tumor cells, *Semin Immunol* 18(3), 151–158, 2006.

117. Algarra, I., Garcia-Lora, A., Cabrera, T., Ruiz-Cabello, F., and Garrido, F., The selection of tumor variants with altered expression of classical and nonclassical MHC class I molecules: Implications for tumor immune escape, *Cancer Immunol Immunother* 53(10), 904–910, 2004.

118. Molling, J. W., Kolgen, W., van der Vliet, H. J., Boomsma, M. F., Kruizenga, H., Smorenburg, C. H., Molenkamp, B. G., Langendijk, J. A., Leemans, C. R., von Blomberg, B. M., Scheper, R. J., and van den Eertwegh, A. J., Peripheral blood IFN-gamma-secreting Valpha24+Vbeta11+ NKT cell numbers are decreased in cancer patients independent of tumor type or tumor load, *Int J Cancer* 116(1), 87–93, 2005.

119. Motohashi, S., Kobayashi, S., Ito, T., Magara, K. K., Mikuni, O., Kamada, N., Iizasa, T., Nakayama, T., Fujisawa, T., and Taniguchi, M., Preserved IFN-alpha production of circulating Valpha24 NKT cells in primary lung cancer patients, *Int J Cancer* 102(2), 159–165, 2002.

120. Dhodapkar, M. V., Geller, M. D., Chang, D. H., Shimizu, K., Fujii, S., Dhodapkar, K. M., and Krasovsky, J., A reversible defect in natural killer T cell function characterizes the progression of premalignant to malignant multiple myeloma, *J Exp Med* 197(12), 1667–1676, 2003.

121. Tahir, S. M., Cheng, O., Shaulov, A., Koezuka, Y., Bubley, G. J., Wilson, S. B., Balk, S. P., and Exley, M. A., Loss of IFN-gamma production by invariant NK T cells in advanced cancer, *J Immunol* 167(7), 4046–4050, 2001.

122. Galli, G., Nuti, S., Tavarini, S., Galli-Stampino, L., De Lalla, C., Casorati, G., Dellabona, P., and Abrignani, S., Innate immune responses support adaptive immunity: NKT cells induce B cell activation, *Vaccine* 21 Suppl 2, S48–S54, 2003.

123. Dellabona, P., Abrignani, S., and Casorati, G., iNKT-cell help to B cells: A cooperative job between innate and adaptive immune responses, *Eur J Immunol* 44(8), 2230–2237, 2014.

124. Zeng, S. G., Ghnewa, Y. G., O'Reilly, V. P., Lyons, V. G., Atzberger, A., Hogan, A. E., Exley, M. A., and Doherty, D. G., Human invariant NKT cell subsets differentially promote differentiation, antibody production, and T cell stimulation by B cells in vitro, *J Immunol* 191(4), 1666–1676, 2013.

125. Rossignol, A., Barra, A., Herbelin, A., Preud'homme, J. L., and Gombert, J. M., Freshly isolated Valpha24+ CD4+ invariant natural killer T cells activated by alpha-galactosylceramide-pulsed B cells promote both IgG and IgE production, *Clin Exp Immunol* 148(3), 555–563, 2007.

126. Kojo, S., Adachi, Y., Keino, H., Taniguchi, M., and Sumida, T., Dysfunction of T cell receptor AV24AJ18+, BV11+ double-negative regulatory natural killer T cells in autoimmune diseases, *Arthritis Rheum* 44(5), 1127–1138, 2001.

127. Green, M. R., Kennell, A. S., Larche, M. J., Seifert, M. H., Isenberg, D. A., and Salaman, M. R., Natural killer T cells in families of patients with systemic lupus erythematosus: Their possible role in regulation of IGG production, *Arthritis Rheum* 56(1), 303–310, 2007.

128. Cho, Y. N., Kee, S. J., Lee, S. J., Seo, S. R., Kim, T. J., Lee, S. S., Kim, M. S., Lee, W. W., Yoo, D. H., Kim, N., and Park, Y. W., Numerical and functional deficiencies of natural killer T cells in systemic lupus erythematosus: Their deficiency related to disease activity, *Rheumatology (Oxford)* 50(6), 1054–1063, 2011.

129. Bosma, A., Abdel-Gadir, A., Isenberg, D. A., Jury, E. C., and Mauri, C., Lipid-antigen presentation by CD1d(+) B cells is essential for the maintenance of invariant natural killer T cells, *Immunity* 36(3), 477–490, 2012.

130. Shen, L., Zhang, H., Caimol, M., Benike, C. J., Chakravarty, E. F., Strober, S., and Engleman, E. G., Invariant natural killer T cells in lupus patients promote IgG and IgG autoantibody production, *Eur J Immunol* 45(2), 612–623, 2015.

131. Gao, Y., Workman, S., Gadola, S., Elliott, T., Grimbacher, B., and Williams, A. P., Common variable immunodeficiency is associated with a functional deficiency of invariant natural killer T cells, *J Allergy Clin Immunol* 133(5), 1420–1428, 1428.e1, 2014.

132. Fulcher, D. A., Avery, D. T., Fewings, N. L., Berglund, L. J., Wong, S., Riminton, D. S., Adelstein, S., and Tangye, S. G., Invariant natural killer (iNK) T cell deficiency in patients with common variable immunodeficiency, *Clin Exp Immunol* 157(3), 365–369, 2009.

133. Carvalho, K. I., Melo, K. M., Bruno, F. R., Snyder-Cappione, J. E., Nixon, D. F., Costa-Carvalho, B. T., and Kallas, E. G., Skewed distribution of circulating activated natural killer T (NKT) cells in patients with common variable immunodeficiency disorders (CVID), *PLoS One* 5(9), 2010.

134. Venken, K., Decruy, T., Aspeslagh, S., Van Calenbergh, S., Lambrecht, B. N., and Elewaut, D., Bacterial CD1d-restricted glycolipids induce IL-10 production by human regulatory T cells upon cross-talk with invariant NKT cells, *J Immunol* 191(5), 2174–2183, 2013.

135. Azuma, T., Takahashi, T., Kunisato, A., Kitamura, T., and Hirai, H., Human CD4+ CD25+ regulatory T cells suppress NKT cell functions, *Cancer Res* 63(15), 4516–4520, 2003.

136. Jiang, S., Game, D. S., Davies, D., Lombardi, G., and Lechler, R. I., Activated CD1d-restricted natural killer T cells secrete IL-2: Innate help for CD4+CD25+ regulatory T cells?, *Eur J Immunol* 35(4), 1193–1200, 2005.

137. Wilson, S. B., Kent, S. C., Patton, K. T., Orban, T., Jackson, R. A., Exley, M., Porcelli, S., Schatz, D. A., Atkinson, M. A., Balk, S. P., Strominger, J. L., and Hafler, D. A., Extreme Th1 bias of invariant Valpha24JalphaQ T cells in type 1 diabetes, *Nature* 391(6663), 177–181, 1998.

138. Wilson, S. B., Kent, S. C., Horton, H. F., Hill, A. A., Bollyky, P. L., Hafler, D. A., Strominger, J. L., and Byrne, M. C., Multiple differences in gene expression in regulatory Valpha 24Jalpha Q T cells from identical twins discordant for type I diabetes, *Proc Natl Acad Sci USA* 97(13), 7411–7416, 2000.

139. Kukreja, A., Cost, G., Marker, J., Zhang, C., Sun, Z., Lin-Su, K., Ten, S., Sanz, M., Exley, M., Wilson, B., Porcelli, S., and Maclaren, N., Multiple immuno-regulatory defects in type-1 diabetes, *J Clin Invest* 109(1), 131–140, 2002.

140. Illes, Z., Kondo, T., Newcombe, J., Oka, N., Tabira, T., and Yamamura, T., Differential expression of NK T cell V alpha 24J alpha Q invariant TCR chain in the lesions of multiple sclerosis and chronic inflammatory demyelinating polyneuropathy, *J Immunol* 164(8), 4375–4381, 2000.

141. Gausling, R., Trollmo, C., and Hafler, D. A., Decreases in interleukin-4 secretion by invariant CD4(–)CD8(–)V alpha 24J alpha Q T cells in peripheral blood of patients with relapsing-remitting multiple sclerosis, *Clin Immunol* 98(1), 11–17, 2001.

142. van der Vliet, H. J., von Blomberg, B. M., Nishi, N., Reijm, M., Voskuyl, A. E., van Bodegraven, A. A., Polman, C. H., Rustemeyer, T., Lips, P., van den Eertwegh, A. J., Giaccone, G., Scheper, R. J., and Pinedo, H. M., Circulating V(alpha24+) Vbeta11+ NKT cell numbers are decreased in a wide variety of diseases that are characterized by autoreactive tissue damage, *Clin Immunol* 100(2), 144–148, 2001.

143. Gigli, G., Caielli, S., Cutuli, D., and Falcone, M., Innate immunity modulates autoimmunity: Type 1 interferon-beta treatment in multiple sclerosis promotes growth and function of regulatory invariant natural killer T cells through dendritic cell maturation, *Immunology* 122(3), 409–417, 2007.

144. Nguyen, K. D., Vanichsarn, C., and Nadeau, K. C., Increased cytotoxicity of CD4+ invariant NKT cells against CD4+CD25hiCD127lo/– regulatory T cells in allergic asthma, *Eur J Immunol* 38(7), 2034–2045, 2008.

145. Schipper, H. S., Rakhshandehroo, M., van de Graaf, S. F., Venken, K., Koppen, A., Stienstra, R., Prop, S., Meerding, J., Hamers, N., Besra, G., Boon, L., Nieuwenhuis, E. E., Elewaut, D., Prakken, B., Kersten, S., Boes, M., and Kalkhoven, E., Natural killer T cells in adipose tissue prevent insulin resistance, *J Clin Invest* 122(9), 3343–3354, 2012.

146. Balato, A., Zhao, Y., Harberts, E., Groleau, P., Liu, J., Fishelevich, R., and Gaspari, A. A., CD1d-dependent, iNKT-cell cytotoxicity against keratinocytes in allergic contact dermatitis, *Exp Dermatol* 21(12), 915–920, 2012.

147. Hofmann, S. C., Bosma, A., Bruckner-Tuderman, L., Vukmanovic-Stejic, M., Jury, E. C., Isenberg, D. A., and Mauri, C., Invariant natural killer T cells are enriched at the site of cutaneous inflammation in lupus erythematosus, *J Dermatol Sci* 71(1), 22–28, 2013.

148. Akbari, O., Faul, J. L., Hoyte, E. G., Berry, G. J., Wahlstrom, J., Kronenberg, M., DeKruyff, R. H., and Umetsu, D. T., CD4+ invariant T-cell-receptor+ natural killer T cells in bronchial asthma, *N Engl J Med* 354(11), 1117–1129, 2006.

149. Hodge, G., Mukaro, V., Holmes, M., Reynolds, P. N., and Hodge, S., Enhanced cytotoxic function of natural killer and natural killer T-like cells associated with decreased CD94 (Kp43) in the chronic obstructive pulmonary disease airway, *Respirology* 18(2), 369–376, 2013.

150. Urbanowicz, R. A., Lamb, J. R., Todd, I., Corne, J. M., and Fairclough, L. C., Enhanced effector function of cytotoxic cells in the induced sputum of COPD patients, *Respir Res* 11, 76, 2010.

151. Wang, J., Urbanowicz, R. A., Tighe, P. J., Todd, I., Corne, J. M., and Fairclough, L. C., Differential activation of killer cells in the circulation and the lung: A study of current smoking status and chronic obstructive pulmonary disease (COPD), *PLoS One* 8(3), e58556, 2013.

152. Pichavant, M., Remy, G., Bekaert, S., Le Rouzic, O., Kervoaze, G., Vilain, E., Just, N., Tillie-Leblond, I., Trottein, F., Cataldo, D., and Gosset, P., Oxidative stress-mediated iNKT-cell activation is involved in COPD pathogenesis, *Mucosal Immunol* 7(3), 568–578, 2014.

153. Fuss, I. J., Joshi, B., Yang, Z., Degheidy, H., Fichtner-Feigl, S., de Souza, H., Rieder, F., Scaldaferri, F., Schirbel, A., Scarpa, M., West, G., Yi, C., Xu, L., Leland, P., Yao, M., Mannon, P., Puri, R. K., Fiocchi, C., and Strober, W., IL-13Ralpha2-bearing, type II NKT cells reactive to sulfatide self-antigen populate the mucosa of ulcerative colitis, *Gut* 63(11), 1728–1736, 2014.

154. Tsuneyama, K., Yasoshima, M., Harada, K., Hiramatsu, K., Gershwin, M. E., and Nakanuma, Y., Increased CD1d expression on small bile duct epithelium and epithelioid granuloma in livers in primary biliary cirrhosis, *Hepatology* 28(3), 620–623, 1998.

155. Durante-Mangoni, E., Wang, R., Shaulov, A., He, Q., Nasser, I., Afdhal, N., Koziel, M. J., and Exley, M. A., Hepatic CD1d expression in hepatitis C virus infection and recognition by resident proinflammatory CD1d-reactive T cells, *J Immunol* 173(3), 2159–2166, 2004.

156. Schrumpf, E., Tan, C., Karlsen, T. H., Sponheim, J., Bjorkstrom, N. K., Sundnes, O., Alfsnes, K., Kaser, A., Jefferson, D. M., Ueno, Y., Eide, T. J., Haraldsen, G., Zeissig, S., Exley, M. A., Blumberg, R. S., and Melum, E., The biliary epithelium presents antigens to and activates natural killer T cells, *Hepatology*, 2015.

157. Araki, M., Kondo, T., Gumperz, J. E., Brenner, M. B., Miyake, S., and Yamamura, T., Th2 bias of CD4+ NKT cells derived from multiple sclerosis in remission, *Int Immunol* 15(2), 279–288, 2003.

158. Yanagihara, Y., Shiozawa, K., Takai, M., Kyogoku, M., and Shiozawa, S., Natural killer (NK) T cells are significantly decreased in the peripheral blood of patients with rheumatoid arthritis (RA), *Clin Exp Immunol* 118(1), 131–136, 1999.

159. Kojo, S., Tsutsumi, A., Goto, D., and Sumida, T., Low expression levels of soluble CD1d gene in patients with rheumatoid arthritis, *J Rheumatol* 30(12), 2524–2528, 2003.

160. Parietti, V., Chifflot, H., Sibilia, J., Muller, S., and Monneaux, F., Rituximab treatment overcomes reduction of regulatory iNKT cells in patients with rheumatoid arthritis, *Clin Immunol* 134(3), 331–339, 2010.

161. Berzins, S. P., Smyth, M. J., and Baxter, A. G., Presumed guilty: Natural killer T cell defects and human disease, *Nat Rev Immunol* 11(2), 131–142, 2011.

162. Iwamura, C., and Nakayama, T., Role of NKT cells in allergic asthma, *Curr Opin Immunol* 22(6), 807–813, 2010.

163. Shim, J. U., and Koh, Y. I., Increased Th2-like invariant natural killer T cells in peripheral blood from patients with asthma, *Allergy Asthma Immunol Res* 6(5), 444–448, 2014.

164. Nickoloff, B. J., Wrone-Smith, T., Bonish, B., and Porcelli, S. A., Response of murine and normal human skin to injection of allogeneic blood-derived psoriatic immuno-cytes: Detection of T cells expressing receptors typically present on natural killer cells, including CD94, CD158, and CD161, *Arch Dermatol* 135(5), 546–552, 1999.
165. Zhao, Y., Fishelevich, R., Petrali, J. P., Zheng, L., Anatolievna, M. A., Deng, A., Eckert, R. L., and Gaspari, A. A., Activation of keratinocyte protein kinase C zeta in psoriasis plaques, *J Invest Dermatol* 128(9), 2190–2197, 2008.
166. Bobryshev, Y. V., and Lord, R. S., Co-accumulation of dendritic cells and natural killer T cells within rupture-prone regions in human atherosclerotic plaques, *J Histochem Cytochem* 53(6), 781–785, 2005.
167. Chan, W. L., Pejnovic, N., Hamilton, H., Liew, T. V., Popadic, D., Poggi, A., and Khan, S. M., Atherosclerotic abdominal aortic aneurysm and the interaction between autolo-gous human plaque-derived vascular smooth muscle cells, type 1 NKT, and helper T cells, *Circ Res* 96(6), 675–683, 2005.
168. Kyriakakis, E., Cavallari, M., Andert, J., Philippova, M., Koella, C., Bochkov, V., Erne, P., Wilson, S. B., Mori, L., Biedermann, B. C., Resink, T. J., and De Libero, G., Invariant natural killer T cells: Linking inflammation and neovascularization in human atherosclerosis, *Eur J Immunol* 40(11), 3268–3279, 2010.
169. Giaccone, G., Punt, C. J., Ando, Y., Ruijter, R., Nishi, N., Peters, M., von Blomberg, B. M., Scheper, R. J., van der Vliet, H. J., van den Eertwegh, A. J., Roelvink, M., Beijnen, J., Zwierzina, H., and Pinedo, H. M., A phase I study of the natural killer T-cell ligand alpha-galactosylceramide (KRN7000) in patients with solid tumors, *Clin Cancer Res* 8(12), 3702–3709, 2002.
170. Ishikawa, A., Motohashi, S., Ishikawa, E., Fuchida, H., Higashino, K., Otsuji, M., Iizasa, T., Nakayama, T., Taniguchi, M., and Fujisawa, T., A phase I study of alpha-galactosylceramide (KRN7000)-pulsed dendritic cells in patients with advanced and recurrent non-small cell lung cancer, *Clin Cancer Res* 11(5), 1910–1917, 2005.
171. Motohashi, S., Nagato, K., Kunii, N., Yamamoto, H., Yamasaki, K., Okita, K., Hanaoka, H., Shimizu, N., Suzuki, M., Yoshino, I., Taniguchi, M., Fujisawa, T., and Nakayama, T., A phase I-II study of alpha-galactosylceramide-pulsed IL-2/GM-CSF-cultured peripheral blood mononuclear cells in patients with advanced and recurrent non-small cell lung cancer, *J Immunol* 182(4), 2492–2501, 2009.
172. Kim, H. Y., Kim, H. J., Min, H. S., Kim, S., Park, W. S., Park, S. H., and Chung, D. H., NKT cells promote antibody-induced joint inflammation by suppressing transforming growth factor beta1 production, *J Exp Med* 201(1), 41–47, 2005.
173. Jahng, A. W., Maricic, I., Pedersen, B., Burdin, N., Naidenko, O., Kronenberg, M., Koezuka, Y., and Kumar, V., Activation of natural killer T cells potentiates or prevents experimental autoimmune encephalomyelitis, *J Exp Med* 194(12), 1789–1799, 2001.
174. Yang, J. Q., Saxena, V., Xu, H., Van Kaer, L., Wang, C. R., and Singh, R. R., Repeated alpha-galactosylceramide administration results in expansion of NK T cells and alle-viates inflammatory dermatitis in MRL-lpr/lpr mice, *J Immunol* 171(8), 4439–4446, 2003.
175. Zeng, D., Liu, Y., Sidobre, S., Kronenberg, M., and Strober, S., Activation of natural killer T cells in NZB/W mice induces Th1-type immune responses exacerbating lupus, *J Clin Invest* 112(8), 1211–1222, 2003.
176. East, J. E., Kennedy, A. J., and Webb, T. J., Raising the roof: The preferential pharma-cological stimulation of Th1 and th2 responses mediated by NKT cells, *Med Res Rev* 34(1), 45–76, 2014.
177. Rayapudi, M., Rajavelu, P., Zhu, X., Kaul, A., Niranjan, R., Dynda, S., Mishra, A., Mattner, J., Zaidi, A., Dutt, P., and Mishra, A., Invariant natural killer T-cell neutral-ization is a possible novel therapy for human eosinophilic esophagitis, *Clin Transl Immunology* 3(1), e9, 2014.

178. Scheuplein, F., Thariath, A., Macdonald, S., Truneh, A., Mashal, R., and Schaub, R., A humanized monoclonal antibody specific for invariant Natural Killer T (iNKT) cells for in vivo depletion, *PLoS One* 8(9), e76692, 2013.

179. Wen, X., Rao, P., Carreno, L. J., Kim, S., Lawrenczyk, A., Porcelli, S. A., Cresswell, P., and Yuan, W., Human CD1d knock-in mouse model demonstrates potent antitumor potential of human CD1d-restricted invariant natural killer T cells, *Proc Natl Acad Sci USA* 110(8), 2963–2968, 2013.

180. Felio, K., Nguyen, H., Dascher, C. C., Choi, H. J., Li, S., Zimmer, M. I., Colmone, A., Moody, D. B., Brenner, M. B., and Wang, C. R., CD1-restricted adaptive immune responses to Mycobacteria in human group 1 CD1 transgenic mice, *J Exp Med* 206(11), 2497–2509, 2009.

181. Lawrenczyk, A., Kim, S., Wen, X., Xiong, R., and Yuan, W., Exploring the therapeutic potentials of iNKT cells for anti-HBV treatment, *Pathogens* 3(3), 563–576, 2014.

182. Jiang, X., Zhang, M., Lai, Q., Huang, X., Li, Y., Sun, J., Abbott, W. G., Ma, S., and Hou, J., Restored circulating invariant NKT cells are associated with viral control in patients with chronic hepatitis B, *PLoS One* 6(12), e28871, 2011.

183. Wen, X., Kim, S., Xiong, R., Li, M., Lawrenczyk, A., Huang, X., Chen, S. Y., Rao, P., Besra, G. S., Dellabona, P., Casorati, G., Porcelli, S. A., Akbari, O., Exley, M. A., and Yuan, W., A subset of CD8alphabeta+ invariant NKT cells in a humanized mouse model, *J Immunol* 195(4), 1459–1469, 2015.

184. Lockridge, J. L., Chen, X., Zhou, Y., Rajesh, D., Roenneburg, D. A., Hegde, S., Gerdts, S., Cheng, T. Y., Anderson, R. J., Painter, G. F., Moody, D. B., Burlingham, W. J., and Gumperz, J. E., Analysis of the CD1 antigen presenting system in humanized SCID mice, *PLoS One* 6(6), e21701, 2011.

185. Germanov, E., Veinotte, L., Cullen, R., Chamberlain, E., Butcher, E. C., and Johnston, B., Critical role for the chemokine receptor CXCR6 in homeostasis and activation of CD1d-restricted NKT cells, *J Immunol* 181(1), 81–91, 2008.

186. Wong, C. H., Jenne, C. N., Lee, W. Y., Leger, C., and Kubes, P., Functional innervation of hepatic iNKT cells is immunosuppressive following stroke, *Science* 334 (6052), 101–105, 2011.

# 9 Transmission of T Cell Receptor-Mediated Signaling via the GRB2 Family of Adaptor Proteins

*Mahmood Y. Bilal and Jon C. D. Houtman*

## CONTENTS

## ABSTRACT

The GRB2 family of proteins consists of three highly related homologs, GRB2, GADS, and GRAP. These proteins have a conserved domain structure with a central SH2 domain, which binds to phosphorylated tyrosine ligands, flanked by two SH3 domains, which interact with proline- or arginine/lysine-rich sequences in numerous signaling proteins. The function of the GRB2 family of proteins is to regulate signaling events by connecting phosphorylated ligands, such as receptor tyrosine kinases and adaptor proteins, to downstream effectors needed for the propagation of intracellular signaling. Because of these functions, the GRB2 family of proteins is critical for the differentiation and function of numerous cell types, as well as for the initiation and progression of a panoply of human diseases. GRB2, GADS, and GRAP are especially vital for the function of T cells. These proteins control signaling complexes at the T cell antigen receptor and the adaptor protein LAT, where they facilitate the activation, differentiation, and function of T cells. This chapter will describe the domain structure of the GRB2 family of proteins, their general role in the initiation of signaling in multiple cell types, and their function in the activation and differentiation of T cells.

## 9.1 INTRODUCTION

Helper CD4+ T cells are critical in the fight against parasitic, bacterial, and viral infections.[1–4] However, dysregulated CD4+ T cell responses lead to multiple autoimmune and pathological disorders such as type I diabetes, asthma, solid and hematopoietic cancers, and cardiovascular disease.[5–10] Upon activation, they mediate cellular immunity through the secretion of cytokines that recruit or activate other immune cells, promotion of humoral-mediated immunity through the "help" of antigen-specific B-cells, optimization of cytotoxic CD8+ T cell functions, and suppression of inappropriate immune cell responses.[11–14]

The transition from naïve to effector CD4+ T cells requires activation through multiple signals. Ligation of the T cell receptor (TCR) by foreign peptide antigen bound to the major histocompatibility complex (MHC) class II on the antigen-presenting cells is the primary signal required for CD4+ T cell activation.[15,16] TCR-induced signals are required for differentiation, proliferation, cytokine release, and cytotoxic killing.[15,16] TCR-induced signals are further complemented and amplified by costimulatory receptors and cytokines derived from antigen-presenting cells and the environment. In this review, we focus on a subset of adaptor proteins, the GRB2 family of adaptors, known to transduce signals from numerous receptors and other signaling pathways.[17–22] GRB2 and its homologs, GADS and GRAP, contribute to TCR-mediated signal transduction.[16,23] However, the full functions of these adaptor proteins and how they positively and negatively regulate TCR-derived signaling complexes have been elusive. We will describe the current paradigm of how the GRB2 family members function and discuss recent experiments describing the mechanisms used by these proteins to transduce signals from the TCR.

## 9.2 STRUCTURE, FUNCTION, AND HOMOLOGY OF THE GRB2 FAMILY PROTEINS

The GRB2 family of adaptor proteins consists of three members: GRB2, GADS, and GRAP. These proteins are composed of a central SH2 domain flanked by two SH3 domains (Figure 9.1). The amino acid sequence homology of the full proteins as well as the specific domains highlights the high degree of similarity between GRB2, GADS, and GRAP (Table 9.1).[24] The SH2 domain facilitates the recognition of specific phospho-tyrosyl sequences on intracellular ligands such as the TCR ζ chains and associated adaptor protein linker for activation of T cells (LAT), RTKs such as EGFR and FGFR, the T cell costimulatory receptor CD28, and CML-associated

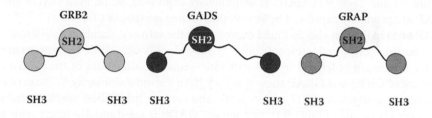

FIGURE 9.1 GRB2 family of adaptor proteins. Diagram of the structure of GRB2, GADS, and GRAP. Shown below each protein are some of the proline-rich and phospho-tyrosine ligands that bind the SH3 and SH2 domains of the GRB2 family members.

---

## TABLE 9.1
## Size, Tissue Expression, and Relative Protein Sequence Homology of GRB2 Family of Adaptors

| Protein | Amino Acids | Tissue Expression | Overall Homology to GRB2 | N-SH3 Homology to GRB2 | SH2 Homology to GRB2 | C-SH3 Homology to GRB2 |
|---|---|---|---|---|---|---|
| GRB2 | 217 | Ubiquitous | 100% | 100% | 100% | 100% |
| GADS | 330 | Hematopoietic and neuroendocrine cells | 50% | 52% | 57% | 52% |
| GRAP | 217 | Hematopoietic, salivary glands, and kidney | 60% | 67% | 62% | 48% |

kinase BCR-Abl.[25-34] In contrast, the SH3 domain allows binding and recruitment of multiple proline-rich ligands, such as the guanine nucleotide exchange factor (GEF) son of sevenless homolog-1 (SOS1) and the SH2 domain-containing leukocyte protein of 76kDa (SLP-76), to active proximal signaling zones initiated by tyrosine kinases (Figure 9.1). Nonetheless, the complete inventory of SH3 or SH2 domain ligands that bind the GRB2 family of adaptors is yet to be determined. These proteins are essential for driving the assembly of various multiprotein signaling complexes by connecting SH2 and SH3 domain ligands. Due to their role in core signaling pathways, dysregulated signaling complexes driven by GRB2 family members have been linked to oncogenesis, autoimmunity, diabetes, and cardiovascular disease.[21,29,30,35-37]

The core differences between GRB2, GADS, and GRAP are tissue expression, size and structure, and affinity of the SH3 and SH2 domain to distinct ligands (Figure 9.1 and Table 9.1). GRB2 is ubiquitously expressed, while both GADS and GRAP are primarily expressed by hematopoietic lineage tissues (Table 9.1).[24,28,38-41] GRAP and GADS can also be found expressed in the salivary glands and differentiated neuroendocrine cells, respectively.[42-44] Second, although all three proteins are similar in domain order and overall structure, sequence alignment of the primary sequence of GRB2 and GRAP show that they have the most similarity.[35] These two proteins are composed of 217 amino acids, and contain the highest single domain and overall homology (Table 9.1). In contrast, GADS is substantially larger with a length of 330 amino acids. The major structural difference is that GADS contains an extended 120 amino acid linker region containing proline-rich stretches between the SH2 domain and the C-terminal SH3 domain.[45] Biochemical experiments have shown that this region is cleaved by caspases, which results in the alteration and regulation of SLP-76/LAT signaling in T cells.[46] These biochemical studies have found that the linker region of GADS contains other negative regulatory sites such as T262 (murine T254) that is phosphorylated by the serine/threonine kinase hematopoietic regulatory kinase 1 (HPK1) upon TCR stimulation. The phosphorylation of this site has been recently shown to control the interaction of GADS/SLP-76 with LAT signaling complexes.[47,48] Interestingly, GRB2 also undergoes posttranslational modifications upon phosphorylation on Tyr209 by kinases such as BCR-Abl and FGFR that lead to reduced affinity of SOS1 binding and release of inhibitory interaction at FGFR, respectively.[29,32,38,49,50] However, the role of tyrosine phosphorylated GRB2 in T cells is unclear. Further studies are required to elucidate the role of posttranslational modifications acquired by the GRB2 family members.

The similarity between GRB2 and GRAP can further be observed in the N-terminus SH3 domain where there is 67% amino acid sequence homology. This explains the ability of their SH3 domains to bind similar proline-rich ligands such as SOS1, SAM68, and Dynamin (Figure 9.1).[51] Binding of the N-terminal SH3 region of GRB2, and theoretically GRAP, to its ligand is thought to be mediated through a consensus of PXXPXR or an atypical PXXXR motif.[26,51-53] In contrast, biochemical and biophysical studies found that the SH3 domains of GADS bind to proline-rich ligands, such as SLP-76 and HPK1, via RXXK motifs.[54-59] However, some studies have suggested that these interactions are not limited to specific adaptor proteins. GRB2 and GADS have the ability to bind SLP-76 and SOS1 in vivo and in vitro, respectively, albeit with a much lower affinity when compared to GRB2/SOS1 or

GADS/SLP-76 interactions.[24,60–62] It is not clear if these interactions have any physiological relevance. In addition, the binding of GRB2 to SLP-76 and other ligands such as GAB1 appears to be driven by the C-terminal SH3 domain via the RXXK motif that is also required for the high-affinity binding between GADS and SLP-76.[58–60,63] This is in contrast to the binding of GRB2 to SOS1 and CBL, which require both SH3 domains of GRB2.[64–66] Another structural difference is the stoichiometry of binding of GRB2 or GADS to its ligands. Biophysical analysis using isothermal titration calorimetry and analytical ultracentrifugation indicated that GRB2 binds to SOS1 and CBL in a 2:1 GRB2:ligand ratio, while GADS interacts with SLP-76 with a 1:1 stoichiometry.[61] Overall, these disparities highlight a potential difference in the ability of GRB2 and GADS to nucleate diverse multiprotein signaling complexes. The SH2 domains of the GRB2 family members have a higher protein homology when compared to the SH3 domains (Table 9.1). This is illustrated in phospho-tyrosyl binding specificities, especially in the case of the adaptor protein LAT. Upon TCR activation, all GRB2 family members are able to bind the same phospho-tyrosyl sequences on LAT with a consensus of pYXNX.[23,27,45,51,67] Additionally, the SH2 domains of GRB2 and GADS can both interact through the same binding motif with CD28, BCR-Abl, SHC, and c-Kit.[21–25,29,30,67]

Overall, members of the GRB2 family of adaptors are highly homologous both in the basic structure and amino acid sequence, but each protein has distinct ligand specificities (i.e., GRB2/SOS1 and GADS/SLP-76). Ligand specificities facilitate the formation of diverse signaling complexes that allow T cells to transduce differential signaling pathways important for T cell effector functions. However, the similarity between these proteins highlights the potential for redundancy or compensation through recruitment of similar ligands to signaling complexes in the absence of other members. For example, GRB2 could potentially bind a larger pool of proline-rich ligands, such as SLP-76, in the absence of GRAP or GADS in hematopoietic and no-hematopoietic tissues. For example, we have observed that GADS expression is regulated between the T cell lines HuT78 (relatively low), and Jurkat (high), as well as primary CD4+ T cells (moderate). In these cases, GRB2 could have increased binding of multiple proline-rich ligands including SLP-76. Additionally, since proteins such as SLP-76 or PLC-γ1 can increase or decrease in expression depending on the subtypes of T cells, this may allow increased binding of these proteins to either GRB2 or GRAP.[68–70]

## 9.3 ROLE OF GRB2 FAMILY MEMBERS IN HUMAN DISEASE

The importance of GRB2 in driving essential cell signaling was first demonstrated in the nematode *Caenorhabditis elegans* where Sem-5 (Nematoda homologue of mammalian GRB2) was required for vulval development.[71] Due to the ubiquitous expression of GRB2 relative to GADS or GRAP, and its role in driving signaling complexes and cellular development, mice that have a complete deficiency in the GRB2 gene do not survive past early embryogenesis.[40] The requirement of GRB2 in development was also demonstrated by multiple studies showing it facilitates signaling from various RTKs, such as EGFR, VEGF, HGF, PDGFR, FGFR, and other protein complexes. Signaling from these receptors, transduced in part by GRB2, leads to critical functions including vasculogenesis, angiogenesis, insulin signaling, metabolism, proliferation, differentiation, development of neointima, coagulation, and cytoskeletal rearrangement.[17–20,72,73] For this reason, GRB2-mediated complexes have been linked to the initiation and progression of multiple disorders such as autoimmune diseases, asthma, cardiovascular disorders, and diabetes. Moreover, GRB2-mediated complexes are involved in the transformation and progression of numerous types of cancers, including solid and hematopoietic tumors, breast, bladder, and pancreatic cancers.[17,29,35,74–78] Interestingly, the transforming ability of GRB2 may not be limited to the cytoplasm as GRB2 is overexpressed in the nucleus of human breast cancer cells.[75,77,79] The presence of GRB2 in the nucleus could potentially be mediating malignant cell transformation by regulating transcriptional pathways. However, the role that GRB2 plays in the nucleus is not completely understood.

In contrast to GRB2-deficient mice that do not survive past early embryogenesis or SLP-76-deficient mice that succumb to severe systemic hemorrhage, GADS-deficient mice are overall healthy.[40,56,80,81] However, similar to GRB2, GADS-mediated signaling complexes have been linked to the progression of hematopoietic malignancies. Specifically, both GRB2 and GADS mediate BCR-Abl-driven myeloid leukemia by propagating BCR-Abl-induced Ras signaling and cytoskeletal rearrangement, respectively.[21,29,30,37] Additionally, GADS regulates the activation of RET receptors in medullary carcinoma cells by inhibiting RET-induced activation of the transcription factor NF-κB.[43,44] Because GRAP has not been well characterized, its role in human disease is still elusive. GRAP-deficient mice are viable and healthy, and there is no apparent pathogenic phenotype that is suggestive of disease.[82] However, GRAP may play a role in driving pathogenesis in patients with Sjögren's syndrome as they express high levels of GRAP in the salivary glands.[42] Additionally, GRAP has been shown to be overexpressed in kidney tubules and is a regulator in TGFβ signaling.[83]

Due to the significance of GRB2 family members, especially GRB2, in the alteration of numerous signaling pathways, these proteins were selected as prominent pharmaceutical targets for cancer therapy.[17,35,36] For example, C90, a small molecule SH2 domain antagonist for GRB2, blocks both in vivo and in vitro tumor growth.[84,85] GRB2 SH3 domain antagonists have also demonstrated antitumor effects in human cancer cells.[86,87] Although considerable progress has occurred in the development of GRB2 domain antagonists during preclinical trials, uncertainty regarding nonspecific side effects of these inhibitors has delayed their advance into clinical trials.[17,88] Alternatively, small molecule inhibitors targeting GRB2 associated proteins, such as the membrane-anchored metalloproteinases ADAMS, are currently undergoing clinical trials in patients with breast cancer.[88,89] ADAMS have been implicated in mediating tumor metastasis by driving angiogenesis and motility through releasing membrane-bound cytokines, growth factors, and receptors.[88] In contrast to protein inhibitors, RNAi-based suppression of GRB2 may be an effective and specific method for the inhibition of GRB2 activity. A current clinical trial is targeting GRB2 utilizing liposomal GRB2 oligonucleotides for the suppression of GRB2 protein levels in patients with myeloid leukemia (Clinical Trial Code: NCT01159028). Overall, members of the GRB2 family of adaptors are critical for the initiation and progression of multiple developmental and disease-related signaling complexes.

## 9.4 REGULATION OF T CELL DEVELOPMENT BY THE GRB2 FAMILY MEMBERS

### 9.4.1 GRB2 IN T CELL DEVELOPMENT

The first insight into the role of GRB2 in T cell development was acquired from genetic studies performed by Gong et al. utilizing mice with GRB2 haplo-deficiency, as GRB2-null mice undergo embryonic developmental arrest.[40,90] Thymocytes and splenocytes obtained from mice with GRB2 haplo-deficiency displayed 60% reduction in total GRB2 protein levels.[90] Analysis of positive and negative selection in thymocytes was performed by crossing GRB2 WT or haplo-deficient mice with mice that express MHC class I-restricted transgenic TCR specific for the male HY antigen. Upon TCR engagement in the thymus, male CD8+ T cells would respond to HY antigens and undergo apoptosis, thereby testing negative selection.[91,92] In contrast, CD8+ T cells in female mice will not differentiate due to the lack of HY antigen expression, allowing for the testing of altered positive selection. Thymocytes with reduced GRB2 expression had no differences in positive selection or proliferative capacity as determined by the percentage of double-positive (DP) and single-positive (SP) thymocytes in females.[90] Similar results were observed in CD4+ T cells when GRB2 haplo-deficient mice were crossed with DO11.10 mice that express a transgenic TCR restricted to MHC class II presenting ovalbumin peptides. Conversely, male thymocytes with reduced GRB2 expression displayed impaired negative selection as seen in the increased numbers of DP and SP CD8+ thymocytes.[90] These results demonstrated that negatively selecting signals are attenuated in the absence of GRB2. In agreement with the negative selection analysis, T cell depletion experiments utilizing the injection of soluble anti-CD3 antibodies to nonspecifically stimulate the TCR demonstrated a resistance to deletion in the DP compartment in thymocytes with reduced expression of GRB2.[90]

One caveat in the study by Gong et al. was that 40% GRB2 expression was not enough to completely analyze positive selection in the thymus. Although negative selection is more sensitive and requires GRB2 protein levels of >40% of wild-type levels, positive selection may be impacted when the GRB2 gene is completely disrupted. To address this knowledge gap, Jang et al. produced thymocytes that are deficient for the GRB2 gene by crossing mice with floxed GRB2 alleles with LCK-Cre transgenic mice.[93] Although LCK is expressed at the DN2 stage, this method begins to remove GRB2 alleles in thymocytes at DN3, thereby conditionally ablating GRB2 in the T cell lineage.[93] Similar to thymocytes with GRB2 haplo-deficiency, GRB2-deficient thymocytes exhibit impairment in negative selection as seen in the increased numbers of DP and CD8+ T cells in GRB2-deficient mice expressing the HY transgenic TCR.[93] However, these cells were unable to undergo positive selection as a developmental block at the DP stage, as evidenced by the significant reductions in the numbers of SP CD4+ and CD8+ in the thymus and periphery.[93] Overall, these studies demonstrate that thymocytes require GRB2 for optimal development.

The developmental defects in GRB2-deficient thymocytes stemmed in part from impaired activation of the MAP kinases p38 and JNK but not ERK1/ERK2.[93] These cells also had reduction of activation in proximal signaling proteins associated with the TCR with a normal to slight reduction in TCR-induced calcium influx.[90,93] Moreover, part of the developmental defects stems from the inability of GRB2 partners to be recruited to membrane signaling complexes, such as the complexes formed by the adaptor protein LAT, as well as at the immunological synapse. LAT is an important adaptor protein in TCR-mediated signaling that binds the SH2 domain of GRB2 and is critical for T cell development.[23,94,95] Similarly, thymocyte-expressed molecule involved in selection (THEMIS), a protein essential for T cell positive and negative selection, was observed to bind to the SH3 domain of GRB2 in vitro and in vivo, and this binding is required for its recruitment to LAT and the immunological synapse.[96,97] Additional defects that lead to impaired development are observed in GRB2-deficient CD4+ T cells that fail to upregulate CD3 and downmodulate heat-stable antigen (HSA), a costimulatory molecule required for optimal T cell development.[93,98] The failure to express molecules critical for TCR signaling during development may explain some of the signaling deficiency displayed by GRB2-deficient thymocytes.

### 9.4.2 GADS in T Cell Development

The first study to examine the role of GADS in T cell development was completed by Yoder et al. utilizing GADS-deficient mice.[56] Similar to GRB2, GADS-deficient mice had severe impairment in T cell development.[45,56] Examination of both CD4+ and CD8+ T cell populations from GADS-deficient mice revealed impairment in both positive and negative selection.[45,56,99,100] As determined by flow cytometry, there was an increase in the DN thymocyte fraction, but a significant decrease in the numbers of SP thymocytes with the greatest defects occurring in numbers of CD4+ T cells.[56] However, in contrast to GRB2-deficient thymocytes, the total number of GADS-deficient mature T cells in the periphery accumulated over time, indicating that GADS is not essential to the survival of mature T cells.[56,93,99] The aforementioned studies were confirmed by using transgenic mice expressing a dominant negative version of GADS driven by the LCK proximal promoter.[101] Expression of dominant negative GADS resulted in impaired T cell development and reduction of total numbers of DP and SP thymocytes.[101] The impaired development in the absence of GADS was due in part to defects in recruitment of the GADS-binding protein SLP-76 to LAT, subsequent PLC-γ1 phosphorylation, and calcium mobilization.[56,99] Similar to the role of LAT, SLP-76 has been demonstrated to be required for T cell development.[102,103] This observation highlights LAT as a central hub for the GRB2 family of adaptors in mediating T cell development by recruiting effector molecules to the plasma membrane.

## 9.5 FUNCTION OF THE GRB2 FAMILY AT TCR-DRIVEN SIGNALING COMPLEXES

The TCR is composed of multimeric subunits with no intrinsic enzymatic activity (Figure 9.2). The main function of these units is to detect specific peptide/MHC

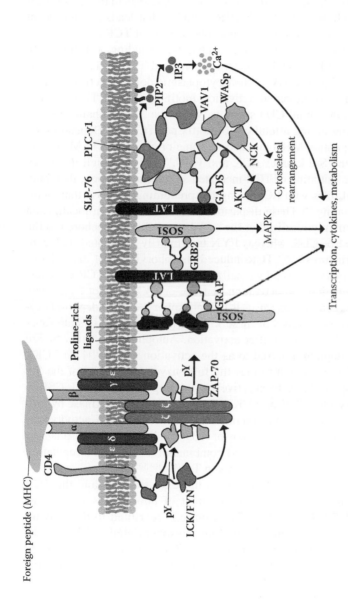

**FIGURE 9.2** T cell receptor signal transduction. Ligation of the TCR by peptide/MHC complexes induces the phosphorylation and activation of Src kinase LCK. Activated LCK phosphorylates the ITAM residues present in the TCR complex, including the CD3 subunits and the TCR ζ chains. ZAP-70 is then recruited to the TCR complex by binding phosphorylated ITAMS on the CD3 subunits and the TCR ζ chains via its SH2 domains. LCK phosphorylates ZAP-70, enhancing its enzymatic activity. Activation of ZAP-70 results in the phosphorylation of the LAT complex, which facilitates the recruitment of the GRB2 members, GRB2, GADS, and GRAP. These proteins nucleate the LAT signalosome by recruiting multiple proline-rich ligands via SH3 domain-mediated interactions while binding phosphorylated LAT molecules through the SH2 domain. Proline-rich ligands and other associated proteins, such as SOS1, PLC-γ1, NCK, and WASp, mediate multiple signaling pathways important for T cell function, such as MAP kinase activation, calcium mobilization, and actin polymerization.

complexes with varying affinity and/or avidity to produce changes in the structure of the TCR that initiate signal transduction. It is still unclear how the TCR subunits sense and transfer related messages from APCs carrying peptide/MHC. In addition to the role of peptide/MHC complexes, the CD4 coreceptor assists in the recruitment of signaling kinases, such as LCK, to the TCR complex that leads to subsequent initiation of TCR signal transduction via the phosphorylation of TCR subunits.[15,16]

A single TCR is composed of eight total subunits, including a TCRαTCRβ homodimer, CD3εCD3δ and CD3εCD3γ heterodimers, and TCR ζ chain homodimer (Figure 9.2). These subunits carry, in their cytoplasmic domain, activation sites termed immunoreceptor tyrosine-based activation motifs (ITAMs). These motifs are present as a single copy on all CD3 subunits, while the TCR ζ chains contain three copies each, thereby giving a total of 10 motifs per TCR.[15,16] Depending on the affinity of the TCR-peptide/MHC interaction, some or all of the ITAMs motifs are phosphorylated by the Src kinases LCK and FYN.[15,16,104] Although these two kinases contribute to the initiating of TCR-mediated signal transduction, they both seem to play differential roles with LCK being the predominant kinase that mediates proximal TCR signaling.[104] These disparities partially stem from localization differences, as LCK is mainly localized at the cellular membrane, anchored to the cytoplasmic tails of CD4 or CD8, whereas FYN is primarily associated with centrosomal and mitotic structures.[104,105] TCR-induced phosphorylation of the ζ chains and CD3 subunits is primarily driven by localization of LCK to the TCR complex. The mechanisms of regulation and activation of LCK or FYN are not completely understood, however, the general consensus is that phosphorylation or dephosphorylation of regulatory tyrosine residues of these kinases regulates steady state function and downregulates their activity after activation. Autophosphorylation by Src kinases in the catalytic domain, referred to as the activation site (Y394 on LCK), promotes an open conformation and primes the full enzymatic activity of the Src kinase. In contrast, phosphorylation by negatively regulating kinases such as CSK in the C-terminal inhibitory site (Y505 on LCK) results in a less active kinase by producing a closed conformation and reduction in enzymatic activity. LCK activity may also be mediated through dephosphorylation of Y394 and Y505 by phosphatases such as CD45.[106] In general, the interplay and combination of these two regulatory sites dictates the activation status of Src kinases in T cells.[104,105]

The phosphorylation of the ITAM motifs on the CD3 subunits and the TCR ζ chains by active LCK facilitates the recruitment of another essential TCR-associated kinase, ZAP-70 (Figure 9.2). This kinase is recruited to the proximity of the plasma membrane by directly binding phosphorylated ITAMs on the CD3 and TCR ζ subunits. The interaction between ZAP-70 and phosphorylated ITAMs is facilitated by its two SH2 domains. Upon binding of phosphorylated CD3 subunits or TCR ζ chains, ZAP-70 undergoes conformational changes, which facilitates its enzymatic activation through LCK-mediated phosphorylation. Thus, cascades of multiprotein regulatory mechanisms are present to control the activation and extent of TCR proximal signaling.[106,107] Nonetheless, the complete regulatory mechanisms of how the TCR subunits and associated enzymes are regulated are not entirely clear. However, adaptor proteins that are also recruited to the ζ chains, ZAP-70 or LCK, such as GRB2 or the GRB2-binding protein SHC, appear to directly or indirectly play a role in these events.[33,34,108,109] Overall, TCR subunits and the associated proteins LCK and ZAP-70 are the initiators of TCR-mediated signaling. Alteration or mutation of these proteins produces overactive or defective downstream signaling and can significantly impair T cell function resulting in human immunodeficiencies or autoimmunity.[110–115] Therefore, elucidating how the TCR is regulated is essential to better understand the mechanisms that drive and initiate T cell-mediated pathology, malignancies, and immunosuppression.

### 9.5.1 GRB2 NEGATIVELY CONTROLS THE TCR COMPLEX

We have recently used GRB2-specific microRNAs to suppress the expression of GRB2 in HuT78 T cells. In these studies, we observed substantially enhanced proximal signaling events.[116] Specifically, molecules associated with the TCR complex including LCK, ZAP-70, and the CD3 ζ chains had enhanced phosphorylation in the absence of GRB2 as assessed by site-specific immunoblotting studies. These results were unanticipated due to the bulk of literature suggesting a positive regulatory role for GRB2 in all cell types. Additionally, in murine T cells, disruption of the *GRB2* gene significantly reduces TCR-induced proximal signals, which include enhanced LCK, ZAP-70, CD3 ζ chains, and LAT.[93] Nonetheless, the role of GRB2 as a negative regulator has been previously reported in B cells as it inhibits BCR-induced calcium mobilization.[117] Our results highlight a new role for GRB2 through negatively regulating the TCR complex.

An unanswered question stemming from these observations is how GRB2 is negatively controlling the TCR complex. We propose that GRB2 may control the TCR complex through a combination of two potential mechanisms (Figure 9.3). These mechanisms are due to the ability of GRB2 to bind the CD3 ζ chains and ZAP-70.[33,34,118] The conformation of GRB2 molecules bound to the TCR complex will determine the mechanism of negative regulation. First, similar to the inhibitory effects on FGFR activation by dimeric GRB2, inhibition of the TCR complex may also be mediated by the binding of GRB2 dimers.[32] We propose that the binding of dimeric GRB2 to the TCR complex increases the threshold of activation. To this end, upon TCR activation, LCK or other kinases may phosphorylate dimeric GRB2 molecules that drive the release of the "locking" functions of these proteins. The release of pY-GRB2 molecules in turn allows effector proteins such as ZAP-70 to be recruited to the TCR ζ chains. A second mechanism is derived from the ability of GRB2 to recruit proline-rich negative regulators to the TCR/LAT complexes. In this case, GRB2 may directly bind the TCR complex and recruit the ubiquitin ligase CBL, the negative regulator FAK, and phosphatases such as LYP (Figure 9.3).[107,119,120] Interestingly, compared to GRB2, suppression of GADS expression did not have an effect on proximal TCR signals, with the exception of increased SLP-76 phosphorylation.[121] These observations indicate that out of the GRB2 family of adaptors, GRB2 is essential for the regulation of LCK, the CD3 ζ chains, and ZAP-70. Overall, we have demonstrated a new regulatory role for GRB2 in controlling proximal TCR-mediated signaling, but further studies are needed to examine the mechanism of this novel effect.

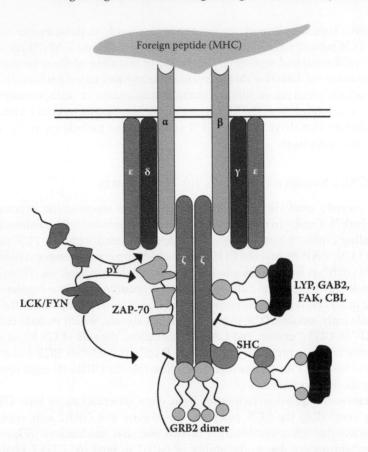

**FIGURE 9.3** Potential mechanisms utilized by GRB2 for negatively controlling the TCR complex. Similar to FGFR, GRB2 homodimers may bind the TCR ζ chains, thereby increasing the threshold of activation. In the absence of dimeric GRB2, the TCR complex becomes sensitive to lower levels of TCR stimulation. GRB2 molecules may also recruit negative regulators or phosphatases such as LYP, GAB2, FAK, or CBL to the TCR complex by directly binding the TCR ζ chains, ZAP-70, or LCK.

## 9.5.2  NEGATIVE REGULATION OF T CELLS BY GRAP

Similar to our observations with GRB2, previous studies reported that GRAP negatively controls TCR-mediated signals. GRAP-deficient mice had increased TCR-induced proliferation, IL-2 production, and c-fos induction.[82] The authors further demonstrate that GRAP deficiency augments T cell activation by enhancing ERK1/ERK2 signaling pathways.[82] It is unclear if GRAP negatively controls ERK1/ERK2 pathways by directly regulating proximal TCR signals or by competing for GRB2 binding with various ligands such as SOS1. First, in contrast to GRB2, it may be that the negative effects of GRAP are directed against pathways downstream of the TCR/LAT complexes such as ERK1/ERK2 activation. In this case, GRAP could be binding SOS1, which forms an inefficient complex relative to GRB2/SOS1 for RAS activation and subsequent ERK1/ERK2 phosphorylation. This could be a mechanism used by T cells to control the extent of MAPK activation in different developmental stages. Second, based on their related homology, both GRB2 and GRAP could bind similar or different negative regulators, thereby generating a synergistic negative effect on the TCR complex. Thus, when one protein is suppressed, a reduced number of negative regulators are recruited to the TCR complex and the TCR signal is therefore increased. Finally, based on potential competitive binding with the TCR, the absence of dimeric GRB2 discussed earlier allows theoretical negatively regulating GRAP dimers to bind the TCR complex, albeit generating a substantially reduced threshold of activation for the TCR compared to GRB2. It is also possible that GRAP in reality recruits positive regulators to the TCR complex but with less efficiency when compared to molecules recruited by GRB2. Therefore, in the absence of competitive GRB2, GRAP recruits more positive regulators, thereby enhancing TCR-induced proximal signaling. Overall, the role of GRAP in mediating TCR-mediated signals and regulating T cell biology is not completely understood and should be addressed in subsequent studies.

## 9.6 THE FUNCTION OF THE GRB2 FAMILY AT THE LAT SIGNALING COMPLEX

### 9.6.1 GRB2 Family Members Bind the Phosphorylated Adaptor Protein LAT

The adaptor protein LAT serves as a TCR signal transduction hub that is essential in relaying the signals driven by TCR-peptide/MHC interactions.[23] Thymocytes from LAT deficient mice do not develop past the DN3 stage, and there are no αβ T cells present in the periphery of these mice.[95] Additionally, mutation of tyrosines on LAT that are important for the recruitment of signaling complexes also impair T cell development.[122,123] LAT is rapidly phosphorylated by ZAP-70 on five tyrosines located in its cytoplasmic tail.[16,124] Four of these tyrosines (Y132, Y171, Y191, and Y226 in humans) have been implicated in driving multiple distinct signaling pathways.[26,67,125]

GRB2, GADS, and GRAP directly bind specific phosphorylated tyrosine residues at LAT. This results in the recruitment of proline-rich ligands that drive different signaling pathways such as MAP kinase activation, calcium mobilization, adhesion, and metabolic regulation (Figure 9.2).[23,126,127] However, the association of GRB2, GRAP, or GADS for LAT phospho-tyrosines are not identical. Although both GRB2 and GADS can bind to phosphorylated Y171 and Y191 in vivo and in vitro, only GRB2 can bind phosphorylated Y226 in vivo.[27,67,125] Additionally, although the GRB2/GADS in vitro binding affinities are not significantly different for phosphorylated Y171 and Y191, GADS has lower in vivo binding for phosphorylated Y171 as determined by isothermal titration calorimetry.[27] The distinct in vivo and in vitro binding affinities for GADS likely stem from cooperative interactions that occur in vivo with other signaling molecules such as GRB2, SLP-76, and PLC-γ1. Similarly, as determined by immunoprecipitation experiments, GRB2 requires two LAT tyrosines in any combination for stable complex formation with LAT.[125,128] Both GRB2 and GADS are unable to bind phosphorylated Y132, the PLC-γ1 binding and activation site on LAT due to reduced binding affinities rather than cooperative interactions.[27,67] GRAP can also bind phospho-LAT, however, relative binding affinities compared to GRB2 or GADS are not well defined.[51] Overall, the LAT and the GRB2 family of adaptors nucleate large multiprotein signaling complexes based on both cooperative and affinity-based interactions.

### 9.6.2  GRB2/LAT-Mediated Signaling through Recruitment of Adaptor Proteins

GRB2 is constitutively bound to several SH3 domain ligands, thereby linking GRB2 complexes to multiple signaling pathways. The full list of GRB2 ligands (proline and tyrosine-based) and the involved pathways is still under investigation, but these ligands can be found in all compartments of the cell. Proline-rich ligands include GEFs SOS1 and SOS2, the RNA-binding protein SAM68, the cyclin-dependent kinase (Cdk) inhibitor p27kip1, the GRB2-associated binding protein 2 (GAB2), the dynamin family of transport proteins, THEMIS, and the E3 ubiquitin ligases CBL and CBLB.[61,65,96,116,129–131] In addition, GRB2 controls the activation of proteins that drive metabolic pathways such as the energy-sensing AMP-activated protein kinase (AMPK).[72] An increase in the AMP/ATP ratio from cellular metabolism and nutrient variation enhances the enzymatic activity of AMPK. Increased binding of AMP to AMPK leads to conformational changes that facilitate the phosphorylation of AMPK by upstream kinases, which significantly increases its enzymatic activity.[132,133] In this case, GRB2 directly binds to AMPK and plays a role in regulating its phosphorylation and subsequent increased enzymatic activity.[72] Activation of AMPK is essential for the induction of metabolic cellular pathways in all cells, including T lymphocytes, that generate ATP and downregulate pathways that consume ATP.[134,135]

One of the most studied signaling pathways downstream of RTK and the TCR is the activation of MAP kinases (ERK1/ERK2) induced by GRB2/SOS1 complexes.[136–142] Following receptor activation, GRB2 recruits SOS1 to the plasma membrane where it facilitates the exchange of GDP for GTP on RAS. In T cells, the interaction of SOS1 with RAS is mediated through GRB2-mediated recruitment of SOS1 to the LAT complex.[143,144] Activation of RAS by SOS1 is thought to propagate a kinase cascade that eventually results in the activation of MAP kinases ERK1/ERK2. The activation of ERK1/ERK2 in T cells through the recruitment of SOS1 by GRB2 has recently come under criticism in studies demonstrating optimal activation of ERK1/ERK2 in the absence of GRB2 or SOS1.[93,145] Other RAS GEFs, such as RASGRP, have been suggested to drive RAS activation in T cells. Although SOS1 is recruited by GRB2 to the T cell membrane, RASGRP is dependent on membrane-bound diacylglycerol (DAG).[142] Loss of RASGRP expression impairs T cell development and differentiation due to its role in controlling ERK1/ERK2 activation.[146]

Our recent microRNA-based studies demonstrated that, similar to GRB2-deficient thymocytes, HuT78 T cells deficient in GRB2 have only modest reduction of ERK1/ERK2 activation.[116] These observations suggest compensatory or alternative mechanisms for the activation of ERK1/ERK2. Since GRAP can bind phosphorylated

LAT as well as SOS1, it is very feasible that GRAP could compensate, albeit to a lower extent, for the absence of GRB2 in mediating ERK1/ERK2 activation.[51] It is also possible that another ERK1/ERK2-inducing mechanism, such as RASGRP, is mediating their activation in the absence of GRB2. However, because GRB2 deficiency markedly reduces PLC-γ1 activation and subsequent DAG production, we do not suspect a compensatory action of this pathway.[116] GRB2 has also been implicated in the optimal activation of MAP kinases p38 and JNK downstream of the TCR.[90,93] However, the full mechanisms of how these proteins are activated or the role of GRB2 in their activation is unclear. In support of these findings, we also observed that p38 and JNK were moderately reduced in the absence of GRB2 in T cells.[116] Together, these data suggest that GRB2 is required for optimal activation of MAP kinases in T cells, but that other proteins strongly contribute to the induction of these important signaling molecules.

In addition to propagating positive signaling pathways from the LAT complex, GRB2 has also been implicated in facilitating negative signaling to control TCR-derived signals via the recruitment of negative regulators such as the E3 ubiquitin ligase CBL and adaptor protein GAB2 to the LAT complex.[107] Upon recruitment, CBL downregulates signaling by targeting molecules including ZAP-70 and LAT for proteasome-mediated degradation.[107,147,148] T cells that are deficient in CBL have increased phosphorylation of ZAP-70, LAT, and SLP-76 as well as increased TCR surface expression.[107] GRB2 also mediates the phosphorylation of the adaptor protein GAB2 by recruiting it to LAT. Phosphorylated GAB2 can then bind the phosphatase SHP-2, which then dephosphorylates the TCR ζ chains and the adaptor protein SHC.[149]

Dephosphorylation of proximal T cell molecules drives the reduction of transcription factors NFAT and NFκB.[26,149,150] Other negative regulators of T cell signaling such as FAK or the phosphatase LYP also bind and are potentially recruited by GRB2 to the T cell plasma membrane.[119,120,151] Therefore, GRB2 is responsible for driving positive signaling events, as well as negatively controlling the extent of signaling transduced by TCR/LAT complexes.

### 9.6.3 GRB2 Nucleates LAT Microclusters in Human T Cells

The formation of megadalton-sized LAT microclusters is important for driving LAT-mediated signaling. The formation of these signaling complexes requires ligands associated with the GRB2 family of adaptors, such as SOS1, CBL, and possibly SLP-76.[61,152] Mutation of the LAT tyrosines that bind the GRB2 family abolishes the formation of microclusters and subsequent T cell signaling.[61,67,125] Recently, the GRB2-binding protein, SOS1, has been directly shown to be important in oligomerization of LAT microclusters.[61,153] Oligomerization of LAT is thought to occur due to the dimerization of GRB2 via binding SOS1 with a stoichiometry of 2:1.[61] This allows GRB2:SOS1:GRB2 complexes to interact with two different LAT proteins. The ability of an individual LAT to bind to three separate GRB2-containing complexes drives the oligomerization of LAT. Prior to our recent studies described next, it was unclear if other ligands of GRB2 or its homologs, GRAP and GADS, can induce the formation of LAT clusters.[116] For example, a recent study suggested that the GADS-binding ligand, SLP-76, is involved in the induction and stability of LAT microclusters through cooperative interactions and the recruitment of protein complexes at LAT.[154] It is also possible that GRAP/SOS1 binding may produce redundant complexes that can facilitate LAT oligomerization. Therefore, the concept that GRB2 alone drives clustering has not been directly demonstrated. Direct suppression or mutation of GRB2 proteins is required for conclusively demonstrating the role of GRB2 in inducing LAT clusters.

Our results show that GRB2 is absolutely required for the formation of LAT microclusters.[116] Specifically, our data demonstrate that CD4+ T cells require the presence of GRB2 as a nucleation factor for LAT microclusters.[116] We have also found that N-terminal sites on GRB2 (residues 26–29; KVLN→AAAA) are also essential for oligomerization of the LAT complex.[116] This mutant of GRB2 can bind phosphorylated LAT but is unable to oligomerize LAT microclusters due to attenuated SOS1 binding. We could not rule out the possibility that both SH3 domains are required for oligomerizing LAT, but we confirmed that at least the function of the GRB2 N-terminus is essential in mediating LAT-induced signaling. It would be interesting to determine if other high-affinity SH3 domain ligands of GRB2 (i.e., CBL) can compensate partially or completely for the absence of SOS1. We speculate, however, that SOS1 plays a major role in mediating LAT clusters due to previous studies utilizing SOS1 mutants and murine T cells deficient in SOS1 that clearly illustrate its role in inducing LAT cluster formation. Overall, our data confirm GRB2 as the critical player in the formation of LAT-induced signaling clusters. These results clearly illustrate that GRAP and GADS cannot compensate for the absence of GRB2 in aggregating LAT molecules. In fact, our unpublished data illustrate that GADS-deficient HuT78 T cells displayed normal LAT cluster formation.

### 9.6.4 LAT Microcluster-Induced Signaling

One compelling, yet unanswered, question is how does clustering of LAT control downstream signal transduction. Our data reveal that LAT has to be both phosphorylated and oligomerized to create a stable framework for the optimal activation of signaling molecules bound to LAT. GRB2-deficient T cells have normal phosphorylation of LAT tyrosine 226 but are unable to drive clustering of LAT. These cells also have markedly reduced TCR-induced calcium mobilization and subsequent release of IL-2 and IFN-γ.[116] The main defect observed in these cells is the instability of PLC-γ1 at the LAT complex. Interestingly, in these cells, the phosphorylation of Y783 on PLC-γ1 and the PLC-γ1 binding site on LAT Y132 were reduced by ~55%. The reduced phosphorylation is likely due to the increased access of these proteins to the cellular phosphatases. However, reduced phosphorylation on these sites alone cannot explain the absence of PLC-γ1 from the LAT complex, striking reduction of calcium flux, and substantial decrease of cytokine release.[116] Based on these data, we propose that phosphorylation of LAT on GRB2-binding sites (Y171, Y191, Y226) must occur to allow GRB2–LAT binding and subsequent LAT oligomerization. This creates a framework at the cell membrane for the stable binding of signaling proteins such as PLC-γ1. The stable interaction of PLC-γ1 at the cellular membrane drives the induction of optimal calcium mobilization and ultimately cytokine release. In support of this model, previous reports illustrated that mutation of the GRB2 binding sites on LAT inhibited PLC-γ1 recruitment to LAT.[67,125] Overall, our studies demonstrate that GRB2 nucleates TCR-induced LAT oligomerization, a process that is essential for the stability and function of LAT-binding effector molecules.

## 9.6.5 GADS/LAT-MEDIATED SIGNALING

The most studied SH3 domain ligand for GADS is SLP-76. The current model is that recruitment of GADS/SLP-76 complex to LAT facilitates the binding of interleukin-2-inducible T-cell kinase (ITK) and VAV1. These proteins then regulate the recruitment of PLC-γ1 to the LAT signaling complex (Figure 9.2).[16,23,155–158] PLC-γ1 is recruited to the cellular membrane through the binding of Y132 at LAT. This binding catalyzes the formation of inositol 1,4,5-trisphosphate (IP3) and DAG from phosphatidylinositol 4,5-bisphosphate (PIP2). Increased concentration of IP3 and DAG in the cytoplasm induced by the GADS/SLP76 complexes mediates critical functions through the induction of calcium influx, and activation of protein kinase C (PKCθ) and RASGRP, respectively.[56,155,159–161] The importance of GADS in mediating TCR-induced calcium mobilization was first described by Yoder et al., wherein GADS deficient CD4+ thymocytes displayed reduced PLC-γ1 phosphorylation and TCR-induced calcium influx in peripheral CD4+ and CD8+ (Yankee et al.) T cells compared to wild-type controls.[56,99] Interestingly, the absence of GADS did not seem to be required for cytokine production as GADS-deficient CD8+ T cells were still able to produce levels of IFN-γ comparable to wild-type controls.[162] Similarly, Jurkat T cells overexpressing GADS did not have increased TCR-induced NFAT or IL-2 promoter activity unless SLP-76 was synergistically coexpressed.[57] Other studies utilized mutant SLP-76 molecules or dominant negative versions of GADS that inhibited the recruitment of SLP-76 into LAT clusters and subsequent LAT-induced signaling.[55,154,163] Similarly, the disruption of the GADS/SLP-76 complex inhibited the recruitment of SLP-76 into signaling clusters in mast cells, and subsequent calcium mobilizing and cytokine release.[160]

Similar to the aforementioned studies, we have observed that GADS/SLP-76-mediated complexes are also important for optimal stability of PLC-γ1 at LAT.[121] Our recent work in HuT78 T cells demonstrated that suppression of GADS expression using GADS-specific microRNAs substantially attenuates SLP-76 recruitment to LAT.[121] The absence of SLP-76 from the LAT complex destabilizes LAT/PLC-γ1 interaction and subsequent induced calcium mobilization, IL-2, and IFN-γ production.[121] These results indicate that in addition to the framework created by GRB2-induced LAT clustering, SLP-76 is important in the optimal stability of PLC-γ1 at the cellular membrane.[116,121] These results match well with the previously described role of SLP-76 in mediating PLC-γ1 stability in the LAT complex.[23,157,158,161] It should be noted, however, that the decreased levels of PLC-γ1–LAT interaction and subsequent calcium mobilization in the absence of GADS is not as marked compared to the reduction observed in the absence of GRB2.[116,121] Our data collectively highlights the interplay between GRB2 and GADS in the formation of fine-tuned signaling complexes at the LAT signalosome.

## 9.6.6　GADS/SLP-76 COMPLEXES ARE DISPENSABLE FOR TCR-INDUCED ACTIN POLYMERIZATION

TCR activation drives extensive cytoskeletal rearrangement including actin polymerization.[164,165] The actin cytoskeleton controls T cell motility, adhesion, and morphology, all of which are critical for moving into secondary lymphoid organs, stabilizing interactions with antigen-presenting cells, and migrating to sites of infection.[164,165] The role of the LAT signaling complex in driving TCR-induced actin polymerization has been examined in Jurkat T cells deficient in LAT that were unable to recruit proteins associated with the actin cytoskeleton to the T cell plasma membrane.[166] As a result, these cells were unable to undergo TCR-induced spreading due to deficiency in actin polymerization.[167] These signaling defects were rescued by reconstitution of wild-type LAT, which stabilized NCK recruitment to signaling clusters.[166] However, reconstitution with LAT proteins lacking tyrosines important for recruitment of SLP-76 via GADS were unable to rescue these defects.[166] These results were not surprising as SLP-76 has been demonstrated to bind to actin-regulatory molecules such as NCK and WASp protein at the LAT complex (Figure 9.2).[166,168–171] Thus, SLP-76 appears to serve as a docking site for NCK, WASp, ITK, and VAV1, which form interactions essential for LAT-mediated cytoskeletal organization (Figure 9.2).

Based on these studies, we were surprised to observe increased actin polymerization and cellular adhesion in GADS-deficient HuT78 T cells that are unable to recruit SLP-76 to the LAT complex.[121] Similarly, we observed increased antigen-induced cellular adhesion in GADS-deficient primary CD8+ T cells compared to wild-type controls.[121] Our data demonstrate that the recruitment of SLP-76 to LAT by GADS is dispensable for TCR-mediated adhesion. One explanation for this is that SLP-76 can drive actin polymerization and subsequent adhesion utilizing different receptors other than LAT such as CD6.[172,173] Increased recruitment of SLP-76 to CD6 or other receptors may also explain the increased SLP-76 phosphorylation observed in the absence of GADS.[121] Another explanation may be that compensation by GRB2 or GRAP may occur in the absence of GADS. For example, GRB2 or GRAP may recruit low but sufficient levels of SLP-76 to the LAT complex, but this interaction may be too weak to be observed in immunoprecipitation assays. Similarly, previous studies demonstrated that GRB2 can drive actin polymerization in different cell types through binding and mediating the activation of WASp.[20,73,174] Although we have not performed direct experiments to test the role for GRB2 in cellular adhesion, the fact that GRB2-deficient cells adhere normally to anti-CD3 antibody-coated glass chambers suggests a nonessential role for GRB2 in TCR-mediated cellular adhesion.[116] The last possibility is that SLP-76 is not required for T cell adhesion. In support of this model, other studies have demonstrated that SLP-76-deficient Jurkat T cells were able to form normal TCR-induced actin rings, which indicates redundant pathways at the LAT complex.[166] Moreover, recent studies demonstrated that NCK and VAV1 could interact and induce actin polymerization independent of SLP-76.[169,170] Interestingly, a recent study by our group demonstrated that proline-rich tyrosine kinase 2 (PYK2), and not LAT, is needed for a second phase of TCR-induced actin rearrangement and T cell adhesion.[175] Overall, individual members of the GRB2 adaptors do not seem to be essential in driving TCR-induced adhesion due to compensation and redundancy by other proteins and signaling molecules.

## 9.7 THE REQUIREMENT FOR THREE DIFFERENT GRB2 FAMILY HOMOLOGS IN T CELLS

It is not clear why T cells would need three different adaptor proteins that can potentially produce the same signaling complexes. We speculate that multiple homologous adaptors are expressed to fine-tune numerous signals received by the T cell from the environment. These signals may be derived from the TCR, costimulatory receptors such as CD28, cytokine, or adhesion receptors. Additionally, since the signals appear in different strengths and combinations, it is important that they are integrated carefully by the cellular signal transduction system. T cells may need to express different levels of GRB2 family members in distinct developmental stages. When this occurs, GRB2, GRAP, or GADS may compensate for the loss of other family members, albeit with differing strengths and kinetics. For example, TCR-induced activation of MAP kinases ERK1/ERK2 is preferentially triggered in naïve compared to antigen-experienced T cells.[176] Conversely, activation of p38 is more prominent in antigen-experienced compared to naïve T cells.[176] In this case, potential reduced expression of GRAP in naïve T cells may allow more GRB2 to bind and recruit SOS1 to the membrane, thereby enhancing ERK1/ERK2 activation. Conversely, antigen-experienced cells may contain increased relative GRAP expression, which is not optimal in TCR-induced ERK1/ERK2 activation compared to GRB2. Similarly, the GRB2 family of adaptors may play distinct roles in driving CD4+ and CD8+ T cell lineage commitment based on their ability to differentially activate MAP kinases or other signaling pathways.[177] In this case, GRB2 may be more important for CD4+ lineage, as these cells need optimal ERK1/ERK2 activation, while increased GRAP expression may determine CD8 lineage commitment. Interestingly, we have observed differential regulation of GRB2, GRAP, or GADS in T cells. Leukemic Jurkat T cells express substantially more GADS when compared to lymphoma-derived HuT78 (low) and primary CD4+ T cells (moderate). These differences may considerably shift GADS/SLP-76-mediated signaling at LAT and other signaling complexes, and drive differential T cell outputs. It also potentially explains why deficiency of GADS produces different phenotypes between CD4+ and CD8+ T cells.[99,178]

## 9.8 CONCLUSION

Overall, the GRB2 family of adaptor proteins are critical players in driving TCR-mediated signaling. Our work and others using a multitude of complementary techniques has revealed that GRB2 and GADS have distinct roles at the TCR and LAT complexes. GRB2 performs differential functions through negatively controlling TCR proximal signals, and by inducing calcium mobilization through aggregation of LAT signaling complexes. In contrast, GADS regulates TCR-induced calcium influx through the recruitment of SLP-76 to LAT but not through LAT-induced clustering

**TABLE 9.2**

**Comparison of TCR-Induced Signals between Cells Deficient in the GRB2 Family Members**

| | GRB2 | GADS | GRAP |
|---|---|---|---|
| Total pY | +++ | NC | NT |
| pY394 LCK | +++ | NT | NT |
| pY142 ζ | +++ | NT | NT |
| pY128 SLP76 | +++ | ++ | NT |
| pAKT | +++ | NC | NT |
| pY226 LAT | + | NT | NT |
| pY132 LAT | – | NT | NT |
| pERK1/ERK2 | – | NT | ++ |
| pp38 | – | NT | NT |
| pJNK | – | NT | NT |
| pY783 PLC-γ1 | – | NC | NT |
| TCR-induced calcium influx | – | – | NT |
| PLC-γ1/LAT stability | – | NT | NT |
| SLP-76/LAT stability | NT | – | NT |
| Formation of LAT microclusters | – | NC | NT |
| Actin polymerization/adhesion | NT | + | NT |
| IL-2 release | – | | ++ |
| IFN-γ release | – | – | NT |

*Source:* Shen, R. et al., *Molecular and Cellular Biology* **22**, 3230–3236 (2002); Bilal, M. Y., and Houtman, J. C., *Frontiers in Immunology* **6**, 141 (2015); Bilal, M. Y. et al., *Cellular Signalling* **27**, 841–850 (2015).

*Note:* +, increased; –, decreased; NC, no change; NT, not tested.

(Table 9.2). The work discussed here suggests the therapeutic potential for targeting T-cell mediated pathological disorders such as autoimmune disorders, GRB2/GADS-mediated leukemias, and other hematopoietic malignancies.[17,21]

## REFERENCES

1. Romagnani, S. Regulation of the T cell response. *Clinical and Experimental Allergy: Journal of the British Society for Allergy and Clinical Immunology* **36**, 1357–1366 (2006).
2. Mobs, C., Slotosch, C., Loffler, H., Pfutzner, W., and Hertl, M. Cellular and humoral mechanisms of immune tolerance in immediate-type allergy induced by specific immunotherapy. *International Archives of Allergy and Immunology* **147**, 171–178 (2008).
3. Romani, L. Immunity to fungal infections. *Nature Reviews Immunology* **11**, 275–288 (2011).
4. Magombedze, G., Reddy, P. B., Eda, S., and Ganusov, V. V. Cellular and population plasticity of helper CD4(+) T cell responses. *Frontiers in Physiology* **4**, 206 (2013).
5. Dobrzanski, M. J. Expanding roles for CD4 T cells and their subpopulations in tumor immunity and therapy. *Frontiers in Oncology* **3**, 63 (2013).

6. Jutel, M., and Akdis, C. A. T-cell subset regulation in atopy. *Current Allergy and Asthma Reports* **11**, 139–145 (2011).
7. Podojil, J. R., and Miller, S. D. Molecular mechanisms of T-cell receptor and costimulatory molecule ligation/blockade in autoimmune disease therapy. *Immunological Reviews* **229**, 337–355 (2009).
8. Datta, S., and Milner, J. D. Altered T-cell receptor signaling in the pathogenesis of allergic disease. *The Journal of Allergy and Clinical Immunology* **127**, 351–354 (2011).
9. Hansson, G. K., and Hermansson, A. The immune system in atherosclerosis. *Nature Immunology* **12**, 204–212 (2011).
10. Gilboa, E. The promise of cancer vaccines. *Nature Reviews Cancer* **4**, 401–411 (2004).
11. Hamilton, S. E., Prlic, M., and Jameson, S. C. Environmental conservation: Bystander CD4 T cells keep CD8 memories fresh. *Nature Immunology* **5**, 873–874 (2004).
12. Hammarlund, E. et al. Duration of antiviral immunity after smallpox vaccination. *Nature Medicine* **9**, 1131–1137 (2003).
13. Bevan, M. J. Helping the CD8(+) T-cell response. *Nature Reviews Immunology* **4**, 595–602 (2004).
14. Schmidt, A., Oberle, N., and Krammer, P.H. Molecular mechanisms of treg-mediated T cell suppression. *Frontiers in Immunology* **3**, 51 (2012).
15. Nel, A. E. T-cell activation through the antigen receptor. Part 1: Signaling components, signaling pathways, and signal integration at the T-cell antigen receptor synapse. *The Journal of Allergy and Clinical Immunology* **109**, 758–770 (2002).
16. Smith-Garvin, J. E., Koretzky, G. A., and Jordan, M. S. T cell activation. *Annual Review of Immunology* **27**, 591–619 (2009).
17. Dharmawardana, P. G., Peruzzi, B., Giubellino, A., Burke, T. R., Jr., and Bottaro, D. P. Molecular targeting of growth factor receptor-bound 2 (Grb2) as an anti-cancer strategy. *Anti-Cancer Drugs* **17**, 13–20 (2006).
18. Dutting, S. et al. Grb2 contributes to (hem)ITAM-mediated signaling in platelets. *Circulation Research* (2013).
19. Zhang, S. et al. Grb2 is required for the development of neointima in response to vascular injury. *Arteriosclerosis, Thrombosis, and Vascular Biology* **23**, 1788–1793 (2003).
20. Carlier, M. F. et al. GRB2 links signaling to actin assembly by enhancing interaction of neural Wiskott-Aldrich syndrome protein (N-WASp) with actin-related protein (ARP2/3) complex. *Journal of Biological Chemistry* **275**, 21946–21952 (2000).
21. Gillis, L. C., Berry, D. M., Minden, M. D., McGlade, C. J., and Barber, D. L. Gads (Grb2-related adaptor downstream of Shc) is required for BCR-ABL-mediated lymphoid leukemia. *Leukemia* **27**, 1666–1676 (2013).
22. Watanabe, R. et al. Grb2 and Gads exhibit different interactions with CD28 and play distinct roles in CD28-mediated costimulation. *Journal of Immunology* **177**, 1085–1091 (2006).
23. Balagopalan, L., Coussens, N. P., Sherman, E., Samelson, L. E., and Sommers, C. L. The LAT story: A tale of cooperativity, coordination, and choreography. *Cold Spring Harbor Perspectives in Biology* **2**, a005512 (2010).
24. Liu, S. K., and McGlade, C. J. Gads is a novel SH2 and SH3 domain-containing adaptor protein that binds to tyrosine-phosphorylated Shc. *Oncogene* **17**, 3073–3082 (1998).
25. Takeda, K. et al. CD28 stimulation triggers NF-kappaB activation through the CARMA1-PKCtheta-Grb2/Gads axis. *International Immunology* **20**, 1507–1515 (2008).
26. Bartelt, R. R., and Houtman, J. C. The adaptor protein LAT serves as an integration node for signaling pathways that drive T cell activation. *Wiley Interdisciplinary Reviews. Systems Biology and Medicine* **5**, 101–110 (2013).

27. Houtman, J. C. et al. Binding specificity of multiprotein signaling complexes is determined by both cooperative interactions and affinity preferences. *Biochemistry* **43**, 4170–4178 (2004).

28. Feng, G. S. et al. Grap is a novel SH3-SH2-SH3 adaptor protein that couples tyrosine kinases to the Ras pathway. *Journal of Biological Chemistry* **271**, 12129–12132 (1996).

29. Li, S., Couvillon, A. D., Brasher, B. B., and Van Etten, R. A. Tyrosine phosphorylation of Grb2 by Bcr/Abl and epidermal growth factor receptor: A novel regulatory mechanism for tyrosine kinase signaling. *EMBO Journal* **20**, 6793–6804 (2001).

30. Preisinger, C., and Kolch, W. The Bcr-Abl kinase regulates the actin cytoskeleton via a GADS/Slp-76/Nck1 adaptor protein pathway. *Cellular Signalling* **22**, 848–856 (2010).

31. Bazley, L. A., and Gullick, W. J. The epidermal growth factor receptor family. *Endocrine-Related Cancer* **12 Suppl 1**, S17–S27 (2005).

32. Lin, C. C. et al. Inhibition of basal FGF receptor signaling by dimeric Grb2. *Cell* **149**, 1514–1524 (2012).

33. Osman, N., Lucas, S. C., Turner, H., and Cantrell, D. A comparison of the interaction of Shc and the tyrosine kinase ZAP-70 with the T cell antigen receptor zeta chain tyrosine-based activation motif. *Journal of Biological Chemistry* **270**, 13981–13986 (1995).

34. Pacini, S. et al. Tyrosine 474 of ZAP-70 is required for association with the Shc adaptor and for T-cell antigen receptor-dependent gene activation. *Journal of Biological Chemistry* **273**, 20487–20493 (1998).

35. Feller, S. M., and Lewitzky, M. Potential disease targets for drugs that disrupt protein–protein interactions of Grb2 and Crk family adaptors. *Current Pharmaceutical Design* **12**, 529–548 (2006).

36. Lung, F. D., and Tsai, J. Y. Grb2 SH2 domain-binding peptide analogs as potential anticancer agents. *Biopolymers* **71**, 132–140 (2003).

37. Brehme, M. et al. Charting the molecular network of the drug target Bcr-Abl. *Proceedings of the National Academy of Sciences of the United States of America* **106**, 7414–7419 (2009).

38. Lowenstein, E. J. et al. The SH2 and SH3 domain-containing protein GRB2 links receptor tyrosine kinases to ras signaling. *Cell* **70**, 431–442 (1992).

39. Suen, K. L., Bustelo, X. R., Pawson, T., and Barbacid, M. Molecular cloning of the mouse Grb2 gene: Differential interaction of the Grb2 adaptor protein with epidermal growth factor and nerve growth factor receptors. *Molecular and Cellular Biology* **13**, 5500–5512 (1993).

40. Cheng, A. M. et al. Mammalian Grb2 regulates multiple steps in embryonic development and malignant transformation. *Cell* **95**, 793–803 (1998).

41. Bourette, R. P. et al. Mona, a novel hematopoietic-specific adaptor interacting with the macrophage colony-stimulating factor receptor, is implicated in monocyte/macrophage development. *EMBO Journal* **17**, 7273–7281 (1998).

42. Shiraiwa, H. et al. Detection of Grb-2-related adaptor protein gene (GRAP) and peptide molecule in salivary glands of MRL/lpr mice and patients with Sjogren's syndrome. *Journal of International Medical Research* **32**, 284–291 (2004).

43. Ludwig, L. et al. Grap-2, a novel RET binding protein, is involved in RET mitogenic signaling. *Oncogene* **22**, 5362–5366 (2003).

44. Ludwig, L. et al. Expression of the Grb2-related RET adapter protein Grap-2 in human medullary thyroid carcinoma. *Cancer Letters* **275**, 194–197 (2009).

45. Liu, S. K., Berry, D. M., and McGlade, C. J. The role of Gads in hematopoietic cell signalling. *Oncogene* **20**, 6284–6290 (2001).

46. Berry, D. M., Benn, S. J., Cheng, A. M., and McGlade, C. J. Caspase-dependent cleavage of the hematopoietic specific adaptor protein Gads alters signalling from the T cell receptor. *Oncogene* **20**, 1203–1211 (2001).

47. Lugassy, J. et al. Modulation of TCR responsiveness by the Grb2-family adaptor, Gads. *Cellular Signalling* **27**, 125–134 (2015).
48. Lasserre, R. et al. Release of serine/threonine-phosphorylated adaptors from signaling microclusters down-regulates T cell activation. *J Cell Biol* **195**, 839–853 (2011).
49. Ghosh, J., and Miller, R. A. Rapid tyrosine phosphorylation of Grb2 and Shc in T cells exposed to anti-CD3, anti-CD4, and anti-CD45 stimuli: Differential effects of aging. *Mechanisms of Ageing and Development* **80**, 171–187 (1995).
50. Haines, E. et al. Tyrosine phosphorylation of Grb2: Role in prolactin/epidermal growth factor cross talk in mammary epithelial cell growth and differentiation. *Molecular and Cellular Biology* **29**, 2505–2520 (2009).
51. Trub, T., Frantz, J. D., Miyazaki, M., Band, H., and Shoelson, S. E. The role of a lymphoid-restricted, Grb2-like SH3-SH2-SH3 protein in T cell receptor signaling. *J Biol Chem* **272**, 894–902 (1997).
52. Simon, J. A., and Schreiber, S.L. Grb2 SH3 binding to peptides from Sos: Evaluation of a general model for SH3-ligand interactions. *Chemistry and Biology* **2**, 53–60 (1995).
53. Lock, L. S., Royal, I., Naujokas, M. A., and Park, M. Identification of an atypical Grb2 carboxyl-terminal SH3 domain binding site in Gab docking proteins reveals Grb2-dependent and -independent recruitment of Gab1 to receptor tyrosine kinases. *Journal of Biological Chemistry* **275**, 31536–31545 (2000).
54. Liu, S. K., Smith, C. A., Arnold, R., Kiefer, F., and McGlade, C. J. The adaptor protein Gads (Grb2-related adaptor downstream of Shc) is implicated in coupling hemopoietic progenitor kinase-1 to the activated TCR. *Journal of Immunology* **165**, 1417–1426 (2000).
55. Jordan, M. S. et al. In vivo disruption of T cell development by expression of a dominant-negative polypeptide designed to abolish the SLP-76/Gads interaction. *European Journal of Immunology* **37**, 2961–2972 (2007).
56. Yoder, J. et al. Requirement for the SLP-76 adaptor GADS in T cell development. *Science* **291**, 1987–1991 (2001).
57. Liu, S. K., Fang, N., Koretzky, G. A., and McGlade, C. J. The hematopoietic-specific adaptor protein gads functions in T-cell signaling via interactions with the SLP-76 and LAT adaptors. *Current Biology* **9**, 67–75 (1999).
58. Liu, Q. et al. Structural basis for specific binding of the Gads SH3 domain to an RxxK motif-containing SLP-76 peptide: A novel mode of peptide recognition. *Molecular Cell* **11**, 471–481 (2003).
59. Seet, B. T. et al. Efficient T-cell receptor signaling requires a high-affinity interaction between the Gads C-SH3 domain and the SLP-76 RxxK motif. *EMBO Journal* **26**, 678–689 (2007).
60. Lewitzky, M. et al. The C-terminal SH3 domain of the adapter protein Grb2 binds with high affinity to sequences in Gab1 and SLP-76 which lack the SH3-typical P-x-x-P core motif. *Oncogene* **20**, 1052–1062 (2001).
61. Houtman, J. C. et al. Oligomerization of signaling complexes by the multipoint binding of GRB2 to both LAT and SOS1. *Nature Structural and Molecular Biology* **13**, 798–805 (2006).
62. Motto, D. G., Ross, S. E., Wu, J., Hendricks-Taylor, L. R., and Koretzky, G. A. Implication of the GRB2-associated phosphoprotein SLP-76 in T cell receptor-mediated interleukin 2 production. *Journal of Experimental Medicine* **183**, 1937–1943 (1996).
63. Harkiolaki, M. et al. Structural basis for SH3 domain-mediated high-affinity binding between Mona/Gads and SLP-76. *EMBO Journal* **22**, 2571–2582 (2003).
64. Sastry, L. et al. Quantitative analysis of Grb2-Sos1 interaction: The N-terminal SH3 domain of Grb2 mediates affinity. *Oncogene* **11**, 1107–1112 (1995).

65. Buday, L., Khwaja, A., Sipeki, S., Farago, A., and Downward, J. Interactions of Cbl with two adapter proteins, Grb2 and Crk, upon T cell activation. *Journal of Biological Chemistry* **271**, 6159–6163 (1996).
66. Meisner, H., Conway, B. R., Hartley, D., and Czech, M. P. Interactions of Cbl with Grb2 and phosphatidylinositol 3′-kinase in activated Jurkat cells. *Molecular and Cellular Biology* **15**, 3571–3578 (1995).
67. Zhang, W. et al. Association of Grb2, Gads, and phospholipase C-gamma 1 with phosphorylated LAT tyrosine residues. Effect of LAT tyrosine mutations on T cell antigen receptor-mediated signaling. *Journal of Biological Chemistry* **275**, 23355–23361 (2000).
68. Hussain, S. F., Anderson, C. F., and Farber, D. L. Differential SLP-76 expression and TCR-mediated signaling in effector and memory CD4 T cells. *Journal of Immunology* **168**, 1557–1565 (2002).
69. von Essen, M. R. et al. Vitamin D controls T cell antigen receptor signaling and activation of human T cells. *Nature Immunology* **11**, 344–349 (2010).
70. Clements, J. L., Ross-Barta, S. E., Tygrett, L. T., Waldschmidt, T. J., and Koretzky, G. A. SLP-76 expression is restricted to hemopoietic cells of monocyte, granulocyte, and T lymphocyte lineage and is regulated during T cell maturation and activation. *Journal of Immunology* **161**, 3880–3889 (1998).
71. Clark, S. G., Stern, M. J., and Horvitz, H. R. *C. elegans* cell-signalling gene sem-5 encodes a protein with SH2 and SH3 domains. *Nature* **356**, 340–344 (1992).
72. Pan, Z. et al. The function study on the interaction between Grb2 and AMPK. *Molecular and Cellular Biochemistry* **307**, 121–127 (2008).
73. Kantonen, S. et al. A novel phospholipase D2-Grb2-WASp heterotrimer regulates leukocyte phagocytosis in a two-step mechanism. *Molecular and Cellular Biology* **31**, 4524–4537 (2011).
74. Misra, U. K., and Pizzo, S. V. Potentiation of signal transduction mitogenesis and cellular proliferation upon binding of receptor-recognized forms of alpha2-macroglobulin to 1-LN prostate cancer cells. *Cellular Signalling* **16**, 487–496 (2004).
75. Daly, R. J., Binder, M. D., and Sutherland, R. L. Overexpression of the Grb2 gene in human breast cancer cell lines. *Oncogene* **9**, 2723–2727 (1994).
76. Watanabe, T. et al. Significance of the Grb2 and son of sevenless (Sos) proteins in human bladder cancer cell lines. *IUBMB Life* **49**, 317–320 (2000).
77. Verbeek, B. S., Adriaansen-Slot, S. S., Rijksen, G., and Vroom, T. M. Grb2 overexpression in nuclei and cytoplasm of human breast cells: A histochemical and biochemical study of normal and neoplastic mammary tissue specimens. *Journal of Pathology* **183**, 195–203 (1997).
78. Morlacchi, P., Robertson, F. M., Klostergaard, J., and McMurray, J. S. Targeting SH2 domains in breast cancer. *Future Medicinal Chemistry* **6**, 1909–1926 (2014).
79. Romero, F. et al. Grb2 and its apoptotic isoform Grb3-3 associate with heterogeneous nuclear ribonucleoprotein C, and these interactions are modulated by poly(U) RNA. *Journal of Biological Chemistry* **273**, 7776–7781 (1998).
80. Pivniouk, V. et al. Impaired viability and profound block in thymocyte development in mice lacking the adaptor protein SLP-76. *Cell* **94**, 229–238 (1998).
81. Clements, J. L. et al. Fetal hemorrhage and platelet dysfunction in SLP-76-deficient mice. *Journal of Clinical Investigation* **103**, 19–25 (1999).
82. Shen, R. et al. Grap negatively regulates T-cell receptor-elicited lymphocyte proliferation and interleukin-2 induction. *Molecular and Cellular Biology* **22**, 3230–3236 (2002).
83. Cummins, T. D. et al. Quantitative mass spectrometry of diabetic kidney tubules identifies GRAP as a novel regulator of TGF-beta signaling. *Biochimica et Biophysica Acta* **1804**, 653–661 (2010).

84. Giubellino, A. et al. Inhibition of tumor metastasis by a growth factor receptor bound protein 2 Src homology 2 domain-binding antagonist. *Cancer Research* **67**, 6012–6016 (2007).

85. Atabey, N. et al. Potent blockade of hepatocyte growth factor-stimulated cell motility, matrix invasion and branching morphogenesis by antagonists of Grb2 Src homology 2 domain interactions. *Journal of Biological Chemistry* **276**, 14308–14314 (2001).

86. Gril, B. et al. Grb2-SH3 ligand inhibits the growth of HER2+ cancer cells and has antitumor effects in human cancer xenografts alone and in combination with docetaxel. *International Journal of Cancer* **121**, 407–415 (2007).

87. Kardinal, C. et al. Chronic myelogenous leukemia blast cell proliferation is inhibited by peptides that disrupt Grb2-SoS complexes. *Blood* **98**, 1773–1781 (2001).

88. Giubellino, A., Burke, T. R., Jr., and Bottaro, D. P. Grb2 signaling in cell motility and cancer. *Expert Opinion on Therapeutic Targets* **12**, 1021–1033 (2008).

89. Duffy, M. J. et al. The ADAMs family of proteases: New biomarkers and therapeutic targets for cancer? *Clinical Proteomics* **8**, 9 (2011).

90. Gong, Q. et al. Disruption of T cell signaling networks and development by Grb2 haploid insufficiency. *Nature Immunology* **2**, 29–36 (2001).

91. Kisielow, P., Bluthmann, H., Staerz, U. D., Steinmetz, M., and von Boehmer, H. Tolerance in T-cell-receptor transgenic mice involves deletion of nonmature CD4+8+ thymocytes. *Nature* **333**, 742–746 (1988).

92. Kisielow, P., Teh, H. S., Bluthmann, H., and von Boehmer, H. Positive selection of antigen-specific T cells in thymus by restricting MHC molecules. *Nature* **335**, 730–733 (1988).

93. Jang, I. K. et al. Grb2 functions at the top of the T-cell antigen receptor-induced tyrosine kinase cascade to control thymic selection. *Proceedings of the National Academy of Sciences of the United States of America* **107**, 10620–10625 (2010).

94. Shen, S., Zhu, M., Lau, J., Chuck, M., and Zhang, W. The essential role of LAT in thymocyte development during transition from the double-positive to single-positive stage. *Journal of Immunology* **182**, 5596–5604 (2009).

95. Zhang, W. et al. Essential role of LAT in T cell development. *Immunity* **10**, 323–332 (1999).

96. Paster, W. et al. GRB2-mediated recruitment of THEMIS to LAT is essential for thymocyte development. *Journal of Immunology* **190**, 3749–3756 (2013).

97. Fu, G. et al. Themis controls thymocyte selection through regulation of T cell antigen receptor-mediated signaling. *Nature Immunology* **10**, 848–856 (2009).

98. Hough, M. R., Takei, F., Humphries, R. K., and Kay, R. Defective development of thymocytes overexpressing the costimulatory molecule, heat-stable antigen. *Journal of Experimental Medicine* **179**, 177–184 (1994).

99. Yankee, T. M. et al. The Gads (GrpL) adaptor protein regulates T cell homeostasis. *Journal of Immunology* **173**, 1711–1720 (2004).

100. Dalheimer, S. L. et al. Gads-deficient thymocytes are blocked at the transitional single positive CD4+ stage. *European Journal of Immunology* **39**, 1395–1404 (2009).

101. Kikuchi, K. et al. Suppression of thymic development by the dominant-negative form of Gads. *International Immunology* **13**, 777–783 (2001).

102. Clements, J. L. et al. Requirement for the leukocyte-specific adapter protein SLP-76 for normal T cell development. *Science* **281**, 416–419 (1998).

103. Burns, J. C. et al. The SLP-76 Src homology 2 domain is required for T cell development and activation. *Journal of Immunology* **187**, 4459–4466 (2011).

104. Salmond, R. J., Filby, A., Qureshi, I., Caserta, S., and Zamoyska, R. T-cell receptor proximal signaling via the Src-family kinases, Lck and Fyn, influences T-cell activation, differentiation, and tolerance. *Immunological Reviews* **228**, 9–22 (2009).

105. Palacios, E. H., and Weiss, A. Function of the Src-family kinases, Lck and Fyn, in T-cell development and activation. *Oncogene* **23**, 7990–8000 (2004).
106. Brownlie, R. J., and Zamoyska, R. T cell receptor signalling networks: Branched, diversified and bounded. *Nature Reviews Immunology* **13**, 257–269 (2013).
107. Koretzky, G. A., and Myung, P. S. Positive and negative regulation of T-cell activation by adaptor proteins. *Nature Reviews Immunology* **1**, 95–107 (2001).
108. Ravichandran, K. S. et al. Interaction of Shc with the zeta chain of the T cell receptor upon T cell activation. *Science* **262**, 902–905 (1993).
109. Fukushima, A. et al. Lck couples Shc to TCR signaling. *Cellular Signalling* **18**, 1182–1189 (2006).
110. Picard, C. et al. Hypomorphic mutation of ZAP70 in human results in a late onset immunodeficiency and no autoimmunity. *European Journal of Immunology* **39**, 1966–1976 (2009).
111. Hauck, F. et al. Primary T-cell immunodeficiency with immunodysregulation caused by autosomal recessive LCK deficiency. *Journal of Allergy and Clinical Immunology* **130**, 1144–1152.e1111 (2012).
112. Hubert, P. et al. Defective p56Lck activity in T cells from an adult patient with idiopathic CD4+ lymphocytopenia. *International Immunology* **12**, 449–457 (2000).
113. Sakaguchi, N. et al. Altered thymic T-cell selection due to a mutation of the ZAP-70 gene causes autoimmune arthritis in mice. *Nature* **426**, 454–460 (2003).
114. Sawabe, T. et al. Defect of lck in a patient with common variable immunodeficiency. *International Journal of Molecular Medicine* **7**, 609–614 (2001).
115. Yu, C. C., Mamchak, A. A., and DeFranco, A. L. Signaling mutations and autoimmunity. *Current Directions in Autoimmunity* **6**, 61–88 (2003).
116. Bilal, M. Y., and Houtman, J. C. GRB2 nucleates T cell receptor-mediated LAT clusters that control PLC-gamma1 activation and cytokine production. *Frontiers in Immunology* **6**, 141 (2015).
117. Ackermann, J. A., Radtke, D., Maurberger, A., Winkler, T. H., and Nitschke, L. Grb2 regulates B-cell maturation, B-cell memory responses and inhibits B-cell Ca2+ signalling. *EMBO Journal* **30**, 1621–1633 (2011).
118. Crites, T. J. et al. TCR Microclusters pre-exist and contain molecules necessary for TCR signal transduction. *Journal of Immunology* **193**, 56–67 (2014).
119. Hill, R. J. et al. The lymphoid protein tyrosine phosphatase Lyp interacts with the adaptor molecule Grb2 and functions as a negative regulator of T-cell activation. *Experimental Hematology* **30**, 237–244 (2002).
120. Chapman, N. M., Connolly, S. F., Reinl, E. L., and Houtman, J. C. Focal adhesion kinase negatively regulates Lck function downstream of the T cell antigen receptor. *Journal of Immunology* **191**, 6208–6221 (2013).
121. Bilal, M. Y. et al. GADS is required for TCR-mediated calcium influx and cytokine release, but not cellular adhesion, in human T cells. *Cellular Signalling* **27**, 841–850 (2015).
122. Sommers, C. L. et al. A LAT mutation that inhibits T cell development yet induces lymphoproliferation. *Science* **296**, 2040–2043 (2002).
123. Sommers, C. L. et al. Mutation of the phospholipase C-gamma1-binding site of LAT affects both positive and negative thymocyte selection. *Journal of Experimental Medicine* **201**, 1125–1134 (2005).
124. Paz, P. E. et al. Mapping the Zap-70 phosphorylation sites on LAT (linker for activation of T cells) required for recruitment and activation of signalling proteins in T cells. *Biochemical Journal* **356**, 461–471 (2001).
125. Zhu, M., Janssen, E., and Zhang, W. Minimal requirement of tyrosine residues of linker for activation of T cells in TCR signaling and thymocyte development. *Journal of Immunology* **170**, 325–333 (2003).

126. Samelson, L. E. Signal transduction mediated by the T cell antigen receptor: The role of adapter proteins. *Annual Review of Immunology* **20**, 371–394 (2002).
127. Zhang, W., Irvin, B. J., Trible, R. P., Abraham, R. T., and Samelson, L. E. Functional analysis of LAT in TCR-mediated signaling pathways using a LAT-deficient Jurkat cell line. *International Immunology* **11**, 943–950 (1999).
128. Houtman, J. C., Barda-Saad, M., and Samelson, L. E. Examining multiprotein signaling complexes from all angles. *FEBS Journal* **272**, 5426–5435 (2005).
129. Sugiyama, Y. et al. Direct binding of the signal-transducing adaptor Grb2 facilitates down-regulation of the cyclin-dependent kinase inhibitor p27Kip1. *Journal of Biological Chemistry* **276**, 12084–12090 (2001).
130. Najib, S., and Sanchez-Margalet, V. Sam68 associates with the SH3 domains of Grb2 recruiting GAP to the Grb2-SOS complex in insulin receptor signaling. *Journal of Cellular Biochemistry* **86**, 99–106 (2002).
131. Ando, A. et al. A complex of GRB2-dynamin binds to tyrosine-phosphorylated insulin receptor substrate-1 after insulin treatment. *EMBO Journal* **13**, 3033–3038 (1994).
132. Carling, D. The AMP-activated protein kinase cascade—A unifying system for energy control. *Trends in Biochemical Sciences* **29**, 18–24 (2004).
133. Woods, A. et al. LKB1 is the upstream kinase in the AMP-activated protein kinase cascade. *Current Biology* **13**, 2004–2008 (2003).
134. Ruderman, N., and Prentki, M. AMP kinase and malonyl-CoA: Targets for therapy of the metabolic syndrome. *Nature Reviews Drug Discovery* **3**, 340–351 (2004).
135. Blagih, J., Krawczyk, C. M., and Jones, R. G. LKB1 and AMPK: Central regulators of lymphocyte metabolism and function. *Immunological Reviews* **249**, 59–71 (2012).
136. Roberts, P. J., and Der, C. J. Targeting the Raf-MEK-ERK mitogen-activated protein kinase cascade for the treatment of cancer. *Oncogene* **26**, 3291–3310 (2007).
137. Kortum, R. L., Rouquette-Jazdanian, A. K., and Samelson, L. E. Ras and extracellular signal-regulated kinase signaling in thymocytes and T cells. *Trends in Immunology* **34**, 259–268 (2013).
138. McKay, M. M., and Morrison, D. K. Integrating signals from RTKs to ERK/MAPK. *Oncogene* **26**, 3113–3121 (2007).
139. Poltorak, M., Meinert, I., Stone, J. C., Schraven, B., and Simeoni, L. Sos1 regulates sustained TCR- mediated Erk activation. *European Journal of Immunology* **44**, 1535–1540 (2014).
140. Rincon, M. MAP-kinase signaling pathways in T cells. *Current Opinion in Immunology* **13**, 339–345 (2001).
141. Das, J. et al. Digital signaling and hysteresis characterize ras activation in lymphoid cells. *Cell* **136**, 337–351 (2009).
142. Jun, J. E., Rubio, I., and Roose, J. P. Regulation of ras exchange factors and cellular localization of ras activation by lipid messengers in T cells. *Frontiers in Immunology* **4**, 239 (2013).
143. Buday, L., Egan, S. E., Rodriguez Viciana, P., Cantrell, D. A., and Downward, J. A complex of Grb2 adaptor protein, Sos exchange factor, and a 36-kDa membrane-bound tyrosine phosphoprotein is implicated in ras activation in T cells. *Journal of Biological Chemistry* **269**, 9019–9023 (1994).
144. Salojin, K. V., Zhang, J., Meagher, C., and Delovitch, T. L. ZAP-70 is essential for the T cell antigen receptor-induced plasma membrane targeting of SOS and Vav in T cells. *Journal of Biological Chemistry* **275**, 5966–5975 (2000).
145. Warnecke, N. et al. TCR-mediated Erk activation does not depend on Sos and Grb2 in peripheral human T cells. *EMBO Reports* **13**, 386–391 (2012).
146. Dower, N.A. et al. RasGRP is essential for mouse thymocyte differentiation and TCR signaling. *Nature Immunology* **1**, 317–321 (2000).

147. Rao, N., Dodge, I., and Band, H. The Cbl family of ubiquitin ligases: Critical negative regulators of tyrosine kinase signaling in the immune system. *Journal of Leukocyte Biology* **71**, 753–763 (2002).

148. Balagopalan, L. et al. c-Cbl-mediated regulation of LAT-nucleated signaling complexes. *Molecular and Cellular Biology* **27**, 8622–8636 (2007).

149. Yamasaki, S. et al. Docking protein Gab2 is phosphorylated by ZAP-70 and negatively regulates T cell receptor signaling by recruitment of inhibitory molecules. *Journal of Biological Chemistry* **276**, 45175–45183 (2001).

150. Yamasaki, S. et al. Gads/Grb2-mediated association with LAT is critical for the inhibitory function of Gab2 in T cells. *Molecular and Cellular Biology* **23**, 2515–2529 (2003).

151. Schlaepfer, D. D., and Hunter, T. Evidence for in vivo phosphorylation of the Grb2 SH2-domain binding site on focal adhesion kinase by Src-family protein-tyrosine kinases. *Molecular and Cellular Biology* **16**, 5623–5633 (1996).

152. Coussens, N. P. et al. Multipoint binding of the SLP-76 SH2 domain to ADAP is critical for oligomerization of SLP-76 signaling complexes in stimulated T cells. *Molecular and Cellular Biology* **33**, 4140–4151 (2013).

153. Kortum, R. L. et al. The ability of Sos1 to oligomerize the adaptor protein LAT is separable from its guanine nucleotide exchange activity in vivo. *Science Signaling* **6**, ra99 (2013).

154. Bunnell, S. C. et al. Persistence of cooperatively stabilized signaling clusters drives T-cell activation. *Molecular and Cellular Biology* **26**, 7155–7166 (2006).

155. Bogin, Y., Ainey, C., Beach, D., and Yablonski, D. SLP-76 mediates and maintains activation of the Tec family kinase ITK via the T cell antigen receptor-induced association between SLP-76 and ITK. *Proceedings of the National Academy of Sciences of the United States of America* **104**, 6638–6643 (2007).

156. Sela, M. et al. Sequential phosphorylation of SLP-76 at tyrosine 173 is required for activation of T and mast cells. *EMBO Journal* **30**, 3160–3172 (2011).

157. Yablonski, D., Kadlecek, T., and Weiss, A. Identification of a phospholipase C-gamma1 (PLC-gamma1) SH3 domain-binding site in SLP-76 required for T-cell receptor-mediated activation of PLC-gamma1 and NFAT. *Molecular and Cellular Biology* **21**, 4208–4218 (2001).

158. Gonen, R., Beach, D., Ainey, C., and Yablonski, D. T cell receptor-induced activation of phospholipase C-gamma1 depends on a sequence-independent function of the P-I region of SLP- 76. *Journal of Biological Chemistry* **280**, 8364–8370 (2005).

159. Irvin, B. J., Williams, B. L., Nilson, A. E., Maynor, H. O., and Abraham, R. T. Pleiotropic contributions of phospholipase C-gamma1 (PLC-gamma1) to T-cell antigen receptor-mediated signaling: Reconstitution studies of a PLC-gamma1-deficient Jurkat T-cell line. *Molecular and Cellular Biology* **20**, 9149–9161 (2000).

160. Silverman, M. A., Shoag, J., Wu, J., and Koretzky, G. A. Disruption of SLP-76 interaction with Gads inhibits dynamic clustering of SLP-76 and FcepsilonRI signaling in mast cells. *Molecular and Cellular Biology* **26**, 1826–1838 (2006).

161. Beach, D., Gonen, R., Bogin, Y., Reischl, I. G., and Yablonski, D. Dual role of SLP-76 in mediating T cell receptor-induced activation of phospholipase C-gamma1. *Journal of Biological Chemistry* **282**, 2937–2946 (2007).

162. Zhang, E. Y., Parker, B. L., and Yankee, T. M. Gads regulates the expansion phase of CD8+ T cell-mediated immunity. *Journal of Immunology* **186**, 4579–4589 (2011).

163. Singer, A. L. et al. Roles of the proline-rich domain in SLP-76 subcellular localization and T cell function. *Journal of Biological Chemistry* **279**, 15481–15490 (2004).

164. Burkhardt, J. K., Carrizosa, E., and Shaffer, M.H. The actin cytoskeleton in T cell activation. *Annual Review of Immunology* **26**, 233–259 (2008).

165. Gomez, T. S., and Billadeau, D. D. T cell activation and the cytoskeleton: You can't have one without the other. *Advances in Immunology* **97**, 1–64 (2008).

166. Barda-Saad, M. et al. Dynamic molecular interactions linking the T cell antigen receptor to the actin cytoskeleton. *Nature Immunology* **6**, 80–89 (2005).
167. Bunnell, S. C., Kapoor, V., Trible, R. P., Zhang, W., and Samelson, L. E. Dynamic actin polymerization drives T cell receptor-induced spreading: A role for the signal transduction adaptor LAT. *Immunity* **14**, 315–329 (2001).
168. Zeng, R. et al. SLP-76 coordinates Nck-dependent Wiskott-Aldrich syndrome protein recruitment with Vav-1/Cdc42-dependent Wiskott-Aldrich syndrome protein activation at the T cell-APC contact site. *Journal of Immunology* **171**, 1360–1368 (2003).
169. Barda-Saad, M. et al. Cooperative interactions at the SLP-76 complex are critical for actin polymerization. *EMBO Journal* **29**, 2315–2328 (2010).
170. Reicher, B., and Barda-Saad, M. Multiple pathways leading from the T-cell antigen receptor to the actin cytoskeleton network. *FEBS Letters* **584**, 4858–4864 (2010).
171. Pauker, M. H., Reicher, B., Fried, S., Perl, O., and Barda-Saad, M. Functional cooperation between the proteins Nck and ADAP is fundamental for actin reorganization. *Molecular and Cellular Biology* **31**, 2653–2666 (2011).
172. Hassan, N. J. et al. CD6 regulates T-cell responses through activation-dependent recruitment of the positive regulator SLP-76. *Molecular and Cellular Biology* **26**, 6727–6738 (2006).
173. Roncagalli, R. et al. Quantitative proteomics analysis of signalosome dynamics in primary T cells identifies the surface receptor CD6 as a Lat adaptor-independent TCR signaling hub. *Nature Immunology* **15**, 384–392 (2014).
174. Weisswange, I., Newsome, T. P., Schleich, S., and Way, M. The rate of N-WASP exchange limits the extent of ARP2/3-complex-dependent actin-based motility. *Nature* **458**, 87–91 (2009).
175. Chapman, N. M., Yoder, A. N., and Houtman, J. C. Non-catalytic functions of Pyk2 and Fyn regulate late stage adhesion in human T cells. *PloS One* **7**, e53011 (2012).
176. Adachi, K., and Davis, M. M. T-cell receptor ligation induces distinct signaling pathways in naïve vs. antigen-experienced T cells. *Proceedings of the National Academy of Sciences of the United States of America* **108**, 1549–1554 (2011).
177. Sharp, L. L., Schwarz, D. A., Bott, C. M., Marshall, C. J., and Hedrick, S. M. The influence of the MAPK pathway on T cell lineage commitment. *Immunity* **7**, 609–618 (1997).
178. Zeng, L., Dalheimer, S. L., and Yankee, T.M. Gads–/– mice reveal functionally distinct subsets of TCRbeta+ CD4–CD8– double-negative thymocytes. *Journal of Immunology* **179**, 1013–1021 (2007).

# 10 Defining the Roles of Ca²⁺ Signals during T Cell Activation

*Elsie Samakai,\* Christina Go,\**
*and Jonathan Soboloff*

## CONTENTS

\* Indicates shared first authorship.

## ABSTRACT

$Ca^{2+}$ signaling is critical to T cell activation as a means of rapidly activating and integrating numerous signaling pathways to generate widespread changes in T cell gene expression and function. $Ca^{2+}$ is a second messenger that functions by binding to and altering the function of key proteins leading to pleiotropic changes in cell function. Study of this unique signaling realm has been facilitated by numerous specialized tools including electrophysiological techniques, chemical and genetically designed $Ca^{2+}$ indicators, $Ca^{2+}$ sensors for downstream signaling, and genetically modified mouse models defective in specific $Ca^{2+}$ signaling pathways. Here, we discuss how these techniques have and can be used for the study of T cell responses.

## 10.1   INTRODUCTION

The importance of $Ca^{2+}$ as a key intracellular second messenger has long been established and is evidenced by its use in multiple cellular pathways and events. Binding of antigen to the T cell antigen receptor (TCR) triggers the mobilization of $Ca^{2+}$ and is needed for effective T cell activation, anergy, gene expression, motility, synapse formation, cytotoxicity, development, and differentiation.[1] Although the importance of these signals was recognized as early as the 1960s in which mutant T cells with defective $Ca^{2+}$ influx displayed defects in proliferation and cytokine production,[2,3] the development of cell-permeable esterified forms of organic synthetic $Ca^{2+}$ indicators such as Fura-2 allowed for the advancement of understanding the significance of these $Ca^{2+}$ signals. Today, with the help of current technologies and accompanying procedures used to identify and measure $Ca^{2+}$ signals in single cells, whole tissue, and animals, great strides have been made in not only identifying key molecular components in the generation of these $Ca^{2+}$ signals, but also how these signals are modulated to give rise to specific spatial and temporal $Ca^{2+}$ signatures that determine cell fate and function. In this chapter, we discuss the current understanding of the generation of these complex $Ca^{2+}$ signals, and cellular consequences of these signals, as well as current experimental measurements, methods, and limitations.

## 10.2   CA²⁺ ENTRY IN T CELLS

After it was recognized that $Ca^{2+}$ signals were important for the activation of T cells, it was discovered that stimulation of the TCR with antibodies induced the release of $Ca^{2+}$ from intracellular $Ca^{2+}$ stores.[4] Moreover, stimulation of T cells with either antibodies or the $Ca^{2+}$ ionophore ionomycin led to the appearance of the same new phosphoproteins, suggesting a link between TCR stimulation and the $Ca^{2+}$ pathway.[4-6] It is now generally accepted that successful activation of the T cell requires positive engagement of an antigen, presented by an antigen-presenting cell (APC) onto the TCR followed by increases in cytosolic $Ca^{2+}$ levels.

The TCR is a complex of integral membrane proteins composed of a ligand-sensing subunit, TCRαβ (or to a lesser extent, TCRγδ) and three signaling subunits: CD3εδ, εγ, and ζζ. The core of its antigen-sensing function is a heterodimer composed of TCRα and TCRβ chains that detect MHC-bound antigen in a highly specific manner. Antigen binding triggers phosphorylation of immunoreceptor tyrosine-based activation motifs (ITAMs) by lymphocyte protein tyrosine kinase (Lck) on the cytosolic side of the TCR/CD3 complex. This leads to further recruitment and activation of a series of kinases and their substrates, ultimately leading to the ligation and thus activation of phospholipase C-γ (PLC-γ). PLC-γ activation is a key physiological mediator of receptor-operated $Ca^{2+}$ signaling and an early step in a process that ultimately leads to the initiation of store-operated $Ca^{2+}$ entry (SOCE), which is the main mode of $Ca^{2+}$ entry in T cells.

The mechanism and key molecular components of SOCE have only been elucidated in the last couple of decades. It was long observed that activation of PLC-coupled receptors in virtually all cells leads to a characteristic biphasic $Ca^{2+}$ signal in which a rapid and transient release of $Ca^{2+}$ from the endoplasmic reticulum (ER) is followed by large sustained $Ca^{2+}$ entry via the plasma membrane.[7] This led to James Putney proposing the "capacitative model" in which depletion of intracellular $Ca^{2+}$ stores somehow signaled plasma membrane $Ca^{2+}$ channels to open, thereby permitting refilling of ER $Ca^{2+}$ stores.[8] Subsequent publications using pharmacological inhibitors of the sarco/endoplasmic reticulum $Ca^{2+}$/ATPase (SERCA) to deplete intracellular $Ca^{2+}$ stores independent of receptor activation induced a sustained increase in cytoplasmic $Ca^{2+}$.[9,10] Moreover, an electrophysiological current activated by $Ca^{2+}$ store depletion in mast and T cells named $Ca^{2+}$ release-activated $Ca^{2+}$ current ($I_{CRAC}$) provided quantitative evidence of a relationship between store depletion and agonist-mediated $Ca^{2+}$ influx.[11,12] This capacitative $Ca^{2+}$ entry (CCE), now commonly referred to as SOCE, describes $Ca^{2+}$ entry via the plasma membrane (PM) as a direct consequence of ER store depletion upon receptor activation. This model was further solidified by subsequent discoveries showing that close interactions between the ER and plasma membranes were required for coupling to occur,[13,14] and, finally, the identification of stromal interacting molecule (STIM) family members as ER $Ca^{2+}$ sensors and SOCE regulators,[15,16] and members of the Orai family as $Ca^{2+}$ channels[17–19] using RNAi screens. We now know that SOCE is the key $Ca^{2+}$ entry pathway in T cells and is vital to T cell function as evidenced by severe combined immunodeficiency observed in both STIM/Orai-null mouse mice[20,21] and human patients[22] exhibiting mutations in SOCE components.[22]

The activation of PLC leads to the hydrolysis of phosphatidylinositol 4,5-bisphosphate ($PIP_2$) into inositol-1,4,5-trisphosphate ($InsP_3$) and diacylglycerol (DAG).[23,24] DAG production leads to activation of the Ras/Raf-1/MEK/ERK and nuclear factor kappa-light-chain-enhancer of activated B cell (NF-κB) pathways, which are crucial for T cell activation and function (discussed in later sections). Simultaneously, $InsP_3$ binds to and activates ER-localized $InsP_3$ receptors ($InsP_3Rs$), $Ca^{2+}$ channels that mediate ER $Ca^{2+}$ release. The ER serves as a major $Ca^{2+}$ storage compartment with a $Ca^{2+}$ concentration range estimate of 100–700 μM.[25,26] TCR-mediated $InsP_3R$ activation leads to transient increases in intracellular $Ca^{2+}$ levels to ~500 nM.[1,27] The concomitant decrease in ER $Ca^{2+}$ content is "sensed" by the single pass ER transmembrane protein, STIM1 (Figure 10.1a) via its luminal canonical EF hand (cEF).[28,29] Interestingly, a second non-$Ca^{2+}$-binding "hidden" EF hand (hEF) was later identified as critical to the transduction of $Ca^{2+}$-dependent conformational change to the sterile alpha motif (SAM) that initiates the process of oligomerization.[29] This oligomerization seems to be necessary and sufficient to drive membrane

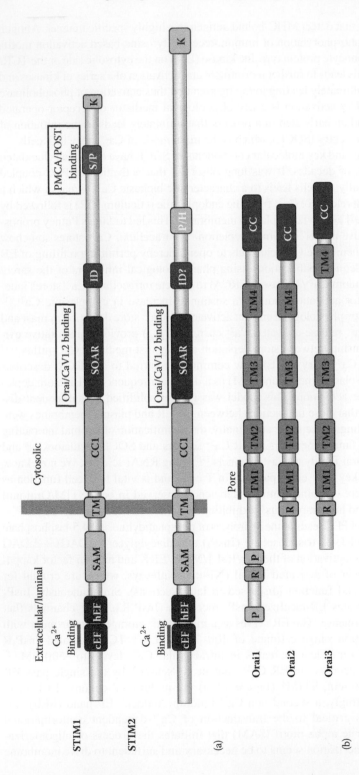

**FIGURE 10.1** Ca²⁺ signaling proteins. (a) Domains of STIM isoforms showing canonical EF-hand (cEF), hidden EF-hand (hEF), sterile alpha motif (SAM), transmembrane (TM), coiled-coil (CC1), STIM1 Orai1 activation region (SOAR), inactivation domain (ID), serine/proline-rich region (S/P), proline/histidine-rich region (P/H), lysine-rich region (K). (b) Domains of Orai isoforms including proline-rich region (P), arginine-rich region (R), transmembrane domains (TM), and coiled-coil domain (CC). Within cells, Orai subunits form hexamers with a Ca²⁺-permeable pore created between the 6 TM1 domains.

localization and Orai1 activation.[30] Interestingly, although SAM domains initiate the process of oligomerization, domains located on the cytosolic side of STIM1 are the primary mediators of STIM–STIM interactions.[31] It has since been shown that it is the STIM-Orai activating region (SOAR[32]; also known as CAD[33]) that mediates oligomerization[34]; significant, in that SOAR is also the minimal required component of STIM1 required for Orai1 activation.[32,33] Hence, loss of Ca²⁺ from the STIM1 EF hand drives association between SAM domains, relieving autoinhibition of SOAR by coiled-coil 1 (CC1)[35,36] to drive SOAR-mediated aggregation, thereby facilitating interaction/gating of Orai1 by SOAR.[37–39] Interactions between c-terminal lysine-rich domains and negatively charged phospholipids in the PM also support STIM–Orai interactions,[28] although this does not seem to be required for this process.

STIM2 is a close homolog of STIM1 (see Figure 10.1a), is also expressed in T cells, and contributes to Ca²⁺ homeostasis and T cell activation. Although STIM2 is an ER Ca²⁺ sensor and Orai1 activator similar to STIM1, it is activated at higher ER Ca²⁺ concentrations[40] with slower kinetics[29,41,42] and decreased efficiency.[42,43] These features of STIM2 likely contribute to its inability to compensate for loss of STIM1, since the presence of STIM2 fails to prevent severe immunodeficiency in both STIM1-null mice[21] and human patients exhibiting STIM1 mutations.[22] However, this does not mean that STIM2 serves no role in Ca²⁺ signaling in T cells. Indeed, STIM2-null mice exhibit defects in T cell activation[21] and STIM1/STIM2 double knockout mice develop complex autoimmunity[21,44,45] not observed in STIM1-null animals.

Members of the Orai family serve as required pore forming units in the process of store-operated Ca²⁺ entry (Figure 10.1b).[17–19] These are highly Ca²⁺-selective ion channels gated almost exclusively by STIM1 and STIM2.[37,39,46] Interestingly, although all 3 Orai channels can be activated by STIM proteins when overexpressed,[47,48] loss of Orai1 seems to eliminate SOCE in most but not all cell types. This includes human, but not murine T cells, which seem to be able to utilize Orai2 in the absence of Orai1.[49] While a role for Orai3 in T cells has not been defined, several studies have implicated Orai3-containing Orai as lipid- rather than store-operated Ca²⁺ channels.[50,51] While future investigations may shed new light on the roles of these Orai homologs in Ca²⁺ signaling, Orai1 seems to be the primary mediator of endogenous SOCE, particularly in T cells, the subject of the current chapter.

## 10.3  MEASURING CA²⁺ SIGNALING IN T CELLS

The availability of diverse experimental approaches to monitor intracellular Ca²⁺ and measure Ca²⁺ currents in real time has significantly contributed to our current understanding of SOCE, and the cellular processes that regulate or are regulated by it. In this section, we review the two main approaches utilized in the study of SOCE: use of fluorescent indicators and electrophysiological techniques. Here, we will outline these approaches, highlighting their key advantages and limitations.

### 10.3.1  CHEMICAL CA²⁺ INDICATORS

Much of our current understanding of Ca²⁺ signaling in T cells is owed to continuous advancements in the development of Ca²⁺-dependent dyes such as Fluo-4, Indo-1, and Fura-2.[52,53] These chemical indicators are charged $Ca^{2+}$-sensitive dyes that can be introduced to the cytoplasm either by microinjection,[54] electroporation,[55] or, more commonly, chemically masked with an acetoxymethyl ester group (AM) enabling them to cross the plasma membrane and enter the cytoplasm, where the AM groups are cleaved off by constitutively active nonspecific esterases.[56,57] The ease with which this can be accomplished has made these fluorophores the first choice by most investigators interested in measuring $Ca^{2+}$ signals. However, recent developments have made fluorophores available with different binding affinities and spectral properties that provide unique advantages depending upon imaging equipment and what precisely is being measured.

### 10.3.1.1    Spectral Properties

When considering spectral properties, the importance of accurate quantification of $Ca^{2+}$ signals to your analysis should be considered. Ratiometric dyes shift either their excitation or emission wavelengths upon $Ca^{2+}$ binding.[53] Although the acquisition and data manipulation is more complex due to the need to acquire multiple fluorescent wavelengths, ratiometric imaging corrects for confounding factors such as uneven loading and cell size differences. Fura-2 and Indo-1 are the most common ratiometric indicators by far. They have similar $Ca^{2+}$ binding affinities, but ratio quite differently. Hence, Fura-2 always emits at the same wavelength but exhibits a shift in excitation wavelength upon $Ca^{2+}$ binding, while the opposite is true for Indo-1.[53,56] This makes Fura-2 the preferred $Ca^{2+}$ indicator for fluorescence microscopy, whereas Indo-1 is usually only used for $Ca^{2+}$ measurement by flow cytometry.[56]

A limitation of Fura-2 and Indo-1 is that they require excitation by ultraviolet light. Since not all confocal microscopes and flow cytometers have ultraviolet (UV) lasers, many investigators use single wavelength dyes such as Fluo-4. Although advantages associated with ratiometric imaging are lost, Fluo-4 exhibits a large dynamic range and is heavily used. Typically, loading differences are discounted by normalizing data to the first acquisition from the beginning of the experiment.[56,58]

### 10.3.1.2    Alternative Indicators with Specialized Properties

Due to their binding affinity for $Ca^{2+}$ at or near resting cytosolic $Ca^{2+}$ levels, Fura-2, Indo-1, and Fluo-4 are by far the most common indicators used, however, alternatives do exist with distinct binding affinity and intracellular location. Although we cannot realistically list all $Ca^{2+}$ indicators used for all situations, we have included several examples with specific and useful properties. Rhod-2 is a $Ca^{2+}$-sensitive dye that is sequestered into mitochondria due to its net positive charge and is used primarily to measure $Ca^{2+}$ levels in this intracellular compartment.[52,59] Alternatively, Mag-Fura-2 (also known as furaptra) exhibits a very low $Ca^{2+}$ affinity and is used primarily to measure ER $Ca^{2+}$ content in permeabilized cells.[60] Recently, mid-affinity dyes[56] such as Fura-5F have gained in popularity; although less sensitive to small changes in cytosolic $Ca^{2+}$ content, the decreased $Ca^{2+}$ binding improves resolution at peak $Ca^{2+}$ levels.

### 10.3.2    GENETICALLY ENCODED FLUORESCENT PROTEINS (GEFPs)

Over the last 20 years, considerable progress has been made in the development of genetically encoded fluorescent proteins (GEFPs). In all GEFPs we are aware of, Ca²⁺ sensitivity is conferred by modified forms of calmodulin, however, several different strategies to transduce Ca²⁺ binding into fluorescence change are outlined next.

### 10.3.2.1   Cameleons

The first GEFPs designed were the Cameleons, genetically encoded indicators consisting of CFP and YFP separated by calmodulin.[61] Upon Ca²⁺-binding, the CFP and YFP moieties are forced together, enabling Förster resonance energy transfer (FRET), in which fluorescence emissions from CFP lead to concomitant excitation of YFP. There have since been several generations of modifications of these proteins to improve their dynamic range, fine-tune Ca²⁺ sensitivity, and decrease pH sensitivity. Although Cameleons can be used for cytosolic Ca²⁺ measurement, the most common Cameleon used, D1ER, is targeted to the ER and exhibits very low Ca²⁺ affinity optimized for the measurement of ER Ca²⁺ content.[62–64]

### 10.3.2.2   Circular Permutated Fluorescent Proteins (cpFPs)

Several different families of Ca²⁺ indicators based on *circular permutated fluorescent proteins* (cpFPs) have been designed. In essence, these synthetic proteins consist of a "broken" nonfluorescent GFP or YFP moiety linked to a calmodulin Ca²⁺-binding domain and a calmodulin-binding domain.[64,65] When Ca²⁺ binds, interactions between calmodulin and its binding protein force the broken GFP or YFP into a fluorescent conformation.[65] As with Cameleons, multiple generations of these proteins have improved their dynamic range and specificity for Ca²⁺. Commonly used cpFPs include Pericams, GCaMPs, and Camagaroos.[66] The first utilized cpFP family, Camagaroos, utilize a YFP chromophore.[66] Although their design was an important conceptual breakthrough, their relatively high pH sensitivity and poor visibility in cells at basal Ca²⁺ levels has limited their usefulness. Pericams also use YFP variants as chromophores, but are preferred, as they contain mutations within their YFP and calmodulin Ca²⁺-binding domain that enhance both fluorescence and Ca²⁺ affinity.[67] Additionally, a "ratiometric-pericam" is also available, allowing for the advantage of imaging and analysis that corrects for confounding factors such as uneven loading and cell size differences.[65] Experimentally, ratiometric-pericams have been shown to be the most reliable to measure changes in mitochondrial Ca²⁺ levels.[64,68] GCaMPs use circularly permutated GFP rather than YFP, resulting in substantially improved fluorescence properties; the most recently designed GCaMPs are up to 40 times brighter than their counterparts, while their relatively rapid response kinetics are ideal for resolving rapid calcium fluctuations.[65]

Newer family members of *cpFPs* continue to be introduced,[65] which now include RFP-based *cpFPs*[69] allowing for even more versatility in assessing Ca²⁺ signals.

### 10.3.2.3   Advantages and Disadvantages

The major advantages of GEFPs over chemical indicators are highly specific organelle targeting and tunable Ca²⁺ sensitivity via modification of the EF hands introduced. In addition, fluorescent proteins are also highly stable, ideal for studies lasting longer than 5 hours.[70] The major disadvantages are the relatively limited dynamic range and need for introduction by transfection.

### 10.3.3    Ca²⁺ Signaling Assays

To mimic the effects of antigen presentation to the T cell and subsequent activation, monoclonal antibodies to the CD3 chains (anti-CD3) or to the TCRαβ subunit (anti-TCR) are often used.[6] As with antigen presentation, stimulation with the agonistic antibodies will induce TCR-mediated $Ca^{2+}$ signals that can be assessed in *in vitro* studies. Nevertheless, since TCR stimulation activates multiple pathways that can confound interpretation, direct measurement of SOCE is more commonly done using pharmacological SERCA inhibitors or $Ca^{2+}$ ionophores.[71] SERCA pumps residing on the ER membrane sequester $Ca^{2+}$ ions from the cytoplasm. This serves to counter a poorly defined "$Ca^{2+}$ leak," enabling the continuous maintenance of high $Ca^{2+}$ levels in the ER.[72,73] Hence, when SERCA is inhibited, a passive $Ca^{2+}$ leak leads to ER $Ca^{2+}$ depletion. Although there are several alternatives, thapsigargin (TG) and cyclopiazonic acid (CPA) are the most commonly used SERCA inhibitors. TG is a highly selective and irreversible SERCA inhibitor,[10] whereas CPA is selective but reversible and can be washed out of cells.[74] An alternative approach to ER $Ca^{2+}$ depletion is to use $Ca^{2+}$ ionophores such as ionomycin or Br-A23187, which actively transport $Ca^{2+}$ across cellular membranes. Although they permeabilize all membranes at higher concentrations, for reasons unknown, at lower levels they selectively permeabilize the ER to $Ca^{2+}$, leading to rapid and selective ER $Ca^{2+}$ depletion.[75,76]

### 10.3.4    "Ca²⁺ Add Back" Method

Since receptor stimulation induces a biphasic $Ca^{2+}$ signal, the $Ca^{2+}$ add-back protocol is heavily utilized to facilitate quantification of ER $Ca^{2+}$ release and subsequent SOCE separately. Essentially, ER $Ca^{2+}$ depletion via either TCR stimulation or pharmacological depletion is induced in the absence of extracellular $Ca^{2+}$. $Ca^{2+}$ levels are then increased to normal physiological levels after the ER $Ca^{2+}$ efflux is completed, facilitating separate measurement of ER $Ca^{2+}$ content and SOCE (Figure 10.2a).

### 10.3.5    Measuring Ca²⁺ Clearance

Recent investigations have led to the recognition that activated T cells exhibit defective $Ca^{2+}$ clearance.[77,78] $Ca^{2+}$ clearance can be measured via a simple modification of the $Ca^{2+}$ add back protocol.[77–80] Briefly, stores are depleted using a SERCA inhibitor in the absence of extracellular $Ca^{2+}$. Extracellular $Ca^{2+}$ is then added briefly followed by its removal once $Ca^{2+}$ levels reach their peak. Since SERCA function is blocked, $Ca^{2+}$ removal under these conditions is a direct reflection of plasma membrane $Ca^{2+}$ ATPase (PMCA) activity. PMCA activity is assessed by analysis of the rate of decay (Figure 10.2a).

### 10.3.6    Measuring Ca²⁺ Oscillations

T cells are able to tightly control cytosolic $Ca^{2+}$ into oscillations through a network of channels, transporters, and receptors.[81,82] $Ca^{2+}$ oscillations in response to T cell activation can be observed by combining the use of monoclonal antibodies αCD3/CD28 and $Ca^{2+}$ indicator dyes.[83] The effect of TCR ligation itself is made measurable

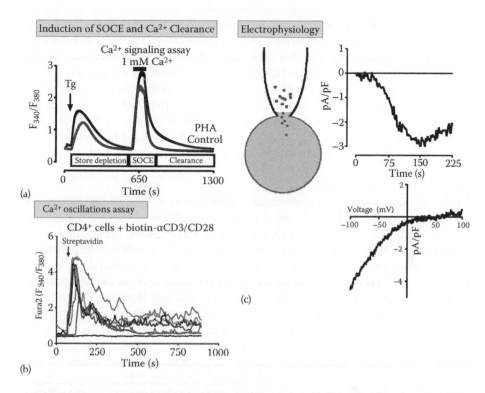

**FIGURE 10.2** Approaches for studying Ca²⁺ signals in T cells. (a) Induction of SOCE and Ca²⁺ clearance. Cells were adhered to polylysine-coated coverslips in the presence or absence of PHA (2 hours) and loaded with Fura-2AM. Stores were depleted via the addition of the SERCA inhibitor, thapsigargin (Tg). The subsequent addition of Ca²⁺ leads to SOCE. The rate at which cytosolic Ca²⁺ levels return to baseline following the subsequent removal of Ca²⁺ reflects PMCA activity. (Data from Ritchie, M. F., Samakai, E., and Soboloff, J., *Embo J* 31(5), 1123–1133, 2012.) (b) Ca²⁺ oscillations are observed in Fura-2-loaded cells adhered to polylysine-coated coverslips. Cells are incubated with biotinylated anti-CD3/CD28. The addition of streptavidin crosslinks the antibodies, leading to TCR activation and Ca²⁺ oscillations (unpublished data). (c) Using whole cell patch clamp, Orai1-mediated currents can be induced and measured via passive store depletion. (Data from Zhou, Y., Lewis, T. L., Robinson, L. J., Brundage, K. M., Schafer, R., Martin, K. H., Blair, H. C., Soboloff, J., and Barnett, J. B., *J Cell Physiol* 226(4), 1082–1089, 2011.) As depicted, currents typically develop over ~2 minutes and exhibit both inward rectification and a positive reversal potential, with essentially no outward current.

in real time by using biotin-tagged monoclonal antibodies that become cross-linked with the addition of streptavidin (Figure 10.2b).

## 10.3.7 Electrophysiology CRAC Recordings

As mentioned in the Introduction, the first unambiguous demonstration of store operated Ca²⁺ currents were in immune cells. The current was termed Ca²⁺ release-activated Ca²⁺ current ($I_{CRAC}$) and is distinguishable by its prominent inward rectification and high selectivity for $Ca^{2+}$ ions.[11,12] Although fluorescent imaging of $Ca^{2+}$ flux is a highly popular and useful method of studying $Ca^{2+}$ flux, electrophysiology recordings of $Ca^{2+}$ currents are advantageous in that they offer specific measurement channel activity and as such, have and continue to provide insight into fundamental channel properties. Electrophysiological measurement of $I_{CRAC}$ can only be carried out in whole-cell patch-clamp mode (sample data[153] depicted in Figure 10.2c), as the single channel conductance is below the level of detection by current methods. A detailed protocol for $I_{CRAC}$ recording has been reviewed previously[84] and will not be repeated here, but we have briefly outlined the general protocol along with major considerations next.

### 10.3.7.1 General Protocol

Whole cell patch clamp first requires the establishment of a tight seal between a recording glass pipette electrode and the surface membrane of the cell. This is followed by the rupture of the membrane establishing electrical and chemical continuity between the pipette and the cytoplasm. Every 1 to 2 seconds a voltage ramp is applied ranging from −100 mV up to +100 mV to reveal currents that develop upon store depletion. This can be achieved actively by including $InsP_3$ in the pipette or via the addition of a $Ca^{2+}$ ionophore to the external bath solution. However, the introduction of a $Ca^{2+}$ chelator such as BAPTA or EGTA to the pipette solution serves to passively activate CRAC over several minutes via lack of availability of $Ca^{2+}$ for store refilling. Both BAPTA and EGTA can be used, although the relatively faster on-rate of BAPTA is generally preferred to avoid $Ca^{2+}$-dependent Orai1 inactivation.[11,85] It is also critical to add 1 millimolar of $Mg^{2+}$/Mg-ATP in the recording pipette to block nonselective TRPM7 channels that otherwise confound this type of recording.[86] One of the defining characteristics of CRAC channels is loss of $Ca^{2+}$ selectivity when all extracellular divalent cations are removed. Hence, while simply removing extracellular $Ca^{2+}$ eliminates all current by removing the charge carrier, the introduction of extracellular $Ca^{2+}$ chelators creates a divalent free (DVF) condition that facilitates $Na^+$ permeability ($Na^+$-$I_{CRAC}$). $Na^+$-$I_{CRAC}$ densities measured are much larger primarily due to a much higher extracellular $Na^+$ concentration, providing a useful tool to measure the very low CRAC activity found in some cell types.[87–89] Further, since Orai1 channels close relatively quickly in the absence of $Ca^{2+}$ (a process known as $Ca^{2+}$ potentiation),[90] this approach provides an additional characteristic facilitating confirmation that the channel of interest is, in fact, CRAC.

## 10.4 DOWNSTREAM EFFECTS OF CA²⁺ SIGNALS

$Ca^{2+}$ signals have been shown to be crucial to both early and late downstream signaling events such as activation, gene expression, differentiation, and effector function events. Discussed next are some key $Ca^{2+}$-mediated processes at various stages post-TCR engagement.

### 10.4.1 ROLE IN EARLY TCR ACTIVATION

As mentioned earlier, TCR engagement leads to ITAM phosphorylation, primarily by Lck.[91] In resting T cells, interactions between positively charged ITAMs and negatively charged phospholipids on the plasma membrane lead to ITAM insertion into the hydrophobic core of the PM, thereby minimizing potential contact with Lck and avoiding spontaneous phosphorylation.[92–94] Recently, it was shown that $Ca^{2+}$ elevation relieves this phenomenon via electrostatic interactions, thereby amplifying proximal TCR signaling at the earliest possible stage.[95] Considering that elevated $Ca^{2+}$ levels are sustained for hours after TCR engagement, this positive feedback likely facilitates sustained TCR engagement by continually maintaining the availability of ITAMs for phosphorylation and signaling.

## 10.4.2 IMMUNOLOGICAL SYNAPSE FORMATION

An immunological synapse (IS) is formed at the interface of the T cell and antigen-presenting cell. TCR engagement induces actin cytoskeleton, organelle reorganization, and redistribution of membrane and intracellular signaling proteins.[96] Interestingly, $Ca^{2+}$ elevation directly facilitates gelsolin interactions with actin filaments, contributing to the rapid cytoskeletal changes occurring upon TCR engagement.[97] However, while increases in cytosolic $Ca^{2+}$ are necessary for F-actin rearrangement, there is also evidence that cytoskeletal rearrangements may also shape downstream $Ca^{2+}$ signals.[98,99] Interruptions to actomyosin retrograde flow leads to loss of PLC-γ phosphorylation by Itk, halting DAG and $InsP_3$ production.[100] Thus, T cell polarization both influences and is influenced by $Ca^{2+}$ signals.

Experimentally, the IS can be visualized by immunostaining or fluorescently tagging proteins that tend to accumulate at the interface of the T cell and the APC, including actin and the CD3 chains.[101] The same approach can be used to identify proteins that may accumulate and are involved with signaling at the IS. Interestingly, STIM1 and Orai1 are both rapidly recruited to the IS after activation, resulting in enhanced localized $Ca^{2+}$ influx at the T cell-APC interface.[102,103] Interestingly, STIM1 and Orai1 also accumulate in "distal caps" at the opposite side of the cell.[102] These studies were followed several years later with the observation that PMCA localization also becomes polarized during IS formation.[77,78] Interestingly, both studies demonstrate that PMCA function is inhibited within the IS, providing a secondary mechanism for local elevation of $Ca^{2+}$ levels at the IS, although different mechanisms were proposed for this phenomenon in these studies. Hence, Ritchie et al. finds that STIM1 associated with PMCA via conformational coupling, leads to decreased function and local $Ca^{2+}$ elevation.[78] In contrast, Quintana et al. attributed this slowing of $Ca^{2+}$ clearance rates to localized sequestration of $Ca^{2+}$ by mitochondria within the IS, thereby decreasing PMCA pumping activity via loss of $Ca^{2+}$ availability.[77] Despite the different mechanisms proffered in these two studies, common to these two studies is that sustained $Ca^{2+}$ signals at the IS are maintained via increased localization of multiple $Ca^{2+}$ signaling molecules and altered functions. Given the importance of sustained TCR signals for T cell activation, it seems likely that these numerous changes in the localization and function of $Ca^{2+}$ signaling molecules all contribute to the maintenance of the IS and TCR signaling; the extent to which this is true is an area of active investigation by us and others.

### 10.4.3    NUCLEAR FACTOR OF ACTIVATED T CELLS (NFAT)

The key family of transcription factors to T cell activation is comprised of five nuclear factor of activated T cell (NFAT) proteins.[104,105] They share a highly conserved DNA binding domain, the Rel homology region (RHR) as well as 14 phosphorylation sites distributed among a transactivation domain (TAD), serine-rich region (SRR), serine/proline (SP) motif, and lysine/threonine/serine (KTS) motif (Figure 10.3a).[106] As evidenced by immunodeficiencies arising from STIM and Orai mutations, T cells are dependent upon SOCE-induced increases in cytosolic $Ca^{2+}$ content, which drive NFAT activity during T cell activation. Inactive NFAT is heavily phosphorylated by the dual-specificity tyrosine-phosphorylation regulated kinases (DYRK), casein kinase (CK), and glycogen synthase kinase 3 (GSK3), and resides in the cytoplasm.[107–111] Upon $Ca^{2+}$ elevation, $Ca^{2+}$/calmodulin activates the serine/threonine protein phosphatase calcineurin, which dephosphorylates NFAT, allowing it to enter the nucleus and act as a transcription factor.[112–115] Critical for the success of this process is the persistence of the $Ca^{2+}$ signal[116] due to the relatively weak binding of NFAT to DNA, and to counteract the activity of kinases that regulate its export from the nucleus.[117,118] Hence, NFAT phosphorylation leads to export from the nucleus by allowing its nuclear export sequence (NES) to bind exportin protein chromosome maintenance 1 (Crm1). Conversely, NFAT dephosphorylation by calcineurin blocks the NES and allows transport into the nucleus via importins.[107]

Although NFAT5 is $Ca^{2+}$-insensitive, NFATc1-c4 depend on $Ca^{2+}$ signals for nuclear localization. The degree to which these different isoforms couple to the $Ca^{2+}$ signal exhibits marked differences. Hence, NFATc3 import and export occurs ~5 to 10 times the rate of NFATc2.[119] As such, NFATc3 activity would tend to be turned on much more quickly than NFATc2, but might also tend to shut off quickly while NFATc2 activity would tend to persist through relatively short decreases in cytosolic $Ca^{2+}$ levels. Recently, differences in the sensitivity of NFATc2 and NFATc3 to local versus global $Ca^{2+}$ content were observed, with NFATc2 translocation to the nucleus being closely matched to CRAC activity, and NFATc3 relying more on increases in both local $Ca^{2+}$ and within the nucleus itself.[120] Future investigations may provide further insight into subtype-dependent differences between these and other NFAT isoforms in the process of T cell activation.

#### 10.4.3.1    Key Assays of NFAT Activity

##### 10.4.3.1.1    Nuclear Localization of NFAT

NFAT undergoes a dramatic movement into the nucleus with activation. This can be observed by expressing a fluorescent fusion protein-tagged NFAT, such as GFP, and visualizing change in localization in response to activation or other stimulus triggering change in cytosolic calcium $Ca^{2+}$. Visualizing NFAT-GFP, particularly when coupled with other methods of measuring $Ca^{2+}$ signals, has given valuable insight into how different NFAT isoforms shape the dynamics of NFAT signaling.

##### 10.4.3.1.2    Binding/Reporter Assays

Although various $Ca^{2+}$-dependent TFs were described earlier, assessing whether they are actually active in various cellular models or under different conditions requires

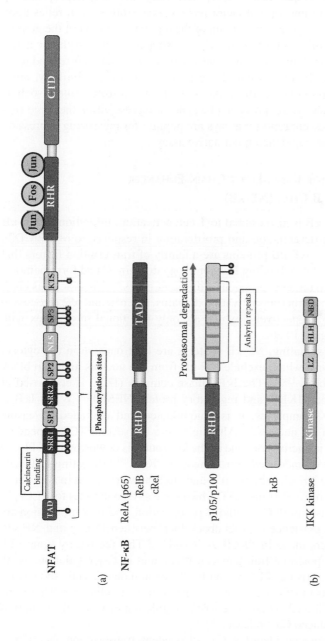

**FIGURE 10.3** Domain structures of NFAT and NF-κB. (a) Domains of Ca²⁺-sensitive (NFATc1–4) isoforms of transcription factor NFAT; transactivation domain (TAD), serine-rich region (SRR), serine/proline motif (SP), lysine/threonine/serine motif (KTS), Rel homology region (RHR), C-terminal domain (CTD). (b) NF-κB signaling molecules. Rel proteins are composed of a Rel homology domain (RHD) and transactivation domain (TAD). NF-κB proteins p105/p100 are comprised of an RHD, ankyrin repeats, and two phosphorylation sites in the C-terminal domain important for proteasomal processing to the functional isoforms p50/p52, respectively. IκB contains several ankyrin repeats. IKK kinases have a kinase domain, leucine zipper (LZ), helix-loop-helix domain (HLH), and NEMO-binding domain (NBD).

approaches to measure their function. Primary assays are ChIP and EMSA. ChIP is advantageous in that it measures endogenous binding, while EMSA can be used to directly measure interactions at specific sites as well as sequence manipulation. However, binding does not equal function; reporter assays are more appropriate for measuring what effects a transcription factor has on gene expression. It relies upon the cloning of the luciferase gene downstream of the regulatory region of the gene of interest and introducing this vector to cells. Upon lysing cells, the enzymatic activity of luciferase is measured, and this enzymatic activity is directly correlated to its expression and therefore the expression of the gene of interest.[121] Similarly, other fluorescent reporter assays combine the expression of a fluorescent protein such as GFP or β-galactosidase with the expression of a gene of interest within the same vector. Luciferase and fluorescent reporter assays are popular for measuring expression because they are sensitive, rapid, and quantitative assays.

### 10.4.4  NUCLEAR FACTOR KAPPA-LIGHT-CHAIN-ENHANCER OF ACTIVATED B CELLS (NF-κB)

Transcription factor NF-κB is also critical to T cell activation. Inhibition of NF-κB is able to block T cell differentiation and proliferation in response to αCD3/CD28 stimulation of the TCR.[122] NF-κB proteins are a family of transcription factors that share a highly conserved DNA-binding Rel homology domain,[123] and form homodimers and heterodimers to target expression of different genes.[124] This includes cRel, RelA, and RelB, which contain transactivation domains, and p50 and p52 (processed forms of p100 and p105, respectively), which are only functional in complex with cRel, RelA, or RelB (Figure 10.3b).

In canonical NF-κB signaling, NF-κB proteins are kept inactive in the cytosol by inhibitory IκB proteins, which exclude NF-κB from the nucleus and also block NF-κB DNA binding ability.[124,125] The IκB kinase complex (IKK) is composed of kinase subunits IKKα and IKKβ, and regulatory protein NEMO.[126] When IκB is phosphorylated by IKK complexes, it is ubiquitinated and degraded, releasing NF-κB to translocate to the nucleus.[127] Although there are several TCR-dependent pathways leading to NF-κB activation, one of the key routes is downstream of PLC-γ via $Ca^{2+}$-independent DAG-mediated recruitment of PKCθ.[128] Interestingly, T cells lacking expression of the NF-κB subunits p50 and cRel show decreases in activation-induced phosphorylation of ZAP70 and LAT, which serve as docking sites for PLC-γ.[129] Correspondingly, decreases in PLC-γ activity, proliferation, and IL-2 production are observed, revealing the existence of a feedback loop between PLC-γ and NF-κB. Similar observations were made in *PKCθ*−/− T cells.[129] The Tec family kinase Itk has been proposed as a potential link between PKCθ and PLC-γ regulation, as Itk overexpression strongly activates PLC-γ, and Itk-deficient mouse T cells show similar $Ca^{2+}$ signaling defects to *PKCθ*−/−, p50−/−, and cRel−/− T cells.[130–132] Therefore, NF-κB may play an additional constitutive role in regulating PLC-γ activation, thus regulating activation-induced $Ca^{2+}$ release.[129,133]

NF-κB has long been considered a $Ca^{2+}$-dependent transcription factor,[116,134] although the mechanisms whereby $Ca^{2+}$ activates NF-κB have never been completely clear. This is at least in part due to the fact that, unlike NFAT, $Ca^{2+}$ is not obligatory for NF-κB activity. Rather, there are Ca²⁺-dependent pathways that intersect with NF-κB signaling which can lead to NF-κB activation. Indeed, unlike for NFAT, a Ca²⁺ signal is insufficient to activate NF-κB, generally requiring coactivation of PKC via the introduction of PMA.[134] Further, unlike NFAT, which requires sustained Ca²⁺ elevation for activation, NF-κB is fully engaged by transient Ca²⁺ responses.[116] There are at least 2 Ca²⁺-dependent pathways that have been identified as leading to IκBα degradation and NF-κB activation. Calcineurin has been shown to activate c-rel[134] via inactivation of IκBβ.[105] More recently, Liu et al. found that Ca²⁺ influx via SOCE controls the nuclear localization and transcriptional activity of NF-κB protein p65 via phosphorylation by PKCα.[135] Irrespective of which pathway, it is clear that the Ca²⁺ and NF-κB signaling pathways crosstalk in multiple ways downstream of TCR engagements.

Future studies will require the use of genetically modified model systems combined with precise control of Ca²⁺ signals. In particular, future investigations using STIM1 and Orai knockout mice may provide new insights into Ca²⁺-mediated control of NF-κB signaling. Patients with STIM1 or Orai1 deficiencies share symptoms with patients who have mutations affecting NF-κB activation. Immunodeficiency and ectodermal dysplasia was observed in patients with mutations in NEMO and IκBα that impaired activation of NF-κB transcription.[136–139] This may suggest closer relationships between SOCE and NF-κB signaling than is currently recognized.

## 10.4.5 CREB-AP1 PATHWAY

Although NFAT and NF-κB are the best-studied Ca²⁺-dependent transcription factor in T cell activation, Ca²⁺ signaling modulates numerous other critical pathways. cAMP response element binding protein (CREB) is best known as downstream of cAMP signaling, however, Ca²⁺ signals are also known to regulate its nuclear localization via calmodulin-dependent kinase IV (CaMKIV)-mediated phosphorylation.[140] CaMKIV binding to calmodulin is mutually exclusive from PP2A binding, therefore, CaMKIV is kept inhibited and bound to PP2A until TCR activation leads to increases in Ca²⁺/calmodulin that displace PP2A and activate CaMKIV.[141] There, CREB binds to cAMP response elements to regulate the transcription of several genes including c-Fos, which forms part of transcription factor heterodimer AP-1.[140,142] AP-1 is critical for T cell activation, forming a complex with NFAT to induce the expression of distinct subsets of genes.[118,143] Interestingly, over the long term, Ca²⁺ signals also stimulate inhibitory pathways via the CREB modulator CREM, which negatively regulate IL-2 production when phosphorylated by CAMKIV.[144,145]

Utilizing novel approaches to differentiate global and local changes in Ca²⁺ signals, Di Capite et al. revealed a "Ca²⁺ signature" for local control of c-fos activation.[146] Hence, whereas the fast on-rate of EGTA blocks Ca²⁺-dependent signals, the relatively slow binding rate of EGTA permits local changes in Ca²⁺ concentration despite equal efficacy in blocking global changes in Ca²⁺ content. Using this approach, it was revealed that changes in Ca²⁺ content near the mouth of the Orai1 pore and not global Ca²⁺ elevation was specifically required for c-fos induction. Using La³⁺ to block both the entry and exit of Ca²⁺ was also used in support of this conclusion, an approach that has been previously used.[147–149]

## 10.5    CONCLUDING STATEMENTS

As outlined in this chapter, advancements in the development of new tools for the study of changes in $Ca^{2+}$ content and downstream signaling events have led to a greater appreciation of how changes in $Ca^{2+}$ concentration modulate the process of T cell activation. See Figure 10.4 for a depiction of the key events in this process, including approaches for the study of $Ca^{2+}$ signals *in vitro*. With the development of

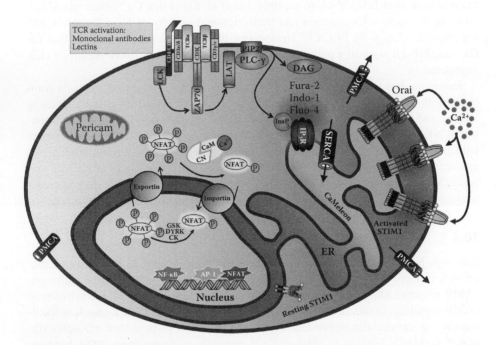

**FIGURE 10.4**   $Ca^{2+}$ signals during T cell activation. Activation in T cells may be studied by ligating the TCR with monoclonal antibodies or lectins. This leads to the recruitment of PLC, production of $InsP_3$, and release of $Ca^{2+}$ from the ER. The corresponding decrease in ER $Ca^{2+}$ content leads to engagement of STIM1, resulting in a conformational change from its resting dimeric state to an extended multimeric form, enabling activation of Orai channels and $Ca^{2+}$ influx. Cytosolic $Ca^{2+}$ is cleared via the combined action of sarco/endoplasmic $Ca^{2+}$/ATPase (SERCA) and the plasma membrane $Ca^{2+}$/ATPase (PMCA). These changes in cytosolic $Ca^{2+}$ content are commonly measured using $Ca^{2+}$ indicator dyes such as Fura-2, Indo-1, and Fluo-4. However, due to their highly negative membrane potential and the presence of channels and transporters on the mitochondrial inner membrane, mitochondrial $Ca^{2+}$ also changes during periods of elevated $Ca^{2+}$ signaling. These events can be measured using the mitochondria-specific $Ca^{2+}$ indicator Pericam or Rhod-2 (not depicted). Changes in ER $[Ca^{2+}]$ are commonly measured using the $Ca^{2+}$ indicator CaMeleon. Also depicted is $Ca^{2+}$-dependent regulation of NFAT. First, increased cytosolic $Ca^{2+}$ leads to calmodulin activation. This leads to activation of the phosphatase calcineurin, which dephosphorylates NFAT, leading to nuclear transloca-tion via importins, where it acts as a transcription factor alongside NF-κB and AP-1, among others. NFAT requires a persistent $Ca^{2+}$ signal to remain in the nucleus due to the activity of kinases such as GSK, DYRK, and CK, which lead to NFAT phosphorylation and the subse-quent transport by exportins to the cytosol.

mouse models expressing genetically encoded $Ca^{2+}$ indicators and/or other genetic modifications combined with recent advances in imaging techniques, it may be possible to shift future studies into the more challenging but ultimately highly rewarding *in vivo* environment. Further, combining tools for the measurement of both $Ca^{2+}$ itself and downstream targets (such as NFAT), it may be possible to examine the downstream consequences of a $Ca^{2+}$ signal, simply by following the changes in location and function of specific cells over time without actually removing them from the animal.

Another longstanding challenge in the field of $Ca^{2+}$ signaling has been linking short-term changes in cytosolic $Ca^{2+}$ content with long-term changes in cell function. Recent investigations from our laboratory have focused on transcriptional mechanisms responsible for the control of STIM1 expression and function. Hence, we have shown that the zinc finger transcription factor early growth response 1 (EGR1) regulates the expression of STIM1.[150] We have further shown that these changes occur as quickly as 2 hours after engagement of the TCR, ultimately leading to surprising differences in both the expression and function of STIM1.[78,151] Interestingly, NF-κB has been shown to regulate both STIM1 and Orai1 expression.[152] How the functions of these different transcription factors are coordinated remain unclear, nor is it certain how many other factors control the expression of these critical $Ca^{2+}$ signaling genes. More global approaches may provide new insights into mechanisms regulating STIM1 and Orai1 transcription, with considerable potential implications to how $Ca^{2+}$ signals regulate the development and function of specific T cell subsets.

## ACKNOWLEDGMENTS

We wish to thank Robert Hooper for the contribution of unpublished data (Figure 10.2b). This work was supported by National Institutes of Health grants R01GM097335 (Jonathan Soboloff), R01GM117907 (Jonathan Soboloff), and F31GM103731 (Elsie Samakai).

## REFERENCES

1. Feske, S., Calcium signalling in lymphocyte activation and disease, *Nat Rev Immunol* 7(9), 690–702, 2007.
2. Whitfield, J. F., Perris, A. D., and Youdale, T., The role of calcium in the mitotic stimulation of rat thymocytes by detergents, agmatine and poly-L-lysine, *Exp Cell Res* 53(1), 155–165, 1968.
3. Asherson, G. L., Davey, M. J., and Goodford, P. J., Increased uptake of calcium by human lymphocytes treated with phytohaemagglutinin, *J Physiol* 206(2), 32P–33P, 1970.
4. Weiss, A., Imboden, J., Shoback, D., and Stobo, J., Role of T3 surface molecules in human T-cell activation: T3-dependent activation results in an increase in cytoplasmic free calcium, *Proc Natl Acad Sci USA* 81(13), 4169–4173, 1984.
5. Imboden, J. B., Weiss, A., and Stobo, J. D., The antigen receptor on a human T cell line initiates activation by increasing cytoplasmic free calcium, *J Immunol* 134(2), 663–665, 1985.
6. Imboden, J. B., and Stobo, J. D., Transmembrane signalling by the T cell antigen receptor. Perturbation of the T3-antigen receptor complex generates inositol phosphates and releases calcium ions from intracellular stores, *J Exp Med* 161(3), 446–456, 1985.

7. Putney, J. W., Jr., Stimulus-permeability coupling: Role of calcium in the receptor regulation of membrane permeability, *Pharmacol Rev* 30(2), 209–245, 1978.

8. Putney, J. W., Jr., A model for receptor-regulated calcium entry, *Cell Calcium* 7(1), 1–12, 1986.

9. Thastrup, O., Cullen, P. J., Drobak, B. K., Hanley, M. R., and Dawson, A. P., Thapsigargin, a tumor promoter, discharges intracellular Ca2+ stores by specific inhibition of the endoplasmic reticulum Ca2(+)-ATPase, *Proc Natl Acad Sci USA* 87(7), 2466–2470, 1990.

10. Takemura, H., Hughes, A. R., Thastrup, O., and Putney, J. W., Jr., Activation of calcium entry by the tumor promoter thapsigargin in parotid acinar cells. Evidence that an intracellular calcium pool and not an inositol phosphate regulates calcium fluxes at the plasma membrane, *J Biol Chem* 264(21), 12266–12271, 1989.

11. Hoth, M., and Penner, R., Depletion of intracellular calcium stores activates a calcium current in mast cells, *Nature* 355(6358), 353–356, 1992.

12. Zweifach, A., and Lewis, R. S., Mitogen-regulated Ca2+ current of T lymphocytes is activated by depletion of intracellular Ca2+ stores, *Proc Natl Acad Sci USA* 90(13), 6295–6299, 1993.

13. Yao, Y., Ferrer-Montiel, A. V., Motal, M., and Tsien, R. Y., Activation of store-operated Ca2+ current in Xenopus oocytes requires SNAP-25 but not a diffusible messenger, *Cell* 98(4), 475–485, 1999.

14. Patterson, R. L., van Rossum, D. B., and Gill, D. L., Store-operated Ca2+ entry: Evidence for a secretion-like coupling model, *Cell* 98(4), 487–499, 1999.

15. Liou, J., Kim, M. L., Heo, W. D., Jones, J. T., Myers, J. W., Ferrell, J. E., Jr., and Meyer, T., STIM is a Ca2+ sensor essential for Ca2+-store-depletion-triggered Ca2+ influx, *Curr Biol* 15(13), 1235–1241, 2005.

16. Roos, J., Digregorio, P. J., Yeromin, A. V., Ohlsen, K., Lioudyno, M., Zhang, S., Safrina, O., Kozak, J. A., Wagner, S. L., Cahalan, M. D., Velicelebi, G., and Stauderman, K. A., STIM1, an essential and conserved component of store-operated Ca2+ channel function, *J Cell Biol* 169(3), 435–445, 2005.

17. Feske, S., Gwack, Y., Prakriya, M., Srikanth, S., Puppel, S. H., Tanasa, B., Hogan, P. G., Lewis, R. S., Daly, M., and Rao, A., A mutation in Orai1 causes immune deficiency by abrogating CRAC channel function, *Nature* 441(7090), 179–185, 2006.

18. Vig, M., Peinelt, C., Beck, A., Koomoa, D. L., Rabah, D., Koblan-Huberson, M., Kraft, S., Turner, H., Fleig, A., Penner, R., and Kinet, J. P., CRACM1 is a plasma membrane protein essential for store-operated Ca2+ entry, *Science* 312(5777), 1220–1223, 2006.

19. Zhang, S. L., Yeromin, A. V., Zhang, X. H., Yu, Y., Safrina, O., Penna, A., Roos, J., Stauderman, K. A., and Cahalan, M. D., Genome-wide RNAi screen of Ca(2+) influx identifies genes that regulate Ca(2+) release-activated Ca(2+) channel activity, *Proc Natl Acad Sci USA* 103(24), 9357–9362, 2006.

20. Gwack, Y., Srikanth, S., Oh-Hora, M., Hogan, P. G., Lamperti, E. D., Yamashita, M., Gelinas, C., Neems, D. S., Sasaki, Y., Feske, S., Prakriya, M., Rajewsky, K., and Rao, A., Hair loss and defective T- and B-cell function in mice lacking ORAI1, *Mol Cell Biol* 28(17), 5209–5222, 2008.

21. Oh-Hora, M., Yamashita, M., Hogan, P. G., Sharma, S., Lamperti, E., Chung, W., Prakriya, M., Feske, S., and Rao, A., Dual functions for the endoplasmic reticulum calcium sensors STIM1 and STIM2 in T cell activation and tolerance, *Nat Immunol* 9(4), 432–443, 2008.

22. Picard, C., McCarl, C. A., Papolos, A., Khalil, S., Luthy, K., Hivroz, C., LeDeist, F., Rieux-Laucat, F., Rechavi, G., Rao, A., Fischer, A., and Feske, S., STIM1 mutation associated with a syndrome of immunodeficiency and autoimmunity, *N Engl J Med* 360(19), 1971–1980, 2009.

23. Dittmar, M. T., McKnight, A., Simmons, G., Clapham, P. R., Weiss, R. A., and Simmonds, P., HIV-1 tropism and co-receptor use, *Nature* 385(6616), 495–496, 1997.
24. van Leeuwen, J. E., and Samelson, L. E., T cell antigen-receptor signal transduction, *Curr Opin Immunol* 11(3), 242–248, 1999.
25. Guse, A. H., Roth, E., and Emmrich, F., Intracellular Ca2+ pools in Jurkat T-lymphocytes, *Biochem J* 291 (Pt 2), 447–451, 1993.
26. Bezprozvanny, I., and Ehrlich, B. E., The inositol 1,4,5-trisphosphate (InsP3) receptor, *J Membr Biol* 145(3), 205–216, 1995.
27. Lewis, R. S., Calcium signaling mechanisms in T lymphocytes, *Annu Rev Immunol* 19, 497–521, 2001.
28. Liou, J., Fivaz, M., Inoue, T., and Meyer, T., Live-cell imaging reveals sequential oligo-merization and local plasma membrane targeting of stromal interaction molecule 1 after Ca2+ store depletion, *Proc Natl Acad Sci USA* 104(22), 9301–9306, 2007.
29. Stathopulos, P. B., Zheng, L., Li, G. Y., Plevin, M. J., and Ikura, M., Structural and mechanistic insights into STIM1-mediated initiation of store-operated calcium entry, *Cell* 135(1), 110–122, 2008.
30. Luik, R. M., Wang, B., Prakriya, M., Wu, M. M., and Lewis, R. S., Oligomerization of STIM1 couples ER calcium depletion to CRAC channel activation, *Nature* 454(7203), 538–542, 2008.
31. Williams, R. T., Senior, P. V., Van Stekelenburg, L., Layton, J. E., Smith, P. J., and Dziadek, M. A., Stromal interaction molecule 1 (STIM1), a transmembrane protein with growth suppressor activity, contains an extracellular SAM domain modified by N-linked glycosylation, *Biochim Biophys Acta* 1596(1), 131–137, 2002.
32. Yuan, J. P., Zeng, W., Dorwart, M. R., Choi, Y. J., Worley, P. F., and Muallem, S., SOAR and the polybasic STIM1 domains gate and regulate Orai channels, *Nat Cell Biol* 11(3), 337–343, 2009.
33. Park, C. Y., Hoover, P. J., Mullins, F. M., Bachhawat, P., Covington, E. D., Raunser, S., Walz, T., Garcia, K. C., Dolmetsch, R. E., and Lewis, R. S., STIM1 clusters and acti-vates CRAC channels via direct binding of a cytosolic domain to Orai1, *Cell* 136(5), 876–890, 2009.
34. Covington, E. D., Wu, M. M., and Lewis, R. S., Essential role for the CRAC activation domain in store-dependent oligomerization of STIM1, *Mol Biol Cell* 21(11), 1897–1907, 2010.
35. Korzeniowski, M. K., Manjarres, I. M., Varnai, P., and Balla, T., Activation of STIM1-Orai1 involves an intramolecular switching mechanism, *Sci Signal* 3(148), ra82, 2010.
36. Zhou, Y., Srinivasan, P., Razavi, S., Seymour, S., Meraner, P., Gudlur, A., Stathopulos, P. B., Ikura, M., Rao, A., and Hogan, P. G., Initial activation of STIM1, the regulator of store-operated calcium entry, *Nat Struct Mol Biol* 20(8), 973–981, 2013.
37. Soboloff, J., Rothberg, B. S., Madesh, M., and Gill, D. L., STIM proteins: Dynamic calcium signal transducers, *Nat Rev Mol Cell Biol* 13(9), 549–565, 2012.
38. Kim, J. Y., and Muallem, S., Unlocking SOAR releases STIM, *EMBO J* 30(9), 1673–1675, 2011.
39. Derler, I., Madl, J., Schutz, G., and Romanin, C., Structure, regulation and biophysics of I(CRAC), STIM/Orai1, *Adv Exp Med Biol* 740, 383–410, 2012.
40. Brandman, O., Liou, J., Park, W. S., and Meyer, T., STIM2 is a feedback regulator that stabilizes basal cytosolic and endoplasmic reticulum Ca2+ levels, *Cell* 131(7), 1327–1339, 2007.
41. Stathopulos, P. B., Zheng, L., and Ikura, M., Stromal interaction molecule (STIM) 1 and STIM2 calcium sensing regions exhibit distinct unfolding and oligomerization kinetics, *J Biol Chem* 284(2), 728–732, 2009.

42. Zhou, Y., Mancarella, S., Wang, Y., Yue, C., Ritchie, M., Gill, D. L., and Soboloff, J., The short N-terminal domains of STIM1 and STIM2 control the activation kinetics of Orai1 channels, *J Biol Chem* 284(29), 19164–19168, 2009.

43. Wang, X., Wang, Y., Zhou, Y., Hendron, E., Mancarella, S., Andrake, M. D., Rothberg, B. S., Soboloff, J., and Gill, D. L., Distinct Orai-coupling domains in STIM1 and STIM2 define the Orai-activating site, *Nat Commun* 5, 3183, 2014.

44. Cheng, K. T., Alevizos, I., Liu, X., Swaim, W. D., Yin, H., Feske, S., Oh-Hora, M., and Ambudkar, I. S., STIM1 and STIM2 protein deficiency in T lymphocytes underlies development of the exocrine gland autoimmune disease, Sjogren's syndrome, *Proc Natl Acad Sci USA* 109(36), 14544–14549, 2012.

45. Ma, J., McCarl, C. A., Khalil, S., Luthy, K., and Feske, S., T-cell-specific deletion of STIM1 and STIM2 protects mice from EAE by impairing the effector functions of Th1 and Th17 cells, *Eur J Immunol* 40(11), 3028–3042, 2010.

46. Prakriya, M., Store-operated Orai channels: Structure and function, *Curr Top Membr* 71, 1–32, 2013.

47. Mercer, J. C., Dehaven, W. I., Smyth, J. T., Wedel, B., Boyles, R. R., Bird, G. S., and Putney, J. W., Jr., Large store-operated calcium selective currents due to co-expression of Orai1 or Orai2 with the intracellular calcium sensor, Stim1, *J Biol Chem* 281(34), 24979–24990, 2006.

48. Lis, A., Peinelt, C., Beck, A., Parvez, S., Monteilh-Zoller, M., Fleig, A., and Penner, R., CRACM1, CRACM2, and CRACM3 are store-operated Ca2+ channels with distinct functional properties, *Curr Biol* 17(9), 794–800, 2007.

49. Shaw, P. J., and Feske, S., Regulation of lymphocyte function by ORAI and STIM proteins in infection and autoimmunity, *J Physiol* 590(Pt 17), 4157–4167, 2012.

50. Gonzalez-Cobos, J. C., Zhang, X., Zhang, W., Ruhle, B., Motiani, R. K., Schindl, R., Muik, M., Spinelli, A. M., Bisaillon, J. M., Shinde, A. V., Fahrner, M., Singer, H. A., Matrougui, K., Barroso, M., Romanin, C., and Trebak, M., Store-independent Orai1/3 channels activated by intracrine leukotrieneC4: Role in neointimal hyperplasia, *Circ Res* 112(7), 1013–1025, 2013.

51. Mignen, O., Thompson, J. L., and Shuttleworth, T. J., Both Orai1 and Orai3 are essential components of the arachidonate-regulated Ca2+-selective (ARC) channels, *J Physiol* 586(1), 185–195, 2008.

52. Minta, A., Kao, J. P., and Tsien, R. Y., Fluorescent indicators for cytosolic calcium based on rhodamine and fluorescein chromophores, *J Biol Chem* 264(14), 8171–8178, 1989.

53. Grynkiewicz, G., Poenie, M., and Tsien, R. Y., A new generation of Ca2+ indicators with greatly improved fluorescence properties, *J Biol Chem* 260(6), 3440–3450, 1985.

54. Glennon, M. C., Bird, G. S., Kwan, C. Y., and Putney, J. W., Jr., Actions of vasopressin and the Ca(2+)-ATPase inhibitor, thapsigargin, on Ca2+ signaling in hepatocytes, *J Biol Chem* 267(12), 8230–8233, 1992.

55. Quintana, A., and Hoth, M., Apparent cytosolic calcium gradients in T-lymphocytes due to fura-2 accumulation in mitochondria, *Cell Calcium* 36(2), 99–109, 2004.

56. Paredes, R. M., Etzler, J. C., Watts, L. T., Zheng, W., and Lechleiter, J. D., Chemical calcium indicators, *Methods* 46(3), 143–151, 2008.

57. Takahashi, A., Camacho, P., Lechleiter, J. D., and Herman, B., Measurement of intracellular calcium, *Physiol Rev* 79(4), 1089–1125, 1999.

58. Gee, K. R., Brown, K. A., Chen, W. N., Bishop-Stewart, J., Gray, D., and Johnson, I., Chemical and physiological characterization of fluo-4 Ca(2+)-indicator dyes, *Cell Calcium* 27(2), 97–106, 2000.

59. Contreras, L., Drago, I., Zampese, E., and Pozzan, T., Mitochondria: The calcium connection, *Biochim Biophys Acta* 1797(6–7), 607–618, 2010.

60. Golovina, V. A., and Blaustein, M. P., Spatially and functionally distinct Ca2+ stores in sarcoplasmic and endoplasmic reticulum, *Science* 275(5306), 1643–1648, 1997.

61. Miyawaki, A., Llopis, J., Heim, R., McCaffery, J. M., Adams, J. A., Ikura, M., and Tsien, R. Y., Fluorescent indicators for Ca2+ based on green fluorescent proteins and calmodulin, *Nature* 388(6645), 882–887, 1997.

62. Graves, T. K., and Hinkle, P. M., Ca(2+)-induced Ca(2+) release in the pancreatic beta-cell: Direct evidence of endoplasmic reticulum Ca(2+) release, *Endocrinology* 144(8), 3565–3574, 2003.

63. Yu, R., and Hinkle, P. M., Rapid turnover of calcium in the endoplasmic reticulum during signaling. Studies with cameleon calcium indicators, *J Biol Chem* 275(31), 23648–23653, 2000.

64. Whitaker, M., Genetically encoded probes for measurement of intracellular calcium, *Methods Cell Biol* 99, 153–182, 2010.

65. Koldenkova, V. P., and Nagai, T., Genetically encoded Ca2+ indicators: Properties and evaluation, *Biochimica Et Biophysica Acta* 1833(7), 1787–1797, 2013.

66. Baird, G. S., Zacharias, D. A., and Tsien, R. Y., Circular permutation and receptor insertion within green fluorescent proteins, *Proc Natl Acad Sci USA* 96(20), 11241–11246, 1999.

67. Nagai, T., Sawano, A., Park, E. S., and Miyawaki, A., Circularly permuted green fluorescent proteins engineered to sense Ca2+, *Proc Natl Acad Sci USA* 98(6), 3197–3202, 2001.

68. Arnaudeau, S., Kelley, W. L., Walsh, J. V., Jr., and Demaurex, N., Mitochondria recycle Ca(2+) to the endoplasmic reticulum and prevent the depletion of neighboring endoplasmic reticulum regions, *J Biol Chem* 276(31), 29430–29439, 2001.

69. Carlson, H. J., and Campbell, R. E., Circular permutated red fluorescent proteins and calcium ion indicators based on mCherry, *Protein Eng Des Sel* 26(12), 763–772, 2013.

70. Palmer, A. E., and Tsien, R. Y., Measuring calcium signaling using genetically targetable fluorescent indicators, *Nat Protoc* 1(3), 1057–1065, 2006.

71. Putney, J. W., Pharmacology of store-operated calcium channels, *Mol Interv* 10(4), 209–218, 2010.

72. Lomax, R. B., Camello, C., Van Coppenolle, F., Petersen, O. H., and Tepikin, A. V., Basal and physiological Ca(2+) leak from the endoplasmic reticulum of pancreatic acinar cells. Second messenger-activated channels and translocons, *J Biol Chem* 277(29), 26479–26485, 2002.

73. Ong, H. L., Liu, X., Sharma, A., Hegde, R. S., and Ambudkar, I. S., Intracellular Ca(2+) release via the ER translocon activates store-operated calcium entry, *Pflugers Arch* 453(6), 797–808, 2007.

74. Sedova, M., Klishin, A., Huser, J., and Blatter, L. A., Capacitative Ca2+ entry is graded with degree of intracellular Ca2+ store depletion in bovine vascular endothelial cells, *J Physiol* 523(Pt 3), 549–559, 2000.

75. Morgan, A. J., and Jacob, R., Ionomycin enhances Ca2+ influx by stimulating store-regulated cation entry and not by a direct action at the plasma membrane, *Biochem J* 300(Pt 3), 665–672, 1994.

76. Deber, C. M., Tom-Kun, J., Mack, E., and Grinstein, S., Bromo-A23187: A nonfluorescent calcium ionophore for use with fluorescent probes, *Anal Biochem* 146(2), 349–352, 1985.

77. Quintana, A., Pasche, M., Junker, C., Al-Ansary, D., Rieger, H., Kummerow, C., Nunez, L., Villalobos, C., Meraner, P., Becherer, U., Rettig, J., Niemeyer, B. A., and Hoth, M., Calcium microdomains at the immunological synapse: How ORAI channels, mitochondria and calcium pumps generate local calcium signals for efficient T-cell activation, *Embo J* 30(19), 3895–3912, 2011.

78. Ritchie, M. F., Samakai, E., and Soboloff, J., STIM1 is required for attenuation of PMCA-mediated Ca2+ clearance during T-cell activation, *Embo J* 31(5), 1123–1133, 2012.

79. Bautista, D. M., Hoth, M., and Lewis, R. S., Enhancement of calcium signalling dynamics and stability by delayed modulation of the plasma-membrane calcium-ATPase in human T cells, *J Physiol* 541(Pt 3), 877–894, 2002.

80. Bautista, D. M., and Lewis, R. S., Modulation of plasma membrane calcium-ATPase activity by local calcium microdomains near CRAC channels in human T cells, *J Physiol* 556(Pt 3), 805–817, 2004.

81. Feske, S., Skolnik, E. Y., and Prakriya, M., Ion channels and transporters in lymphocyte function and immunity, *Nat Rev Immunol* 12(7), 532–547, 2012.

82. Guse, A. H., Berg, I., da Silva, C. P., Potter, B. V., and Mayr, G. W., Ca2+ entry induced by cyclic ADP-ribose in intact T-lymphocytes, *J Biol Chem* 272(13), 8546–8550, 1997.

83. Christo, S. N., Diener, K. R., Nordon, R. E., Brown, M. P., Griesser, H. J., Vasilev, K., Christo, F. C., and Hayball, J. D., Scrutinizing calcium flux oscillations in T lymphocytes to deduce the strength of stimulus, *Sci Rep* 5, 7760, 2015.

84. Parekh, A. B., and Putney, J. W., Jr., Store-operated calcium channels, *Physiol Rev* 85(2), 757–810, 2005.

85. Zweifach, A., and Lewis, R. S., Slow calcium-dependent inactivation of depletion-activated calcium current. Store-dependent and -independent mechanisms, *J Biol Chem* 270(24), 14445–14451, 1995.

86. Hermosura, M. C., Monteilh-Zoller, M. K., Scharenberg, A. M., Penner, R., and Fleig, A., Dissociation of the store-operated calcium current I(CRAC) and the Mg-nucleotide-regulated metal ion current MagNuM, *J Physiol* 539(Pt 2), 445–458, 2002.

87. Shinde, A. V., Motiani, R. K., Zhang, X., Abdullaev, I. F., Adam, A. P., Gonzalez-Cobos, J. C., Zhang, W., Matrougui, K., Vincent, P. A., and Trebak, M., STIM1 controls endothelial barrier function independently of Orai1 and Ca2+ entry, *Sci Signal* 6(267), ra18, 2013.

88. Motiani, R. K., Abdullaev, I. F., and Trebak, M., A novel native store-operated calcium channel encoded by Orai3: Selective requirement of Orai3 versus Orai1 in estrogen receptor-positive versus estrogen receptor-negative breast cancer cells, *J Biol Chem* 285(25), 19173–19183, 2010.

89. Hooper, R., Zhang, X., Webster, M., Go, C., Kedra, J., Marchbank, K., Gill, D. L., Weeraratna, A. T., Trebak, M., and Soboloff, J., Novel protein kinase C-mediated control of Orai1 function in invasive melanoma, *Mol Cell Biol* 35(16), 2790–2798, 2015.

90. Zweifach, A., and Lewis, R. S., Calcium-dependent potentiation of store-operated calcium channels in T lymphocytes, *J Gen Physiol* 107(5), 597–610, 1996.

91. Palacios, E. H., and Weiss, A., Function of the Src-family kinases, Lck and Fyn, in T-cell development and activation, *Oncogene* 23(48), 7990–8000, 2004.

92. Aivazian, D., and Stern, L. J., Phosphorylation of T cell receptor zeta is regulated by a lipid dependent folding transition, *Nat Struct Biol* 7(11), 1023–1026, 2000.

93. Xu, C., Gagnon, E., Call, M. E., Schnell, J. R., Schwieters, C. D., Carman, C. V., Chou, J. J., and Wucherpfennig, K. W., Regulation of T cell receptor activation by dynamic membrane binding of the CD3epsilon cytoplasmic tyrosine-based motif, *Cell* 135(4), 702–713, 2008.

94. DeFord-Watts, L. M., Dougall, D. S., Belkaya, S., Johnson, B. A., Eitson, J. L., Roybal, K. T., Barylko, B., Albanesi, J. P., Wulfing, C., and van Oers, N. S., The CD3 zeta subunit contains a phosphoinositide-binding motif that is required for the stable accumulation of TCR-CD3 complex at the immunological synapse, *J Immunol* 186(12), 6839–6847, 2011.

95. Shi, X., Bi, Y., Yang, W., Guo, X., Jiang, Y., Wan, C., Li, L., Bai, Y., Guo, J., Wang, Y., Chen, X., Wu, B., Sun, H., Liu, W., Wang, J., and Xu, C., Ca2+ regulates T-cell receptor activation by modulating the charge property of lipids, *Nature* 493(7430), 111–115, 2013.

96. Fooksman, D. R., Vardhana, S., Vasiliver-Shamis, G., Liese, J., Blair, D. A., Waite, J., Sacristan, C., Victora, G. D., Zanin-Zhorov, A., and Dustin, M. L., Functional anatomy of T cell activation and synapse formation, *Annu Rev Immunol* 28, 79–105, 2010.

97. Gremm, D., and Wegner, A., Gelsolin as a calcium-regulated actin filament-capping protein, *Eur J Biochem* 267(14), 4339–4345, 2000.

98. Nolz, J. C., Gomez, T. S., Zhu, P., Li, S., Medeiros, R. B., Shimizu, Y., Burkhardt, J. K., Freedman, B. D., and Billadeau, D. D., The WAVE2 complex regulates actin cytoskeletal reorganization and CRAC-mediated calcium entry during T cell activation, *Curr Biol* 16(1), 24–34, 2006.

99. Gomez, T. S., Hamann, M. J., McCarney, S., Savoy, D. N., Lubking, C. M., Heldebrant, M. P., Labno, C. M., McKean, D. J., McNiven, M. A., Burkhardt, J. K., and Billadeau, D. D., Dynamin 2 regulates T cell activation by controlling actin polymerization at the immunological synapse, *Nat Immunol* 6(3), 261–270, 2005.

100. Babich, A., Li, S., O'Connor, R. S., Milone, M. C., Freedman, B. D., and Burkhardt, J. K., F-actin polymerization and retrograde flow drive sustained PLCgamma1 signaling during T cell activation, *J Cell Biol* 197(6), 775–787, 2012.

101. Gascoigne, N. R., Ampudia, J., Clamme, J. P., Fu, G., Lotz, C., Mallaun, M., Niederberger, N., Palmer, E., Rybakin, V., Yachi, P. P., and Zal, T., Visualizing intermolecular interactions in T cells, *Curr Top Microbiol Immunol* 334, 31–46, 2009.

102. Barr, V. A., Bernot, K. M., Srikanth, S., Gwack, Y., Balagopalan, L., Regan, C. K., Helman, D. J., Sommers, C. L., Oh-Hora, M., Rao, A., and Samelson, L. E., Dynamic movement of the calcium sensor STIM1 and the calcium channel Orai1 in activated T-cells: Puncta and distal caps, *Mol Biol Cell* 19(7), 2802–2817, 2008.

103. Lioudyno, M. I., Kozak, J. A., Penna, A., Safrina, O., Zhang, S. L., Sen, D., Roos, J., Stauderman, K. A., and Cahalan, M. D., Orai1 and STIM1 move to the immunological synapse and are up-regulated during T cell activation, *Proc Natl Acad Sci USA* 105(6), 2011–2016, 2008.

104. Crabtree, G. R., and Olson, E. N., NFAT signaling: Choreographing the social lives of cells, *Cell* 109 Suppl, S67–S79, 2002.

105. Biswas, G., Anandatheerthavarada, H. K., Zaidi, M., and Avadhani, N. G., Mitochondria to nucleus stress signaling: A distinctive mechanism of NFkappaB/Rel activation through calcineurin-mediated inactivation of IkappaBbeta, *J Cell Biol* 161(3), 507–519, 2003.

106. Macian, F., NFAT proteins: Key regulators of T-cell development and function, *Nat Rev Immunol* 5(6), 472–484, 2005.

107. Gwack, Y., Sharma, S., Nardone, J., Tanasa, B., Iuga, A., Srikanth, S., Okamura, H., Bolton, D., Feske, S., Hogan, P. G., and Rao, A., A genome-wide Drosophila RNAi screen identifies DYRK-family kinases as regulators of NFAT, *Nature* 441(7093), 646–650, 2006.

108. Beals, C. R., Sheridan, C. M., Turck, C. W., Gardner, P., and Crabtree, G. R., Nuclear export of NF-ATc enhanced by glycogen synthase kinase-3, *Science* 275(5308), 1930–1934, 1997.

109. Okamura, H., Garcia-Rodriguez, C., Martinson, H., Qin, J., Virshup, D. M., and Rao, A., A conserved docking motif for CK1 binding controls the nuclear localization of NFAT1, *Mol Cell Biol* 24(10), 4184–4195, 2004.

110. Zhu, J., Shibasaki, F., Price, R., Guillemot, J. C., Yano, T., Dotsch, V., Wagner, G., Ferrara, P., and McKeon, F., Intramolecular masking of nuclear import signal on NF-AT4 by casein kinase I and MEKK1, *Cell* 93(5), 851–861, 1998.

111. Porter, C. M., Havens, M. A., and Clipstone, N. A., Identification of amino acid residues and protein kinases involved in the regulation of NFATc subcellular localization, *J Biol Chem* 275(5), 3543–3551, 2000.

112. Jain, J., McCaffrey, P. G., Miner, Z., Kerppola, T. K., Lambert, J. N., Verdine, G. L., Curran, T., and Rao, A., The T-cell transcription factor NFATp is a substrate for calcineurin and interacts with Fos and Jun, *Nature* 365(6444), 352–355, 1993.

113. Hogan, P. G., Chen, L., Nardone, J., and Rao, A., Transcriptional regulation by calcium, calcineurin, and NFAT, *Genes Dev* 17(18), 2205–2232, 2003.

114. Muller, M. R., and Rao, A., NFAT, immunity and cancer: A transcription factor comes of age, *Nat Rev Immunol* 10(9), 645–656, 2010.

115. Wu, H., Peisley, A., Graef, I. A., and Crabtree, G. R., NFAT signaling and the invention of vertebrates, *Trends Cell Biol* 17(6), 251–260, 2007.

116. Dolmetsch, R. E., Lewis, R. S., Goodnow, C. C., and Healy, J. I., Differential activation of transcription factors induced by Ca2+ response amplitude and duration, *Nature* 386(6627), 855–858, 1997.

117. Macian, F., Lopez-Rodriguez, C., and Rao, A., Partners in transcription: NFAT and AP-1, *Oncogene* 20(19), 2476–2489, 2001.

118. Chen, L., Glover, J. N., Hogan, P. G., Rao, A., and Harrison, S. C., Structure of the DNA-binding domains from NFAT, Fos and Jun bound specifically to DNA, *Nature* 392(6671), 42–48, 1998.

119. Yissachar, N., Sharar Fischler, T., Cohen, A. A., Reich-Zeliger, S., Russ, D., Shifrut, E., Porat, Z., and Friedman, N., Dynamic response diversity of NFAT isoforms in individual living cells, *Mol Cell* 49(2), 322–330, 2013.

120. Kar, P., and Parekh, A. B., Distinct spatial Ca2+ signatures selectively activate different NFAT transcription factor isoforms, *Mol Cell* 58(2), 232–243, 2015.

121. Fan, F., and Wood, K. V., Bioluminescent assays for high-throughput screening, *Assay Drug Dev Technol* 5(1), 127–136, 2007.

122. Costello, R., Lipcey, C., Algarte, M., Cerdan, C., Baeuerle, P. A., Olive, D., and Imbert, J., Activation of primary human T-lymphocytes through CD2 plus CD28 adhesion molecules induces long-term nuclear expression of NF-kappa B, *Cell Growth Differ* 4(4), 329–339, 1993.

123. Sullivan, J. C., Kalaitzidis, D., Gilmore, T. D., and Finnerty, J. R., Rel homology domain-containing transcription factors in the cnidarian Nematostella vectensis, *Dev Genes Evol* 217(1), 63–72, 2007.

124. Gilmore, T. D., Introduction to NF-kappaB: Players, pathways, perspectives, *Oncogene* 25(51), 6680–6684, 2006.

125. Wegener, E., Oeckinghaus, A., Papadopoulou, N., Lavitas, L., Schmidt-Supprian, M., Ferch, U., Mak, T. W., Ruland, J., Heissmeyer, V., and Krappmann, D., Essential role for IkappaB kinase beta in remodeling Carma1-Bcl10-Malt1 complexes upon T cell activation, *Mol Cell* 23(1), 13–23, 2006.

126. Scheidereit, C., IkappaB kinase complexes: Gateways to NF-kappaB activation and transcription, *Oncogene* 25(51), 6685–6705, 2006.

127. Vallabhapurapu, S., and Karin, M., Regulation and function of NF-kappaB transcription factors in the immune system, *Annu Rev Immunol* 27, 693–733, 2009.

128. Li, Y., Sedwick, C. E., Hu, J., and Altman, A., Role for protein kinase Ctheta (PKCtheta) in TCR/CD28-mediated signaling through the canonical but not the non-canonical pathway for NF-kappaB activation, *J Biol Chem* 280(2), 1217–1223, 2005.

129. Bronk, C. C., Yoder, S., Hopewell, E. L., Yang, S., Celis, E., Yu, X. Z., and Beg, A. A., NF-kappaB is crucial in proximal T-cell signaling for calcium influx and NFAT activation, *Eur J Immunol* 44(12), 3741–3746, 2014.

130. Tomlinson, M. G., Kane, L. P., Su, J., Kadlecek, T. A., Mollenauer, M. N., and Weiss, A., Expression and function of Tec, Itk, and Btk in lymphocytes: Evidence for a unique role for Tec, *Mol Cell Biol* 24(6), 2455–2466, 2004.

131. Liu, R., Zhao, X., Gurney, T. A., and Landau, N. R., Functional analysis of the proximal CCR5 promoter, *AIDS Res Hum Retroviruses* 14(17), 1509–1519, 1998.

132. Sun, Z., Arendt, C. W., Ellmeier, W., Schaeffer, E. M., Sunshine, M. J., Gandhi, L., Annes, J., Petrzilka, D., Kupfer, A., Schwartzberg, P. L., and Littman, D. R., PKC-theta is required for TCR-induced NF-kappaB activation in mature but not immature T lymphocytes, *Nature* 404(6776), 402–407, 2000.

133. Manicassamy, S., Gupta, S., and Sun, Z., Selective function of PKC-theta in T cells, *Cell Mol Immunol* 3(4), 263–270, 2006.

134. Frantz, B., Nordby, E. C., Bren, G., Steffan, N., Paya, C. V., Kincaid, R. L., Tocci, M. J., O'Keefe, S. J., and O'Neill, E. A., Calcineurin acts in synergy with PMA to inactivate I kappa B/MAD3, an inhibitor of NF-kappa B, *EMBO J* 13(4), 861–870, 1994.

135. Liu, X., Berry, C. T., Ruthel, G., Madara, J. J., MacGillivray, K., Gray, C. M., Madge, L. A., McCorkell, K. A., Beiting, D. P., Hershberg, U., May, M. J., and Freedman, B. D., T cell receptor-induced NF-kappaB signaling and transcriptional activation are regulated by STIM1- and Orai1-mediated calcium entry, *J Biol Chem* 291(16) 8440–8452, 2016.

136. Courtois, G., Smahi, A., Reichenbach, J., Doffinger, R., Cancrini, C., Bonnet, M., Puel, A., Chable-Bessia, C., Yamaoka, S., Feinberg, J., Dupuis-Girod, S., Bodemer, C., Livadiotti, S., Novelli, F., Rossi, P., Fischer, A., Israel, A., Munnich, A., Le Deist, F., and Casanova, J. L., A hypermorphic IkappaBalpha mutation is associated with autosomal dominant anhidrotic ectodermal dysplasia and T cell immunodeficiency, *J Clin Invest* 112(7), 1108–1115, 2003.

137. Doffinger, R., Smahi, A., Bessia, C., Geissmann, F., Feinberg, J., Durandy, A., Bodemer, C., Kenwrick, S., Dupuis-Girod, S., Blanche, S., Wood, P., Rabia, S. H., Headon, D. J., Overbeek, P. A., Le Deist, F., Holland, S. M., Belani, K., Kumararatne, D. S., Fischer, A., Shapiro, R., Conley, M. E., Reimund, E., Kalhoff, H., Abinun, M., Munnich, A., Israel, A., Courtois, G., and Casanova, J. L., X-linked anhidrotic ectodermal dysplasia with immunodeficiency is caused by impaired NF-kappaB signaling, *Nat Genet* 27(3), 277–285, 2001.

138. Puel, A., Picard, C., Ku, C. L., Smahi, A., and Casanova, J. L., Inherited disorders of NF-kappaB-mediated immunity in man, *Curr Opin Immunol* 16(1), 34–41, 2004.

139. Feske, S., ORAI1 and STIM1 deficiency in human and mice: Roles of store-operated Ca2+ entry in the immune system and beyond, *Immunol Rev* 231(1), 189–209, 2009.

140. Yu, C. T., Shih, H. M., and Lai, M. Z., Multiple signals required for cyclic AMP-responsive element binding protein (CREB) binding protein interaction induced by CD3/CD28 costimulation, *J Immunol* 166(1), 284–292, 2001.

141. Anderson, K. A., Noeldner, P. K., Reece, K., Wadzinski, B. E., and Means, A. R., Regulation and function of the calcium/calmodulin-dependent protein kinase IV/protein serine/threonine phosphatase 2A signaling complex, *J Biol Chem* 279(30), 31708–31716, 2004.

142. Kuo, C. T., and Leiden, J. M., Transcriptional regulation of T lymphocyte development and function, *Annu Rev Immunol* 17, 149–187, 1999.

143. Macian, F., Garcia-Cozar, F., Im, S. H., Horton, H. F., Byrne, M. C., and Rao, A., Transcriptional mechanisms underlying lymphocyte tolerance, *Cell* 109(6), 719–731, 2002.

144. Solomou, E. E., Juang, Y. T., and Tsokos, G. C., Protein kinase C-theta participates in the activation of cyclic AMP-responsive element-binding protein and its subsequent binding to the -180 site of the IL-2 promoter in normal human T lymphocytes, *J Immunol* 166(9), 5665–5674, 2001.

145. Juang, Y. T., Wang, Y., Solomou, E. E., Li, Y., Mawrin, C., Tenbrock, K., Kyttaris, V. C., and Tsokos, G. C., Systemic lupus erythematosus serum IgG increases CREM binding to the IL-2 promoter and suppresses IL-2 production through CaMKIV, *J Clin Invest* 115(4), 996–1005, 2005.

146. Di Capite, J., Ng, S. W., and Parekh, A. B., Decoding of cytoplasmic Ca(2+) oscillations through the spatial signature drives gene expression, *Curr Biol* 19(10), 853–858, 2009.

147. Bird, G. S., and Putney, J. W., Jr., Capacitative calcium entry supports calcium oscillations in human embryonic kidney cells, *J Physiol* 562(Pt 3), 697–706, 2005.

148. Neher, E., Usefulness and limitations of linear approximations to the understanding of Ca++ signals, *Cell Calcium* 24(5–6), 345–357, 1998.

149. Parekh, A. B., Local Ca2+ influx through CRAC channels activates temporally and spatially distinct cellular responses, *Acta Physiol (Oxf)* 195(1), 29–35, 2009.

150. Ritchie, M. F., Yue, C., Zhou, Y., Houghton, P. J., and Soboloff, J., Wilms tumor suppressor 1 (WT1) and early growth response 1 (EGR1) are regulators of STIM1 expression, *J Biol Chem* 285(14), 10591–10596, 2010.

151. Samakai, E., Hooper, R., Martin, K., M., S., Kappes, D. J., Zhang, Y., Tempera, I., and Soboloff, J., Novel STIM1-dependent control of Ca2+ clearance regulates NFAT activity during T cell activation, Submitted, 2016.

152. Eylenstein, A., Schmidt, S., Gu, S., Yang, W., Schmid, E., Schmidt, E. M., Alesutan, I., Szteyn, K., Regel, I., Shumilina, E., and Lang, F., Transcription factor NF-kappaB regulates expression of pore-forming Ca2+ channel unit, Orai1, and its activator, STIM1, to control Ca2+ entry and affect cellular functions, *J Biol Chem* 287(4), 2719–2730, 2012.

153. Zhou, Y., Lewis, T. L., Robinson, L. J., Brundage, K. M., Schafer, R., Martin, K. H., Blair, H. C., Soboloff, J., and Barnett, J. B., The role of calcium release activated calcium channels in osteoclast differentiation, *J Cell Physiol* 226(4), 1082–1089, 2011.

# 11 Histone Methyltransferases and T Cell Heterogeneity

*Janaki Purushe and Yi Zhang*

## CONTENTS

## ABSTRACT

During the immune response, naïve T cells are activated to produce a multitude of lineages with distinct effector and memory functions. Effector T cells mediate an efficient adaptive immune response to primary antigen encounters, whereas long-lived memory T cells are responsible for the rapid response to subsequent antigen encounters. Distinct subsets of effector T cells, which are primarily characterized by their production of differential cytokines and suppressive molecules, play different roles in eliminating targets and mediating inflammatory responses. Furthermore, different subsets of memory T cells display distinctive features based on their tissue homing capacity, self-renewal capability, and effector recall responsiveness. Recent work has demonstrated the critical role of histone methylation and histone methyltransferases in regulating the generation of heterogeneous subsets of T cells in response to environmental cues. In this chapter, we provide a broad overview of global histone methylation in orchestrating the generation and maintenance of T cell subsets, as well as describe the functional importance of histone methyltransferases in regulating adaptive T cell immunity.

## 11.1 INTRODUCTION

During the immune response, naïve T cells possess a stunning capability to produce distinct subsets of effector cells and memory T cells.[1-7] Effector T cells are characterized by their capacity to efficiently eliminate targets such as pathogens and tumor cells, however, they are short-lived cells that undergo massive apoptotic contraction during late stages of the effector phase.[8-10] Unlike effector T cells, memory T cells are long-lived cells that undergo homeostatic survival in the absence of a specific antigen. Upon reencounter with the specific antigen, memory T cells rapidly acquire effector functions and undergo clonal expansion to produce large numbers of effector T cells, thereby providing protection against secondary infections.[1-3,5,11-15] Thus, effective protective immunity against infection and tumors requires the collective effort of heterogeneous lineages of T cells.

Histone methylation is essential for establishing cell type-specific gene expression patterns.[16-18] Histone methyltransferases (HMTs) catalyze the deposition of these modifications, which result in the mono-, di-, and tri-methylation (me3) of lysine resides at the N-terminal tails of histones.[16-18] The site and degree of methylation is responsible for determining whether each set of modifications will mediate the repression (H3K27me3, H3K9me3, H3K9me2) or activation (H3K4me3, H4K9me3) of associated genes.[18-20] Unlike other histone modifications, particularly acetylation, histone methylation results in transcriptional changes through the recruitment of regulatory proteins to particular genomic regions, rather than through inducing alterations in the charge of the chromatin structure.[18-20]

Global analysis of histone methylation patterns in T cells has revealed a dynamic and complex landscape of modifications proximal to genes critical to T cell activation, expansion, differentiation, and memory formation.[21-23] Methylation can also occur on other residues, particularly arginine; however, the role of this modification and the enzymes that catalyze it are poorly characterized in T cells. This chapter will outline the importance of global histone lysine methylation patterns in T cells and describe roles of particular HMTs in directing effector T cell response and memory formation.

## 11.2 HETEROGENEITY OF ADAPTIVE T CELL RESPONSE

T cells play a critical role in host immune defenses by mediating specific and potent responses against pathogens and cancerous cells. Upon immune stimulation, naïve

T cells activate a multitude of transcriptional programs that direct their vigorous proliferation and subsequent differentiation into effector cells.[5,24,25] CD4+ T cells mediate the clearance of pathogens and differentiate into $T_H1$, $T_H2$, and $T_H17$ effector lineages, each with distinct functions. $T_H1$ cells produce IFN-γ; $T_H2$ cells are characterized by secreting IL-4, IL-5, and IL-13; and $T_H17$ cells produce IL-17 and IL-21.[24–27] Conversely, effector T cell responses can be antagonized by a population of CD4+ T cells termed regulatory T cells ($T_{regs}$), which may develop from naïve precursors ($nT_{regs}$) or be induced in the periphery ($iT_{regs}$).[28] CD8+ T cells, also termed cytotoxic T lymphocytes (CTLs), are essential for the clearance of pathogen-infected and cancerous cells; they produce IFN-γ and directly acting cytotoxic molecules such as granzymes, Fas ligand, and perforins.[5] Upon clearance of an antigen, the vast majority of the T cell population is lost during a period of apoptotic contraction, leaving a residual population of memory cells that can persist throughout the lifetime of an individual.[29–31]

Memory T cells are featured by their capacity to undergo self-renewal in the absence of antigen and their capability to quickly expand and elaborate effector function, thus providing the immune system with long-term protection against secondary antigen encounters. Memory T cells are heterogeneous populations and have distinct capabilities in the context of providing long-term protection against tumor formation. For example, memory CD8+ T cells can be broadly classified into four subsets based on their tissue homing capacity, self-renewal capability, effector recall responsiveness, and surface phenotype: effector memory T cells ($T_{EM}$), central memory T cells ($T_{CM}$), resident memory T cells ($T_{RM}$), and stem cell-like memory T cells ($T_{SCM}$).[3,8,32–38] $T_{EM}$ express low levels of CD62L and CCR7, allowing them to circulate and preferentially home to nonlymphoid tissues, whereas $T_{CM}$ express CD62L and CCR7, restraining their homing to lymphoid tissues. $T_{RM}$ predominantly reside in the local nonlymphoid tissues, such as the brain, mucosa, lung, and skin, where they express cell surface CD69 and CD103, distinguishing them from $T_{EM}$.[7,33,39–42] Finally, $T_{SCM}$ are a memory cell subset expressing a naïve cell-like phenotype of $CD44^{low}CD62L^{high}Sca-1^{high}CD122^{high}Bcl2^{high}$. They possess the ability to differentiate into all subsets of memory CD8 T cells and effector cells, while maintaining self-renewal capabilities.[35,36] The combined functional complexity of effector and memory subsets contributes substantially to the heterogeneity of T cells and ensures a robust response to diverse immune insults.

The maintenance of a precise homeostasis between heterogeneous T cell subsets is essential to balance the beneficial and detrimental effects of T cell immunity. For instance, hosts deficient in one or both of these populations exhibit increased pathology upon exposure to infection. In contrast, intact but dysregulated T cell function can result in autoimmune disease as well as therapeutic complications such as graft rejection and graft-versus-host disease (GVHD), a phenomenon associated with hematopoietic stem cell transplantation.[5,6,14,43–49] For proper T cell homeostasis to be maintained, the transcriptional programs that underlie T cell proliferation and differentiation must be stringently controlled. Indeed, T cell pathologies are associated with widespread alterations in transcription, supporting the possibility of targeting of HMTs, as a strategy for therapeutic immune modulation.[49–51]

## 11.3 GLOBAL HISTONE METHYLATION LANDSCAPE IN EFFECTOR T CELLS

Tightly controlled transcriptional programs guide adaptive T cell responses. Whereas transcription factors are key orchestrators of these programs, histone methylation and its catalyzing HMTs also play a critical role in regulating the abundance and chromatin-localization of these transcription factors in T cells, serving as another essential layer of regulation.[21,22,52–62] Inquiry into histone methylation in T cells has focused on identifying correlations between histone methylation signatures and gene expression. Global chromatin immunoprecipitation (ChIP)-based studies comprehensively characterize genomic regions associated with particular histone modifications. Using antibodies specific to methylation marks such as H3K27me3 and H3K4me3, genomic DNA associated with particular types of histone methyl group modifications may be isolated and subjected to high-throughput sequencing.[63] When combined with gene expression profiling, these studies are able to establish correlations between histone methylation modifications and gene expression. ChIP-based studies of histone methylation in CD4+ and CD8+ T cells have identified a broad association between histone methylation and transcriptional activity within the genomic regions associated with transcription factors and effector molecules. Moreover, these studies highlight the dynamic nature of the histone landscape, in which the presence and degree of methylation remain in flux in naïve, effector, and memory phases.[21–23,52,64,65]

### 11.3.1 HISTONE METHYLATION REGULATES CD4+ T CELL DIFFERENTIATION INTO HELPER SUBSETS

CD4+ T cells can respond to a multitude of pathogen-derived antigens as a result of their capacity to differentiate into diverse subsets. Specialized repertoires of effector cytokines associated with these subsets are able to mediate clearance of varied immune insults. Upon encounter with a particular antigen presented on the surface of an antigen-presenting cell (APC), the surface-bound T cell receptor (TCR) is activated, triggering a complex cascade of transcriptional pathways that results in differentiation into $T_H1$, $T_H2$, $T_H17$, and $T_{reg}$ subsets, depending on various extrinsic and intrinsic stimuli.[24,25]

CD4+ T cells receive three critical signals that precede their differentiation. Stimulation of the TCR triggers activation of nuclear factor of activated T cells (NFATs), along with other transcription factors. Simultaneous costimulation through CD28 induces IL-2 that in turn activates Stat5 through IL-2 receptor signaling. Finally, cytokines present in the surrounding milieu activate particular signaling pathways to induce cell-intrinsic programs specific to distinct helper subsets.[24,25] In parallel to the serial transcriptional changes that occur following T cell stimulation, histone methylation signatures are also reshaped.[22,52,64–67]

CD4+ effector differentiation is critically regulated by transcription factor-mediated changes in gene expression.[24,25] Furthermore, regulatory regions associated with transcription factors critical to differentiation undergo demonstrable alterations in histone methylation signatures upon differentiation into helper subsets

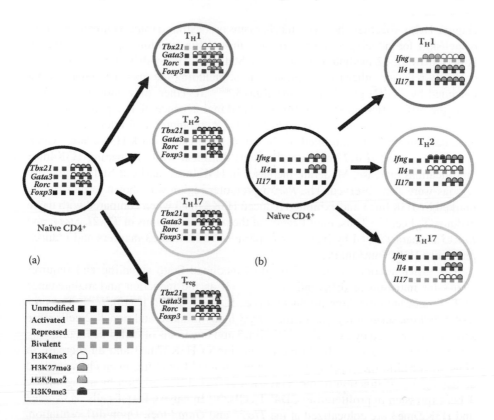

**FIGURE 11.1** Histone methylation signatures for T cell transcription factors. Dynamic histone methylation signatures regulate T effector differentiation. Upon activation, regulatory regions of promoters associated with critical transcription factors undergo alterations that promote the induction and maintenance of distinct effector subsets. (a) Naïve CD4+ T cells possess histone methylation signatures consistent with their quiescent state. Upon stimulation, the $T_H1$-associated *Tbx21* gene regulatory region is marked by increased H3K4me3 and reduced H3K27me3, while non-$T_H2$ regulatory regions gain H3K27me3. The regulatory region of the $T_H2$-associated factor *Gata3* similarly was marked by increased H3K4me3 with a reduction in H3K4me3. In $T_H2$ cells, non-$T_H2$ regulatory regions are repressed by H3K27me3. *Rorc*, which directs $T_H17$ differentiation, undergoes similar changes, although *Foxp3* is not actively silenced by increased H3K27me3. $T_{reg}$ differentiation is accompanied by an increase in H3K4me3 at the *Foxp3* regulatory region and active silencing of $T_H1$ and $T_H17$ differentiation through gains in H3K27me3 at the *Tbx21* and *Rorc* regulatory regions, although *Gata3* exhibits diminished H3K27me3. (b) Naïve CD4+ T cells are marked by H3K27me3 modifications at regulatory regions of the *Ifng* and *Il4* genes, although the *Il17* locus lacks both modifications. Upon activation, nonlineage cytokines are suppressed by gains in H3K27me3 in $T_H1$ cells, $T_H2$, and $T_H17$ lineages. Differentiated $T_H1$ cells were marked by an increase in H3K4me4 at the *Ifng* regulatory region. Unexpectedly this was accompanied by gains in suppressive H3K9me2. Upon $T_H2$ differentiation, H3K4me3 modifications increased at the *Il4* locus. In addition to increases in H3K27me3, the *Ifng* regulatory region also acquired suppressive H3K9me3 in $T_H2$ cells. $T_H17$ differentiation is accompanied by a reduction of H3K27me3 at the *Il17* locus and suppression of *Ifng* and *Il4* by H3K27me3.

(Figure 11.1a). Together, these two mechanisms orchestrate a transcriptional program responsible for inducing and maintaining subset differentiation. $T_H1$ differentiation is driven by IL-12-mediated induction of Stat4 and IFN-$\gamma$-driven Stat1, and also critically relies on other transcription factors, including the master regulator T-bet (encoded by *Tbx21*) plus Eomes and Runx3.[68–72] Together, they cooperate to promote the differentiation, effector function, and maintenance of $T_H1$ populations.[24,73] In naïve and non-$T_H1$ subsets, the *Tbx21* 3′-UTR, promoter, and intergenic regions have been shown to be modified by the repressive histone mark H3K27me3. Upon stimulation, these H3K27me3 modifications diminish while activating H3K4me3 modifications increase. Moderate increases in H3K4me3 marks at Stat4 regulatory regions were also observed. The *Runx3* promoter region was shown to be strongly marked by H3K4me3 and exhibited reduced H3K27me3 when compared with naïve and non-$T_H1$ cells.[64] Thus, modification of the promoter regions of *Tbx21*, *Stat4*, and *Runx3* is characterized by increases in permissive H3K4me3 markers and reduced repressive H3K27me3 markers.

$T_H2$ differentiation is induced by IL-4-mediated Stat6 signaling and requires Gata3 for initiation of differentiation, as well as the induction and maintenance of $T_H2$-associated cytokine production (i.e., IL-4, IL-5, and IL-13).[24,74–77] The promoter and transcribed regions of the *Gata3* locus in naïve and non-$T_H2$ cells were shown to be marked by extensive H3K27me3 and low levels of H3K4me3. Following differentiation, $T_H2$ cells displayed a reduction in H3K27me3 and an increase in H3K4me3 within the regulatory regions of *Gata3*.[64] Gata3 has been characterized as having a dual function in the repression of $T_H1$ differentiation by antagonizing T-bet expression in proliferating CD4$^+$ T cells.[76,77] In naïve CD4$^+$ T cells, H3K4me3 and H3K27me3 are colocalized at the *Tbx21* and *Gata3* loci. Upon differentiation, H3K27me3 is gained at the *Tbx21* locus in non-$T_H1$ cells and at the *Gata3* locus in non-$T_H2$ cells. These changes are consistent with the levels of transcription of both T-bet and Gata3 across $T_H1$ and $T_H2$ subsets.[64] The colocalization of repressive H3K27me3 and H3K4me3 at these loci may present a mechanism by which plasticity between the $T_H1$ and $T_H2$ subsets occurs, however, the functional advantage associated with the reciprocity of these modifications at these loci is not fully understood.[78,79]

In contrast with *Tbx21* and *Gata3* loci, the regulatory region associated with master $T_H17$ regulator, *Rorc* (which encodes ROR$\gamma$t), lacks detectable H3K4me3 in naïve cells, and instead, only H3K27me3 modifications are present. Stat3 initiates $T_H17$ differentiation in response to a diverse set of cytokines, including TGF-$\beta$, IL-6, IL-21, and IL-23, and upon differentiation $T_H17$ cells display a reduction of H3K27me3 and increase of H3K4me3 at *Rorc*.[24,26,64,80] Conversely, $T_H1$ differentiation appears to induce an increase of H3K27me3 at *Rorc*.[64,81] This suggests stringent control of homeostasis between $T_H1$ and $T_H17$ differentiation that is similar to that between $T_H1$ and $T_H2$. Reports of conversion between $T_H1$ and $T_H17$ phenotypes due to cytokine signals have resulted in a revisitation of the concept of "lineage determination," with epigenetic regulation of this plasticity serving as a prevailing hypothesis.[64,67,81–85]

$T_{regs}$ contribute to immune system homeostasis by mediating self-tolerance and modulating inflammation.[86,87] The transcription factor Foxp3 critically mediates the immunosuppressive effects of this subset by directing the expression of IL-10 and TGF-β.[86] Genome-wide analysis identified abundant H3K4me3 at the *Foxp3* locus in T$_{regs}$, and although H3K27me3 was detected within the *Foxp3* regulatory region in T$_H$1 and T$_H$2 cells, it was notably absent in T$_H$17 and naive cells.[64,88–90] Foxp3 expression has been previously reported in T$_H$17 cells, which is consistent with this finding; however, expression has also been noted in T$_H$1 cells.[91,92] Sakaguchi and colleagues have suggested that DNA methylation rather than histone methylation may be more consequential in T$_{reg}$ development and maintenance.[93] Although CpG hypomethylated regions in T$_{regs}$ correlated with H3K4me3-rich regions, H3K4me3 signatures within T$_{reg}$-associated genomic regions were similar between T$_{regs}$ and T$_H$1, T$_H$2, and T$_H$17 subsets. Furthermore, H3K27me3 signatures were similar in naïve and T$_{reg}$ subsets. DNA methylation was found to have a more specific correlation with transcription when compared with histone methylation.[93] Collectively, studies focusing on the regulation of transcription factors within CD4$^+$ T cell subsets by histone methylation have highlighted significant complexity. Further investigation of these mechanisms may facilitate a better understanding of the functional consequences of transcription factor regulation by histone methylation in T cells.

## 11.3.2 HISTONE METHYLATION AND CD4$^+$ T CELL EFFECTOR CYTOKINE PRODUCTION

Naïve T cells are quiescent, having not yet responded to an antigen. These undifferentiated cells display a characteristic histone methylation pattern that reflects their uneducated state and consequential inability to produce effector cytokines. Upon differentiation, promoter regions associated with particular cytokines undergo extensive remodeling in order to repress or promote their expression (Figure 11.1b). Studies have shown naïve CD4$^+$ T cells display moderate H3K27me3 modifications at the 3′UTR of *Ifng* and at the *Il4* gene body, whereas H3K9me3 and H3K4me3 remain undetectable at the *Ifng* or *Il4* loci.[22,52,64] Interestingly, the *Il17* promoter displayed neither H3K27me3 nor H3K4me3. Unlike its cytokine counterparts in T$_H$1 and T$_H$2 cells, this characteristic T$_H$17 cytokine is not actively repressed.[64]

Differentiation triggers remodeling of histone methylation states throughout regulatory regions of subset-specific cytokine loci. At the *Ifng* promoter in T$_H$1 cells, H3K27me3 levels were reduced, while H3K4me3 modifications increased.[52,64] In T$_H$1 cells, T$_H$2 and T$_H$17 cytokines became actively repressed by increased H3K27me3 at the loci of *Il4*, *Il21*, and *Il17*, while H3K4me3 was not detectable.[64] Interestingly, repressive H3K9me2 increased at the *Ifng* locus during T$_H$1 and T$_H$2 differentiation, but was only maintained in T$_H$1 cells. Although repressive H3K9me2 was lost in differentiated T$_H$2 cells, repressive H3K27me3 was gained at the *Ifng* locus in a Gata-3- and Stat6-dependent fashion.[52,94] At the *Il4* locus in T$_H$2 cells, differentiation resulted in a reduction of H3K27me3 and increase of H3K4me3 upon differentiation, with H3K27me3 marking the *Ifng* and *Il17* loci.[64,66] T$_H$17 differentiation was accompanied by the acquisition of H3K4me3 at the *Il17a* and *Il21* promoters, while H3K27me3 marked *Il4* and *Ifng* loci.[64,67]

H3K4me3 was further associated with the *IL21* regulatory region upon $T_H17$ differentiation, while low levels of H3K27me3 were present at the *Il4* locus in $T_H17$ cells. In $T_{regs}$, low levels of H3K27me3 were observed at the *Il4* locus, whereas the *Il17* and *Il21* loci were abundantly marked by H3K27me3. The 3′ untranslated region (UTR) of the *Ifng* gene was marked by H3K27me3 in induced but not natural Tregs.[64] These studies collectively suggest that cytokine effectors of CD4+ T cell subsets are often actively repressed within incorrect lineages, whereas signature cytokine expression is promoted through acquisition of H3K4me3 at gene loci in appropriate lineages.

### 11.3.3 Histone Methylation Regulates CD8+ T Cell Differentiation and Expansion

CD8+ effector T cells, or CTLs, characteristically produce both cytokines and directly cytotoxic molecules. Upon antigen-dependent activation by APCs, naïve CD8+ T cells, like CD4+ T cells, receive TCR and costimulatory signals, along with extrinsic signals from cytokines, which activate transcription programs important for regulating clonal expansion and differentiation. T-bet and Eomes have been shown to function as master regulators for promoting CTL differentiation and function.[95–97] During the immune response, T-bet and Eomes have cooperative and overlapping roles in promoting the differentiation of naïve CD8+ T cells into CTL effector cells capable of secreting cytokines and cytotoxic molecules such as perforin and granzyme B.[95,97] The balance between expression of T-bet and Eomes contributes to the fate of cytotoxic effectors, with high T-bet expression promoting terminal differentiation and high Eomes expression promoting exhaustion.[9,98] In addition to these master regulators, several other transcription factors regulate the expansion, survival, and differentiation of CTLs. For example, IFN regulatory factors 4 (Irf4) and 8 (Irf8) play important roles in late-stage differentiation and expansion.[99–101] Irf4 simultaneously promotes differentiation by enhancing the expression and function of T-bet.[102] Furthermore, Blimp-1 (encoded by *Prdm1*) and Bcl11b serve as enhancers of cytotoxic function.[103–107]

Global ChIP-based studies of H3K4me3 and H3K27me3 modifications in CTLs have identified dynamic alterations within regulatory regions of critical transcription factors (Figure 11.2a) in both polyclonal and antigen-specific CTL populations.[21,23,108] Each identified a correlation of repressive H3K27me3 with transcriptional silencing and activating H3K4me3 with transcriptional activation, as seen in previous studies of CD4+ T cells.[21,23,108] Promoter regions of genes encoding key transcription factors T-bet, Eomes, Irf8, and Irf4 were shown to be bivalently marked by approximately equal degrees of H3K4me3 and H3K27me3 in naïve CD8+ T cells. Upon differentiation, these marks resolved toward a permissive chromatin state through the loss of H3K27me3 at *Tbx21*, *Irf8*, and *Irf4* loci, which correlated with an increase in transcription. The methylation state of the *Eomes* locus was minimally altered during the early stages of activation examined in the study, which is consistent with its late expression in effector cells.[23,108,109] The *Bcl11b* promoter was marked by high H3K4me3 and low H3K27me3 in naïve T cells, and displayed an increase in H3K27me3 after activation.[23]

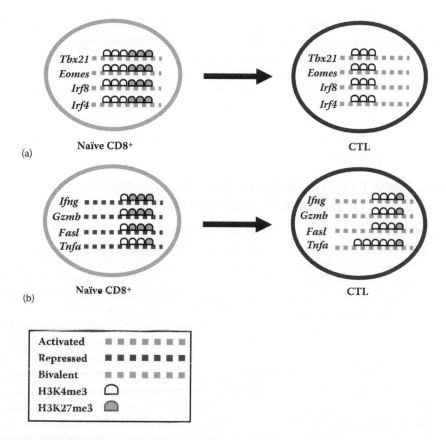

**FIGURE 11.2** Histone methylation signatures for T cell effector cytokines. (a) Critical transcription factor regions associated with CD8⁺ T cell differentiation—*Tbx21, Eomes, Irf8,* and *Ir4*—are bivalently marked by H3K4me3 and H3K4me3. Upon activation, they transition to permissive states through loss of H3K27me3. (b) Critical effector molecules and cytokines are actively repressed in naïve CD8⁺ T cells by H3K27me3 modifications at regulatory regions, with the exception of *Tnfa*, which displays less H3K27me3 and more H3K4me3. Upon differentiation repressive H3K27me3 is diminished and activating H3K4me3 is acquired.

## 11.3.4 HISTONE METHYLATION AND CD8⁺ T CELL EFFECTOR FUNCTION

CTLs mediate direct killing of virally infected and malignant cells through the production of effector cytokines (IFN-γ and TNF-α) and directly cytotoxic molecules (granzymes, perforins, and Fas ligand).[5,6] In naïve CD8 T cells, the *Ifng, FasL,* and *Gzmb* loci are marked by abundant H3K27me3 but low levels of H3K4me3 (Figure 11.2b).[23] Conversely, the naïve *Tnfa* locus displays low H3K27me3 and high H3K4me3.[23,65,69,110] This modification signature is consistent with the observation that TNF-α can be produced more quickly upon activation when compared with *Ifng* and Gzmb.[65,111] Following activation, promoter regions associated with CTL cytokines undergo extensive remodeling in order to repress or promote their expression.

Upon differentiation, the *Tnfa* locus gains additional H3K4me3 in the gene body, while the *Ifng, FasL,* and *Gzmb* loci transition to a permissive chromatin structure through acquisition of H3K4me3 and loss of H3K27me3 throughout the TSS and gene body.[23] Blimp-1 (*Prdm1*) and IL-2Rα (*Cd25*) loci similarly transition from a repressed histone methylation signature in naïve cells to a permissive signature in effector cells through loss of H3K27me3 and gains in H3K4me3.[23,65,110] Genome-wide studies of histone methylation patterns in CD8⁺ effector T cells have established a solid correlation between particular histone signatures and levels of gene transcription. More so, they have introduced the concept of "bivalent" chromatin states, which, when further explored, may critically contribute to the understanding of T cell plasticity and memory formation.

## 11.4    IMPACT OF HISTONE METHYLATION IN MEMORY T CELL DEVELOPMENT AND MAINTENANCE

Memory cells are a critical component of long-term protective immunity. Intrinsic and extrinsic factors that contribute to fate determination in memory development and maintenance have been intensively studied, however, they remain only partially understood.[5,112] Recent studies have suggested that permissive chromatin states at genes critical for effector function are thought to be associated with the ability of memory cells to rapidly respond to secondary antigen encounters.[4,21,25,54,59,110,113,114]

Following resolution of an infection or antitumor response, emerging memory cells express distinct transcriptional profiles.[4] Thus far, the majority of memory studies have been performed in CD8⁺ T cells, however, some mechanisms have been identified in CD4⁺ T cells. For example, Blimp-1 and Bcl-6 have been reported to influence one another as well as effector-versus-memory subset determination in CD4⁺ T cells.[104–106] However, transcriptional control and, by consequence, the role of histone methylation in CD4⁺ T cell memory development and maintenance are poorly understood. Particular challenges that are associated with studying memory formation include the heterogeneity of models used to study antigen-specific response.[112] For example, the composition of memory subsets induced in antigen-specific models may vary dependent on the use of viral versus bacterial pathogens.[115,116] Nevertheless, genome-wide histone methylation patterns in CD8⁺ memory T cells have been explored in both humans and mice, and these studies suggest histone methylation may serve as a precise mechanism for rapid recall response.[21,108,110]

Upon memory formation, promoter regions of the transcription factors critically involved in CD8⁺ memory formation undergo changes in histone methylation patterns. At least three types of dynamic changes in expression of H3K4me3 and H3K27me3 have been reported (Figure 11.3a). The first type is characterized by acquiring permissive H3K4me3 while retaining repressive H3K27me3 modifications. The master transcription factor *Tbx21* has been shown to acquire additional H3K4me3 modifications, when compared with naïve T cells, which displayed abundant H3K27me3 at the gene body and promoter.[108] *Id2*, which plays a role in effector CD8⁺ T cell survival, also displayed an increase in H3K4me3 in $T_{CM}$ cells, despite iats low transcript expression.[21,110,117] The second type is featured by acquisition of repressive H3K27me3 without losing permissive H3K4me3. For example, the transcription factor forkhead-box O 1 (Foxo1) is highly expressed in memory-precursor T cells, and promotes $T_{CM}$ cell response through a mechanism of regulating Transcription factor 7 (Tcf7).[118–121]

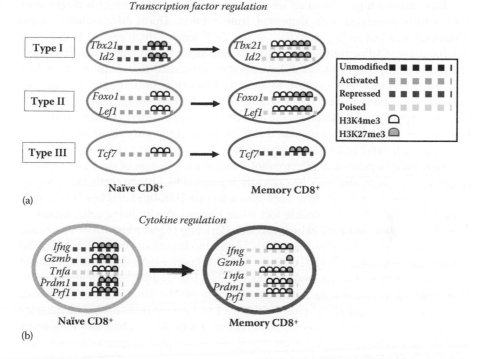

**FIGURE 11.3** Histone methylation signatures for memory CD8+ T cells. Regulatory regions associated with CD8+ transcription factors and effector molecules have distinct histone landscapes in memory populations. This property is thought to contribute to the capacity of memory cells to rapidly elaborate effector function upon antigen reencounter. (a) The regulatory regions of *Tbx21* and *Foxo1* acquire H3K4me3 modification in memory populations when compared with naïve cells. In naïve CD8+ T cells, the *Foxo1* promoter is marked by abundant H3K4me3 but supplemented with H3K27me3 marks in memory populations. Naïve cells exhibited high H3K4me3 levels in the promoter and transcribed region of Tcf7 and *Lef1*. Upon memory formation H3K4me3 is lost at the Tcf7 locus, which acquires H3K27me3 modifications. *Lef1* maintains H3K4me3 in naïve cells and also acquires H3K27me3 in memory cells. (b) Upon memory formation, CD8+ T cells gain H3K4me3 and lose H3K27me3 modifications at *Ifng*, *Prdm1*, and *Prf1* loci when compared with naïve cells. The *Gzmb* regulatory exhibits reduced repressive H3K27me3 modifications but does not acquire H3K4me3 upon memory formation. When compared with naïve cells, H3K4me3 is enriched at the *Tnfa* locus in memory CTLs.

Similar to that for Tbx21, in naïve CD8+ T cells, the *Foxo1* promoter is marked by abundant H3K4me3, while upon formation of $T_{CM}$ and $T_{EM}$, H3K27me3 marks are acquired.[108] The promoter region of *Lef1* was modified by H3K4me3 in naïve CD8+ T cells but supplemented with H3K27me3 in $T_{CM}$ and $T_{EM}$.[108] The final type is characterized by the loss of permissive markers combined with the acquisition of a repressive landscape. For example, Tcf7 was highly expressed in naïve and $T_{CM}$ cells, and poorly expressed in short-lived effector cells (SLECs).[122,123] Naïve CD8+

T cells exhibited high H3K4me3 levels in the promoter and transcribed region of *Tcf7*, which correlated with increased transcription. Upon differentiation, high H3K4me3 was lost in $T_{CM}$ and $T_{EM}$, at the *Tcf7* locus, which gained H3K27me3 modifications. Collectively, these observations suggest that the dynamic change of histone methylation patterns at key memory-associated transcription factor loci may play important roles in naïve to memory transition. Intriguingly, the baseline histone landscape in naïve CD8+ T cells differs in human and murine studies, with the former providing evidence for the existence of bivalent states in naïve cells.[21,108] How differential HMTs are regulated and integrated to modify these histone landscapes in both naïve and antigen-driven CD8+ T cells has yet to be determined.

The rapid upregulation of effector molecules is a key feature of the CD8+ memory response, a capacity that may be critically regulated by histone methylation signatures. CD8+ memory T cells have been shown to gain H3K4me3 and lose H3K27me3 modifications at effector molecule loci when compared with naïve cells, including *Ifng, Fasl, Prdm1*, and *Prf1* (which encodes Perforin) (Figure 11.3b).[21,23] The *Gzmb* locus was shown to lose repressive H3K27me3 modifications, but interestingly did not acquire H3K4me3.[23] This is consistent with lower expression of Granzyme B in memory versus effector populations.[124–126] When compared with naïve cells, H3K4me3 was enriched at the *Tnfa* locus in memory CTLs. Blimp-1, which enhances cytotoxic function, and IL-2Rα lost H3K27me3 and gained H3K4me3 upon memory formation.[23] Collectively, histone methylation, particularly, abundant H3K4me3, may be the mechanism by which an "open" chromatin state is maintained for rapid elaboration of recall response.

Numerous studies demonstrate evidence for open chromatin states mediating recall response in resting CD8+ memory populations.[11,21,54,59,110,113,114] Loci for critical CTL effector cytokines, including *Tnfa, Ifng, Gzmb*, and *Prf1*, display similar histone methylation signatures that are characterized by loss of the repressive modification H3K27me3 with concomitant increases in H3K4me3, as well as another transcriptionally activating modification: acetylation modification H3K9ac.[54,57,65,69,81] The "poised" state of these cytokine loci may provide a mechanism for the observation that when compared with naïve cells, resting memory cells are able to rapidly induce production of these effector cytokines.[4,54,57,65,69,81,114] A similar histone signature was also identified at cytokine loci in $T_H2$ memory populations, facilitating the rapid upregulation of gene expression.[59,113] The highly permissive nature of these histone signatures is consistent with the ability of memory cells to rapidly upregulate cytokine production, however, the negative regulatory mechanisms required to balance this permissive state remain uncharacterized.

## 11.5  HISTONE METHYLTRANSFERASE (HMT)-MEDIATED REGULATION OF T CELL EFFECTOR AND MEMORY FORMATION

HMTs catalyze the transfer of one, two, or three methyl groups to lysine or arginine residues of histone proteins H3 and H4. HMTs that catalyze modifications of lysine residues comprise the majority and better characterized category.[127]

Lysine-modifying HMTs can be classified into two major groups: SET-domain containing and non-SET containing, or DOT1-like, proteins (Table 11.1).[127–129]

Many of these enzymes function as a part of a large multiunit complex and each utilize the cofactor S-adenosylmethionine (SAM) as a methyl group donor.[130] TrxG proteins and PcG proteins comprise epigenetic protein complexes associated with activation and repression of transcription, respectively. MLL family HMTs associate with TrxG family proteins and catalyze the mono-, di-, and tri-methylation of H3K4 in complex with TrxG members menin, WDR5, ASHL2, and RBBP5.[131,132] Ezh2 is a member of the PcG family and catalyzes the trimethylation of H3K27, a mark associated with transcriptional repression.[133–135] Through interactions with Eed and Suz12, it forms the PRC2 complex, which has a demonstrated role in maintenance of transcriptional silencing patterns.[69,136,137] SUV39H1 catalyzes suppressive trimethylation at H3K9, creating a binding site that recruits the adaptor molecule HP1α. In complex with SUV39H1, HP1α facilitates gene silencing and propagation of pericentric heterochromatic subdomains.[138] The HMT G9a catalyzes H3K9me3 with higher activity than SUV39H1 and has the additional capacity to trimethylate H3K27me3. Furthermore, G9a targets transcriptionally active euchromatin, whereas SUV39H1 preferentially methylates repressive pericentric heterochromatin.[139,140] Finally, the HMT SMYD3 catalyzes H3K4 trimethylation through interactions with RNA PolII.[141] Several studies have explored the roles of these particular SET-containing HMTs in the differentiation and function of effector and memory

## TABLE 11.1
## Target Residues of Lysine Histone Methyltransferases

| | Histone Residue | Lysine Histone Methyltransferase | |
| | | me2 | me3 |
| --- | --- | --- | --- |
| | H3K4 | ASH1L, MLL1, MLL2, MLL3, MLL4, SET7/9, SETD1A, SETD1B, SMYD3 | ASH1L, MLL1, MLL2, MLL3, MLL4, PRMD9, SET7/9, SETD1A, SETD1B, SMYD3 |
| | H3K27 | Ezh2, EZH1 | Ezh2, Ezh2 |
| SET domain-containing | H3K9 | EHMT1, G9a, PRDM2, SETDB1, SUV39H1, SUV39H2 | PRDM2, SETDB1, SUV39H1, SUV39H2 |
| | H3K36 | NSD1, NSD2, NSD3, SETD2, SMYD2 | SETD2 |
| | H4K20 | SUV420H1, SUV420H2 | SUV420H1, SUV420H3 |
| DOT1-like | H3K79 | DOT1L | DOT1L |

*Source:* Adapted from Greer, E. L., and Shi, Y., *Nat Rev Genet* 13(5), 343–357, 2012.

*Notes:* Lysine histone methyltransferases (HMTs) catalyze the mono-, di- (me2), and tri-methylation (me3) of lysine resides at the N-terminal tails of histones. The site and degree of methylation is responsible for determining whether each set of modifications will mediate the activation (H3K4me2, H3K4me3, H3K36me2, H3K36me3, H3K79me3, H4K20me2, H4K20me3) or repression (H3K27me2, H3K27me3, H3K79me2, H3K79me3) of associated genes.

T cells with exciting findings, however, the complex composition of HMT multisubunit complexes presents a challenge in characterizing the function of these enzymes.

### 11.5.1 MLL

Four enzymes comprise the MLL family of HMTs, although only MLL1 has begun to be characterized in T cells (Figure 11.4). To date, limited work exploring the role of MLL1 in CD4+ T cells has produced conflicting findings in the context of $T_H1$ and $T_H2$ development, effector function, and memory formation. Using a murine model haploinsufficient for MLL1, Yamashita and colleagues showed that $T_H1$ and $T_H2$ effector subsets differentiate normally in vitro under skewing conditions. Although formation of $T_H1$ memory cells was not impacted, $T_H2$ memory cells lost nearly all capacity to produce IL-4.[59] Conversely, work by Schaller and colleagues using an identical model demonstrated impaired $T_H1$ proliferation, differentiation, and memory formation in vitro under skewing conditions.[142] Moreover, Yamashita et al. reported high expression of MLL1 in naïve CD4+ T cells, followed by a gradual decline in effector populations cultured in the presence of IL-12.[59] In sharp contrast, Schaller et al. described an induction of MLL1 expression by Il-12 signaling, mediated by Stat4.[142] How to reconcile these contradictory findings remains unclear.

The reduction in $T_H2$ memory cells described by Yamashita et al. was accompanied by a decrease in expression of Gata3 and reduced H3K4me3 modifications at $T_H2$ cytokine regulatory regions. In a further characterization of the mechanism behind this phenotype, a complex composed of C-Myb/GATA-3, Menin, and MLL1 was found to be recruited to the Gata3 promoter, where it directly associated with Gata3 and promoted the formation of $T_H2$ memory cells.[59,143] Together, these studies identify a possible role for MLL1 in the maintenance of $T_H2$ memory. However, the function of this enzyme in $T_H1$ subsets remains controversial. Furthermore, the function of MLLs in CD8+ T cells and the function of MLL2, MLL3, and MLL4 in CD4+ T cells remain uncharacterized.

### 11.5.2 Ezh2

Several studies have suggested that Ezh2 plays a crucial role in $T_H1$ and $T_H2$ differentiation. Ezh2 catalyzes H3K27me3 and has been shown to be required to promote the expression of Tbx21 and Stat4, with loss of Ezh2 resulting in impaired differentiation into $T_H1$-subtype IFN-$\gamma$-producing cells.[144,145] Ezh2 also plays an active role in suppressing $T_H2$ cytokine production in $T_H1$ cells through deposition of H3K27me2 at the *IL4* and *Il13* locus of $T_H1$ but not $T_H2$ cells.[56] Ezh2 further modulates *Tbx21* and *Gata3* regulatory regions in differentiating $T_H1$ and $T_H2$ cells through H3K27me3 modifications that maintain ideal expression levels of these transcription factors (Figure 11.4). Deficiency in Ezh2 has been shown to result in spontaneous IFN-$\gamma$ and $T_H2$ cytokine production in nonpolarizing cultures, dependent on the presence of Eomesodermin, which suggests a role for Ezh2 in constraining $T_H1$ and $T_H2$ phenotypes.[145]

Despite these studies, the regulation of T cell survival and CD4+ T helper cell differentiation by Ezh2 is context-dependent.[145–147] In CD4+ T cells, inactivation

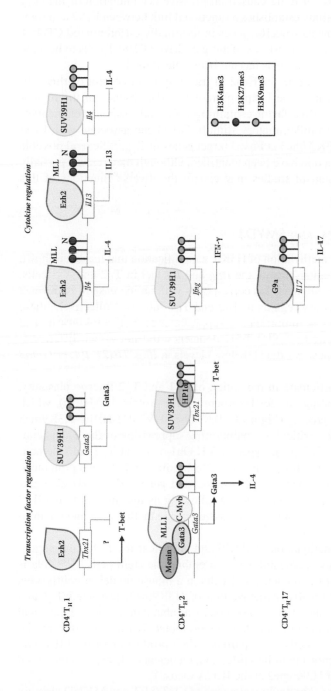

**FIGURE 11.4** Regulation of CD4+ T cell differentiation by histone methyltransferases. Histone methyltransferases are critical regulators of CD4+ T cell-associated transcription factors and effector cytokines. (a) The role of Ezh2 in regulating $Tbx21$ remains controversial. Interaction of this HMT with the $Tbx21$ regulatory region may promote or suppress expression of T-bet. SUV39H1 mediates the suppression $T_H1$ transcription factors in $T_H2$ cells and vice versa through H3K9me3 modifications. HP1α forms a complex with SUV39H1 to further suppress the $T_H1$ phenotype and stabilize the $T_H2$ phenotype in $T_H2$ cells. MLL1 has been shown to interact with key $T_H2$ transcription factors, Gata3, and other components to form a multisubunit complex capable of inducing IL-4 production through interaction with the $Gata3$ promoter. (b) Ezh2 actively suppresses $T_H2$ cytokine production in $T_H1$ cells through deposition of H3K27me2 at the $IL4$ and $Il13$ locus of $T_H1$, but not $T_H2$ cells. SUV39H1 mediates the suppression $T_H1$-associated $Ifng$ in $T_H2$ cells and $T_H2$-associated $Ifl4$ in $T_H1$ through H3K9me3 modifications. Finally, G9a is important for restraining IL-17 production through deposition of H3K9me2 modifications.

of Ezh2 was found to specifically enhance $T_H1$ and $T_H2$ cell differentiation and plasticity, while in vivo loss of Ezh2 caused progressive accumulation of memory phenotype $T_H2$ cells. This study establishes a functional link between Ezh2 and transcriptional regulation of lineage-specific genes in terminally differentiated CD4+ T cells. Whether Ezh2 affected the plasticity of antigen-driven CD8+ T cells to become effector versus memory precursor cells has not been determined.

Studies in $T_{regs}$ also posit a role for Ezh2 in transcriptional control of this subset. For instance, Ezh2 silences the Foxp3 promoter in the absence of the regulatory protein Klf10 (Kruppel-like factor-containing Polycomb response element).[148] Furthermore, in response to inflammatory stimuli, Foxp3 can associate with Ezh2 and direct deposition of H3K27me3 at Foxp3 target genes in $T_{regs}$.[149] Several possible roles for Ezh2 in T cell function have been identified, although they remain partially understood. Further mechanism studies may clarify the diverse functions of this enzyme.

### 11.5.3   SUV39H1, G9A, AND SMYD3

SUV39H1 catalyzes the trimethylation of H3K9, a modification that has been shown to increase at regulatory regions within the *Ifng* and *Il4* loci in $T_H2$ and $T_H1$ cells, respectively, upon differentiation.[52] This corresponds with an increase in H3K9me3 at *Tbx21* and *Gata3* regulatory regions in $T_H2$ and $T_H1$ subsets. Collectively, these findings suggest SUV39H1 is important for suppression of $T_H1$ cytokines in $T_H2$ cells and vice versa. Despite this, SUV39H1 deficiency did not affect differentiation of $T_H1$ and $T_H2$ subtypes, nor the H3K9me3 levels at *Ifng*, *Tbx21*, *Il4*, or *Gata3* regulatory regions.[150]

SUV39H1 may also participate in regulation of $T_H1$ and $T_H2$ lineage plasticity. H3K9me3 serves as a binding site for heterochromatin protein 1α (HP1α), which promotes transcriptional silencing (Figure 11.4).[138] The SUV39H1-HP1α–H3K9me3 interaction was shown to be critical for maintaining suppression of the $T_H1$ phenotype while stabilizing the $T_H2$ phenotype. SUVH39h1-deficient $T_H2$ cells demonstrated a loss of H3K9me3 modification and HP1α binding at $T_H1$ gene promoters, resulting in a reduction in silencing of $T_H1$-associated genes. SUV39H1 also participates in silencing IL-2 cytokine expression following its recruitment to the IL-2 promoter by Smad2 and 3, resulting in suppressive histone methylation and the inhibition of TCR-induced IL-2 expression.[151]

G9a also catalyzes trimethylation of H3K9 and is critical for early embryogenesis and suppression of developmental genes in embryonic stem cells.[139,140] G9a is required for development of pathogenic $T_H1$ cells in a mouse model of colitis. G9a demethylase activity, specifically demethylation of H3K9me2, restricted $T_H1$ and $T_{reg}$ differentiation in vitro and in vivo. H3K9me2 was abundant in naïve T cells and lost after activation. G9a inhibition or deficiency promoted $T_H1$ and $T_{reg}$ differentiation.[152] G9a-deficient CD4+ T cells are impaired in their production of IL-4 IL-5 and Il-13 and do not protect against helminth infection. Deficient cells express increased IL17A as a result of loss of H3K9me2 at the IL17A locus.[153]

Depletion of the H3K4me3 methyltransferase SMYD3 (SET and MYND Domain 3) causes a reduction of H3K4me3 at the promoter of Foxp3 and affected iTreg formation with accompanying dysregulation of IL-17. In an iTreg-dependent RSV (respiratory syncytial virus) in vivo model, SMYD3-deficient mice exhibited exacerbated disease and inflammatory responses within the lung.[154] Preliminary studies suggest SUV39H1, G9a, and SMYD3 HMTs may impact various stages of T cell function. Further investigation of the mechanisms governing their role in transcriptional programs may improve our understanding of T cell heterogeneity.

## 11.6 THERAPEUTIC TARGETING OF HMT TO MODULATE T CELL IMMUNITY

Genome-wide profiles of histone methylation signatures and mechanistic studies of individual histone methyltransferases in T cells significantly contribute to advancing knowledge of basic T cell immunity. Studies outlined in this chapter introduce histone methylation as a compounding layer of regulation that functions in concert with transcription factor-mediated programs. Gaining a better understanding of this novel layer of regulation will not only contribute to our understanding of how T cell function may be dysregulated in disease, but furthermore identify new pathways and molecules to target therapeutically.

Dysregulation of T cell function is responsible for numerous pathologies, including autoimmune disease, graft rejection, and GVHD.[5,6,14,43–49] Several studies have utilized genetic approaches to characterize the roles that HMTs play within T cells in the context of particular pathologies. In vivo models of allergic asthma studies of MLL1 and SUV39H1 function in $T_H2$ memory responses demonstrated an impairment of $T_H2$ memory formation and maintenance upon genetic disruption of these enzymes.[150] Defective MLL1 or SUV39H1 expression was shown to impair memory formation and reduce inflammation within the lung in a murine airway inflammation model.[59] MLL1 was further found to promote the production of $T_H2$ cytokines and mediate the production of IgE and IgG2. Another study characterized MLL1 as playing an important role in $T_H1$-driven responses. MLL1 haploinsufficient (MLL1+/−) mice were infected with *Mycobacterium* and found to develop larger granulomas, when compared with wild-type mice. $T_H1$ effector cell numbers were reduced in the lungs of MLL1-defective mice, although no differences were observed for memory cells. The impaired function of $T_H1$ effectors was suggested to occur as a result of their reduced IFN-γ production.[142] However, whether impaired expansion of effector T cells may contribute to this reduction of IFN-γ-producing cells is not reported.

G9a was found to play a role in mediating $T_H2$ immunity against helminth infection. In a murine model of *Trichuris muris* infection, G9a was shown to be important in promoting $T_H2$ differentiation and suppressing $T_H1$ and $T_H17$. Furthermore, G9a was necessary for mounting an effective response to infection, and deficient mice had impaired capacity to produce signature TH2 cytokines IL-4, IL-5, and IL-13.[153]

Aberrant T cell responses are key drivers of GVHD, a complication arising from hematopoietic stem cell transplantation in which donor T cells mount an immune response against healthy recipient tissues based on histocompatibility disparities.[43,155] A recent study demonstrated that pharmacological inhibition of methyltransferases catalyzing H3K27me3, H3K4me3, and H4K20me3 can selectively ablate alloantigen-specific T cells through induction of apoptosis. Inhibitor treatment improved survival and pathology, and this was suggested to occur through derepression of the proapoptotic molecule *Bim*, which is normally repressed by Ezh2.[156] These studies illustrate the potential for histone methyltransferases as effective therapeutic targets for a diverse array of pathologies, including inflammatory disorders, graft rejection, and chronic infection.

## 11.7 CONCLUSION

Global and mechanistic studies have established a critical role for histone methylation and HMTs in T cell differentiation, effector function, and memory formation. Studies in both CD4+ and CD8+ T cells have demonstrated a broad correlation of histone methylation signatures with gene expression and have uncovered the dynamic nature of these modifications. An important contribution of these studies is the introduction of the concept of an open chromatin state within memory cells that may be responsible for the capacity of these cells to rapidly upregulate effector molecules in response to a reencounter of the antigen. This demonstrated but poorly defined role for histone methylation in the maintenance of plasticity should be more thoroughly explored. A greater understanding of the role of these modifications in maintaining memory and plasticity has the potential to improve antitumor immunotherapies are currently hindered by the poor persistence of therapeutic T cells. Finally, it will be intriguing to determine how individual HMTs are integrated together to regulate effector and memory T cell responses.

## REFERENCES

1. Fearon, D. T., Manders, P., and Wagner, S. D., Arrested differentiation, the self-renewing memory lymphocyte, and vaccination, *Science* 293(5528), 248–250, 2001.
2. Lanzavecchia, A., and Sallusto, F., Dynamics of T lymphocyte responses: Intermediates, effectors, and memory cells, *Science* 290(5489), 92–97, 2000.
3. Lanzavecchia, A., and Sallusto, F., Progressive differentiation and selection of the fittest in the immune response, *Nat Rev Immunol* 2(12), 982–987, 2002.
4. Kaech, S. M., Hemby, S., Kersh, E., and Ahmed, R., Molecular and functional profiling of memory CD8 T cell differentiation, *Cell* 111(6), 837–851, 2002.
5. Kaech, S. M., and Cui, W., Transcriptional control of effector and memory CD8+ T cell differentiation, *Nat Rev Immunol* 12(11), 749–761, 2012.
6. Zhang, N., and Bevan, M. J., CD8(+) T cells: Foot soldiers of the immune system, *Immunity* 35(2), 161–168, 2011.
7. Youngblood, B., Hale, J. S., and Ahmed, R., Memory CD8 T cell transcriptional plasticity, *F1000Prime Rep* 7, 38, 2015.
8. Wherry, E. J., Teichgraber, V., Becker, T. C., Masopust, D., Kaech, S. M., Antia, R., von Andrian, U. H., and Ahmed, R., Lineage relationship and protective immunity of memory CD8 T cell subsets, *Nat Immunol* 4(3), 225–234, 2003.
9. Joshi, N. S., Cui, W., Chandele, A., Lee, H. K., Urso, D. R., Hagman, J., Gapin, L., and Kaech, S. M., Inflammation directs memory precursor and short-lived effector CD8(+) T cell fates via the graded expression of T-bet transcription factor, *Immunity* 27(2), 281–295, 2007.
10. Kaech, S. M., and Wherry, E. J., Heterogeneity and cell-fate decisions in effector and memory CD8+ T cell differentiation during viral infection, *Immunity* 27(3), 393–405, 2007.

11. Kaech, S. M., Wherry, E. J., and Ahmed, R., Effector and memory T-cell differentiation: Implications for vaccine development, *Nat Rev Immunol* 2(4), 251–262, 2002.

12. Wherry, E. J., and Ahmed, R., Memory CD8 T-cell differentiation during viral infection, *J Virol* 78(11), 5535–5545, 2004.

13. Ahmed, R., Bevan, M. J., Reiner, S. L., and Fearon, D. T., The precursors of memory: Models and controversies, *Nat Rev Immunol* 9(9), 662–668, 2009.

14. Fearon, D. T., The expansion and maintenance of antigen-selected CD8(+) T cell clones, *Adv Immunol* 96, 103–139, 2007.

15. Heffner, M., and Fearon, D. T., Loss of T cell receptor-induced Bmi-1 in the KLRG1(+) senescent CD8(+) T lymphocyte, *Proc Natl Acad Sci USA* 104(33), 13414–13419, 2007.

16. Wu, J. I., Lessard, J., and Crabtree, G. R., Understanding the words of chromatin regulation, *Cell* 136(2), 200–206, 2009.

17. Kouzarides, T., Chromatin modifications and their function, *Cell* 128(4), 693–705, 2007.

18. Kelly, T. K., De Carvalho, D. D., and Jones, P. A., Epigenetic modifications as therapeutic targets, *Nat Biotechnol* 28(10), 1069–1078.

19. Jenuwein, T., and Allis, C. D., Translating the histone code, *Science* 293(5532), 1074–1080, 2001.

20. Strahl, B. D., and Allis, C. D., The language of covalent histone modifications, *Nature* 403(6765), 41–45, 2000.

21. Araki, Y., Wang, Z., Zang, C., Wood, W. H., 3rd, Schones, D., Cui, K., Roh, T. Y., Lhotsky, B., Wersto, R. P., Peng, W., Becker, K. G., Zhao, K., and Weng, N. P., Genome-wide analysis of histone methylation reveals chromatin state-based regulation of gene transcription and function of memory CD8+ T cells, *Immunity* 30(6), 912–925, 2009.

22. Roh, T. Y., Cuddapah, S., Cui, K., and Zhao, K., The genomic landscape of histone modifications in human T cells, *Proc Natl Acad Sci USA* 103(43), 15782–15787, 2006.

23. Russ, B. E., Olshanksy, M., Smallwood, H. S., Li, J., Denton, A. E., Prier, J. E., Stock, A. T., Croom, H. A., Cullen, J. G., Nguyen, M. L., Rowe, S., Olson, M. R., Finkelstein, D. B., Kelso, A., Thomas, P. G., Speed, T. P., Rao, S., and Turner, S. J., Distinct epigenetic signatures delineate transcriptional programs during virus-specific CD8(+) T cell differentiation, *Immunity* 41(5), 853–865, 2014.

24. Zhu, J., Yamane, H., and Paul, W. E., Differentiation of effector CD4 T cell populations (*), *Annu Rev Immunol* 28, 445–489, 2010.

25. Zhou, L., Chong, M. M., and Littman, D. R., Plasticity of CD4+ T cell lineage differentiation, *Immunity* 30(5), 646–655, 2009.

26. Stockinger, B., and Veldhoen, M., Differentiation and function of Th17 T cells, *Curr Opin Immunol* 19(3), 281–286, 2007.

27. Kallies, A., Distinct regulation of effector and memory T-cell differentiation, *Immunol Cell Biol* 86(4), 325–332, 2008.

28. Liston, A., and Gray, D. H., Homeostatic control of regulatory T cell diversity, *Nat Rev Immunol* 14(3), 154–165, 2014.

29. Pepper, M., and Jenkins, M. K., Origins of CD4(+) effector and central memory T cells, *Nat Immunol* 12(6), 467–471, 2011.

30. Seder, R. A., and Ahmed, R., Similarities and differences in CD4+ and CD8+ effector and memory T cell generation, *Nat Immunol* 4(9), 835–842, 2003.

31. Taylor, J. J., and Jenkins, M. K., CD4+ memory T cell survival, *Curr Opin Immunol* 23(3), 319–323, 2011.

32. Sallusto, F., Lenig, D., Forster, R., Lipp, M., and Lanzavecchia, A., Two subsets of memory T lymphocytes with distinct homing potentials and effector functions, *Nature* 401(6754), 708–712, 1999.

33. Wakim, L. M., Waithman, J., van Rooijen, N., Heath, W. R., and Carbone, F. R., Dendritic cell-induced memory T cell activation in nonlymphoid tissues, *Science* 319(5860), 198–202, 2008.

34. Zhang, Y., Joe, G., Hexner, E., Zhu, J., and Emerson, S. G., Alloreactive memory T cells are responsible for the persistence of graft-versus-host disease, *J Immunol* 174(5), 3051–3058, 2005.

35. Zhang, Y., Joe, G., Hexner, E., Zhu, J., and Emerson, S. G., Host-reactive CD8+ memory stem cells in graft-versus-host disease, *Nat Med* 11(12), 1299–1305, 2005.

36. Gattinoni, L., Zhong, X. S., Palmer, D. C., Ji, Y., Hinrichs, C. S., Yu, Z., Wrzesinski, C., Boni, A., Cassard, L., Garvin, L. M., Paulos, C. M., Muranski, P., and Restifo, N. P., Wnt signaling arrests effector T cell differentiation and generates CD8+ memory stem cells, *Nat Med* 15(7), 808–813, 2009.

37. Gattinoni, L., Lugli, E., Ji, Y., Pos, Z., Paulos, C. M., Quigley, M. F., Almeida, J. R., Gostick, E., Yu, Z., Carpenito, C., Wang, E., Douek, D. C., Price, D. A., June, C. H., Marincola, F. M., Roederer, M., and Restifo, N. P., A human memory T cell subset with stem cell-like properties, *Nat Med* 17(10), 1290–1297, 2011.

38. Masopust, D., Kaech, S. M., Wherry, E. J., and Ahmed, R., The role of programming in memory T-cell development, *Curr Opin Immunol* 16(2), 217–225, 2004.

39. Masopust, D., Vezys, V., Wherry, E. J., Barber, D. L., and Ahmed, R., Cutting edge: Gut microenvironment promotes differentiation of a unique memory CD8 T cell population, *J Immunol* 176(4), 2079–2083, 2006.

40. Wakim, L. M., Woodward-Davis, A., Liu, R., Hu, Y., Villadangos, J., Smyth, G., and Bevan, M. J., The molecular signature of tissue resident memory CD8 T cells isolated from the brain, *J Immunol* 189(7), 3462–3471, 2012.

41. Wakim, L. M., Smith, J., Caminschi, I., Lahoud, M. H., and Villadangos, J. A., Antibody-targeted vaccination to lung dendritic cells generates tissue-resident memory CD8 T cells that are highly protective against influenza virus infection, *Mucosal Immunol* 8(5), 1060–1071, 2015.

42. Wakim, L. M., Gupta, N., Mintern, J. D., and Villadangos, J. A., Enhanced survival of lung tissue-resident memory CD8(+) T cells during infection with influenza virus due to selective expression of IFITM3, *Nat Immunol* 14(3), 238–245, 2013.

43. Blazar, B. R., Murphy, W. J., and Abedi, M., Advances in graft-versus-host disease biology and therapy, *Nat Rev Immunol* 12(6), 443–458, 2012.

44. Heslop, H. E., How I treat EBV lymphoproliferation, *Blood* 114(19), 4002–4008, 2009.

45. Keir, M. E., Butte, M. J., Freeman, G. J., and Sharpe, A. H., PD-1 and its ligands in tolerance and immunity, *Annu Rev Immunol* 26, 677–704, 2008.

46. Restifo, N. P., Dudley, M. E., and Rosenberg, S. A., Adoptive immunotherapy for cancer: Harnessing the T cell response, *Nat Rev Immunol* 12(4), 269–281, 2012.

47. Rosenberg, S. A., Restifo, N. P., Yang, J. C., Morgan, R. A., and Dudley, M. E., Adoptive cell transfer: A clinical path to effective cancer immunotherapy, *Nat Rev Cancer* 8(4), 299–308, 2008.

48. Wherry, E. J., Ha, S. J., Kaech, S. M., Haining, W. N., Sarkar, S., Kalia, V., Subramaniam, S., Blattman, J. N., Barber, D. L., and Ahmed, R., Molecular signature of CD8+ T cell exhaustion during chronic viral infection, *Immunity* 27(4), 670–684, 2007.

49. O'Shea, J. J. and Paul, W. E., Mechanisms underlying lineage commitment and plasticity of helper CD4+ T cells, *Science* 327(5969), 1098–1102, 2010.

50. Crawford, A., Angelosanto, J. M., Kao, C., Doering, T. A., Odorizzi, P. M., Barnett, B. E., and Wherry, E. J., Molecular and transcriptional basis of CD4(+) T cell dysfunction during chronic infection, *Immunity* 40(2), 289–302, 2014.

51. McKinney, E. F., Lyons, P. A., Carr, E. J., Hollis, J. L., Jayne, D. R., Willcocks, L. C., Koukoulaki, M., Brazma, A., Jovanovic, V., Kemeny, D. M., Pollard, A. J., Macary, P. A., Chaudhry, A. N., and Smith, K. G., A CD8+ T cell transcription signature predicts prognosis in autoimmune disease, *Nat Med* 16(5), 586–591, 1p following 591, 2010.

52. Chang, S., and Aune, T. M., Dynamic changes in histone-methylation "marks" across the locus encoding interferon-gamma during the differentiation of T helper type 2 cells, *Nat Immunol* 8(7), 723–731, 2007.

53. Jacob, E., Hod-Dvorai, R., Schif-Zuck, S., and Avni, O., Unconventional association of the polycomb group proteins with cytokine genes in differentiated T helper cells, *J Biol Chem* 283(19), 13471–13481, 2008.

54. Juelich, T., Sutcliffe, E. L., Denton, A., He, Y., Doherty, P. C., Parish, C. R., Turner, S. J., Tremethick, D. J., and Rao, S., Interplay between chromatin remodeling and epigenetic changes during lineage-specific commitment to granzyme B expression, *J Immunol* 183(11), 7063–7072, 2009.

55. Kimura, M., Koseki, Y., Yamashita, M., Watanabe, N., Shimizu, C., Katsumoto, T., Kitamura, T., Taniguchi, M., Koseki, H., and Nakayama, T., Regulation of Th2 cell differentiation by mel-18, a mammalian polycomb group gene, *Immunity* 15(2), 275–287, 2001.

56. Koyanagi, M., Baguet, A., Martens, J., Margueron, R., Jenuwein, T., and Bix, M., Ezh2 and histone 3 trimethyl lysine 27 associated with Il4 and Il13 gene silencing in Th1 cells, *J Biol Chem* 280(36), 31470–31477, 2005.

57. Northrop, J. K., Wells, A. D., and Shen, H., Cutting edge: Chromatin remodeling as a molecular basis for the enhanced functionality of memory CD8 T cells, *J Immunol* 181(2), 865–868, 2008.

58. Wei, J., Duramad, O., Perng, O. A., Reiner, S. L., Liu, Y. J., and Qin, F. X., Antagonistic nature of T helper 1/2 developmental programs in opposing peripheral induction of Foxp3+ regulatory T cells, *Proc Natl Acad Sci USA* 104(46), 18169–18174, 2007.

59. Yamashita, M., Hirahara, K., Shinnakasu, R., Hosokawa, H., Norikane, S., Kimura, M. Y., Hasegawa, A., and Nakayama, T., Crucial role of MLL for the maintenance of memory T helper type 2 cell responses, *Immunity* 24(5), 611–622, 2006.

60. Yamashita, M., Shinnakasu, R., Nigo, Y., Kimura, M., Hasegawa, A., Taniguchi, M., and Nakayama, T., Interleukin (IL)-4-independent maintenance of histone modification of the IL-4 gene loci in memory Th2 cells, *J Biol Chem* 279(38), 39454–39464, 2004.

61. Ansel, K. M., Lee, D. U., and Rao, A., An epigenetic view of helper T cell differentiation, *Nat Immunol* 4(7), 616–623, 2003.

62. Barski, A., Cuddapah, S., Cui, K., Roh, T. Y., Schones, D. E., Wang, Z., Wei, G., Chepelev, I., and Zhao, K., High-resolution profiling of histone methylations in the human genome, *Cell* 129(4), 823–837, 2007.

63. Raha, D., Hong, M., and Snyder, M., ChIP-Seq: A method for global identification of regulatory elements in the genome, *Curr Protoc Mol Biol* Chapter 21, Unit 21.19.1–14, 2010.

64. Wei, G., Wei, L., Zhu, J., Zang, C., Hu-Li, J., Yao, Z., Cui, K., Kanno, Y., Roh, T. Y., Watford, W. T., Schones, D. E., Peng, W., Sun, H. W., Paul, W. E., O'Shea, J. J., and Zhao, K., Global mapping of H3K4me3 and H3K27me3 reveals specificity and plasticity in lineage fate determination of differentiating CD4+ T cells, *Immunity* 30(1), 155–167, 2009.

65. Denton, A. E., Russ, B. E., Doherty, P. C., Rao, S., and Turner, S. J., Differentiation-dependent functional and epigenetic landscapes for cytokine genes in virus-specific CD8+ T cells, *Proc Natl Acad Sci USA* 108(37), 15306–15311, 2011.

66. Lu, K. T., Kanno, Y., Cannons, J. L., Handon, R., Bible, P., Elkahloun, A. G., Anderson, S. M., Wei, L., Sun, H., O'Shea, J. J., and Schwartzberg, P. L., Functional and epigenetic studies reveal multistep differentiation and plasticity of in vitro-generated and in vivo-derived follicular T helper cells, *Immunity* 35(4), 622–632, 2011.

67. Mukasa, R., Balasubramani, A., Lee, Y. K., Whitley, S. K., Weaver, B. T., Shibata, Y., Crawford, G. E., Hatton, R. D., and Weaver, C. T., Epigenetic instability of cytokine and transcription factor gene loci underlies plasticity of the T helper 17 cell lineage, *Immunity* 32(5), 616–627, 2010.

68. Coccia, E. M., Passini, N., Battistini, A., Pini, C., Sinigaglia, F., and Rogge, L., Interleukin-12 induces expression of interferon regulatory factor-1 via signal transducer and activator of transcription-4 in human T helper type 1 cells, *J Biol Chem* 274(10), 6698–6703, 1999.

69. Afkarian, M., Sedy, J. R., Yang, J., Jacobson, N. G., Cereb, N., Yang, S. Y., Murphy, T. L., and Murphy, K. M., T-bet is a STAT1-induced regulator of IL-12R expression in naive CD4+ T cells, *Nat Immunol* 3(6), 549–557, 2002.

70. Kaplan, M. H., Sun, Y. L., Hoey, T., and Grusby, M. J., Impaired IL-12 responses and enhanced development of Th2 cells in Stat4-deficient mice, *Nature* 382(6587), 174–177, 1996.

71. Lighvani, A. A., Frucht, D. M., Jankovic, D., Yamane, H., Aliberti, J., Hissong, B. D., Nguyen, B. V., Gadina, M., Sher, A., Paul, W. E., and O'Shea, J. J., T-bet is rapidly induced by interferon-gamma in lymphoid and myeloid cells, *Proc Natl Acad Sci USA* 98(26), 15137–15142, 2001.

72. Thierfelder, W. E., van Deursen, J. M., Yamamoto, K., Tripp, R. A., Sarawar, S. R., Carson, R. T., Sangster, M. Y., Vignali, D. A., Doherty, P. C., Grosveld, G. C., and Ihle, J. N., Requirement for Stat4 in interleukin-12-mediated responses of natural killer and T cells, *Nature* 382(6587), 171–174, 1996.

73. Wilson, C. B., Rowell, E., and Sekimata, M., Epigenetic control of T-helper-cell differentiation, *Nat Rev Immunol* 9(2), 91–105, 2009.

74. Lee, D. U., Agarwal, S., and Rao, A., Th2 lineage commitment and efficient IL-4 production involves extended demethylation of the IL-4 gene, *Immunity* 16(5), 649–660, 2002.

75. Fang, T. C., Yashiro-Ohtani, Y., Del Bianco, C., Knoblock, D. M., Blacklow, S. C., and Pear, W. S., Notch directly regulates Gata3 expression during T helper 2 cell differentiation, *Immunity* 27(1), 100–110, 2007.

76. Yagi, R., Zhu, J., and Paul, W. E., An updated view on transcription factor GATA3-mediated regulation of Th1 and Th2 cell differentiation, *Int Immunol* 23(7), 415–420, 2011.

77. Zhu, J., Min, B., Hu-Li, J., Watson, C. J., Grinberg, A., Wang, Q., Killeen, N., Urban, J. F., Jr., Guo, L., and Paul, W. E., Conditional deletion of Gata3 shows its essential function in T(H)1-T(H)2 responses, *Nat Immunol* 5(11), 1157–1165, 2004.

78. Hegazy, A. N., Peine, M., Helmstetter, C., Panse, I., Frohlich, A., Bergthaler, A., Flatz, L., Pinschewer, D. D., Radbruch, A., and Lohning, M., Interferons direct Th2 cell reprogramming to generate a stable GATA-3(+)T-bet(+) cell subset with combined Th2 and Th1 cell functions, *Immunity* 32(1), 116–128, 2010.

79. Peine, M., Rausch, S., Helmstetter, C., Frohlich, A., Hegazy, A. N., Kuhl, A. A., Grevelding, C. G., Hofer, T., Hartmann, S., and Lohning, M., Stable T-bet(+)GATA-3(+) Th1/Th2 hybrid cells arise in vivo, can develop directly from naive precursors, and limit immunopathologic inflammation, *PLoS Biol* 11(8), e1001633, 2013.

80. Weaver, C. T., Hatton, R. D., Mangan, P. R., and Harrington, L. E., IL-17 family cytokines and the expanding diversity of effector T cell lineages, *Annu Rev Immunol* 25, 821–852, 2007.

81. Cohen, C. J., Crome, S. Q., MacDonald, K. G., Dai, E. L., Mager, D. L., and Levings, M. K., Human Th1 and Th17 cells exhibit epigenetic stability at signature cytokine and transcription factor loci, *J Immunol* 187(11), 5615–5626, 2011.

82. Bending, D., De la Pena, H., Veldhoen, M., Phillips, J. M., Uyttenhove, C., Stockinger, B., and Cooke, A., Highly purified Th17 cells from BDC2.5NOD mice convert into Th1-like cells in NOD/SCID recipient mice, *J Clin Invest* 119(3), 565–572, 2009.

83. Nurieva, R., Yang, X. O., Chung, Y., and Dong, C., Cutting edge: In vitro generated Th17 cells maintain their cytokine expression program in normal but not lymphopenic hosts, *J Immunol* 182(5), 2565–2568, 2009.
84. Shi, G., Cox, C. A., Vistica, B. P., Tan, C., Wawrousek, E. F., and Gery, I., Phenotype switching by inflammation-inducing polarized Th17 cells, but not by Th1 cells, *J Immunol* 181(10), 7205–7213, 2008.
85. Oestreich, K. J., and Weinmann, A. S., Master regulators or lineage-specifying? Changing views on CD4+ T cell transcription factors, *Nat Rev Immunol* 12(11), 799–804, 2012.
86. Fontenot, J. D., Gavin, M. A., and Rudensky, A. Y., Foxp3 programs the development and function of CD4+CD25+ regulatory T cells, *Nat Immunol* 4(4), 330–336, 2003.
87. Kitagawa, Y., Ohkura, N., and Sakaguchi, S., Epigenetic control of thymic Treg-cell development, *Eur J Immunol* 45(1), 11–16, 2015.
88. Floess, S., Freyer, J., Siewert, C., Baron, U., Olek, S., Polansky, J., Schlawe, K., Chang, H. D., Bopp, T., Schmitt, E., Klein-Hessling, S., Serfling, E., Hamann, A., and Huehn, J., Epigenetic control of the foxp3 locus in regulatory T cells, *PLoS Biol* 5(2), e38, 2007.
89. Sauer, S., Bruno, L., Hertweck, A., Finlay, D., Leleu, M., Spivakov, M., Knight, Z. A., Cobb, B. S., Cantrell, D., O'Connor, E., Shokat, K. M., Fisher, A. G., and Merkenschlager, M., T cell receptor signaling controls Foxp3 expression via PI3K, Akt, and mTOR, *Proc Natl Acad Sci USA* 105(22), 7797–7802, 2008.
90. Tian, Y., Jia, Z., Wang, J., Huang, Z., Tang, J., Zheng, Y., Tang, Y., Wang, Q., Tian, Z., Yang, D., Zhang, Y., Fu, X., Song, J., Liu, S., van Velkinburgh, J. C., Wu, Y., and Ni, B., Global mapping of H3K4me1 and H3K4me3 reveals the chromatin state-based cell type-specific gene regulation in human Treg cells, *PLoS One* 6(11), e27770, 2011.
91. Stock, P., Akbari, O., Berry, G., Freeman, G. J., Dekruyff, R. H., and Umetsu, D. T., Induction of T helper type 1-like regulatory cells that express Foxp3 and protect against airway hyper-reactivity, *Nat Immunol* 5(11), 1149–1156, 2004.
92. Zhou, L., Lopes, J. E., Chong, M. M., Ivanov, II, Min, R., Victora, G. D., Shen, Y., Du, J., Rubtsov, Y. P., Rudensky, A. Y., Ziegler, S. F., and Littman, D. R., TGF-beta-induced Foxp3 inhibits T(H)17 cell differentiation by antagonizing RORgammat function, *Nature* 453(7192), 236–240, 2008.
93. Ohkura, N., Hamaguchi, M., Morikawa, H., Sugimura, K., Tanaka, A., Ito, Y., Osaki, M., Tanaka, Y., Yamashita, R., Nakano, N., Huehn, J., Fehling, H. J., Sparwasser, T., Nakai, K., and Sakaguchi, S., T cell receptor stimulation-induced epigenetic changes and Foxp3 expression are independent and complementary events required for Treg cell development, *Immunity* 37(5), 785–799, 2012.
94. Schoenborn, J. R., Dorschner, M. O., Sekimata, M., Santer, D. M., Shnyreva, M., Fitzpatrick, D. R., Stamatoyannopoulos, J. A., and Wilson, C. B., Comprehensive epigenetic profiling identifies multiple distal regulatory elements directing transcription of the gene encoding interferon-gamma, *Nat Immunol* 8(7), 732–742, 2007.
95. Pearce, E. L., Mullen, A. C., Martins, G. A., Krawczyk, C. M., Hutchins, A. S., Zediak, V. P., Banica, M., DiCioccio, C. B., Gross, D. A., Mao, C. A., Shen, H., Cereb, N., Yang, S. Y., Lindsten, T., Rossant, J., Hunter, C. A., and Reiner, S. L., Control of effector CD8+ T cell function by the transcription factor Eomesodermin, *Science* 302(5647), 1041–1043, 2003.
96. Intlekofer, A. M., Takemoto, N., Wherry, E. J., Longworth, S. A., Northrup, J. T., Palanivel, V. R., Mullen, A. C., Gasink, C. R., Kaech, S. M., Miller, J. D., Gapin, L., Ryan, K., Russ, A. P., Lindsten, T., Orange, J. S., Goldrath, A. W., Ahmed, R., and Reiner, S. L., Effector and memory CD8+ T cell fate coupled by T-bet and eomesodermin, *Nat Immunol* 6(12), 1236–1244, 2005.
97. Intlekofer, A. M., Takemoto, N., Kao, C., Banerjee, A., Schambach, F., Northrop, J. K., Shen, H., Wherry, E. J., and Reiner, S. L., Requirement for T-bet in the aberrant differentiation of unhelped memory CD8+ T cells, *J Exp Med* 204(9), 2015–2021, 2007.

98. Paley, M. A., Kroy, D. C., Odorizzi, P. M., Johnnidis, J. B., Dolfi, D. V., Barnett, B. E., Bikoff, E. K., Robertson, E. J., Lauer, G. M., Reiner, S. L., and Wherry, E. J., Progenitor and terminal subsets of CD8+ T cells cooperate to contain chronic viral infection, *Science* 338(6111), 1220–1225, 2012.

99. Yao, S., Buzo, B. F., Pham, D., Jiang, L., Taparowsky, E. J., Kaplan, M. H., and Sun, J., Interferon regulatory factor 4 sustains CD8(+) T cell expansion and effector differentiation, *Immunity* 39(5), 833–845, 2013.

100. Nowyhed, H. N., Huynh, T. R., Thomas, G. D., Blatchley, A., and Hedrick, C. C., Cutting edge: The orphan nuclear receptor Nr4a1 regulates CD8+ T cell expansion and effector function through direct repression of Irf4, *J Immunol* 195(8), 3515–3519, 2015.

101. Miyagawa, F., Zhang, H., Terunuma, A., Ozato, K., Tagaya, Y., and Katz, S. I., Interferon regulatory factor 8 integrates T-cell receptor and cytokine-signaling pathways and drives effector differentiation of CD8 T cells, *Proc Natl Acad Sci USA* 109(30), 12123–12128, 2012.

102. Nayar, R., Schutten, E., Jangalwe, S., Durost, P. A., Kenney, L. L., Conley, J. M., Daniels, K., Brehm, M. A., Welsh, R. M., and Berg, L. J., IRF4 regulates the ratio of T-bet to eomesodermin in CD8+ T cells responding to persistent LCMV infection, *PLoS One* 10s(12), e0144826, 2015.

103. Martins, G. A., Cimmino, L., Shapiro-Shelef, M., Szabolcs, M., Herron, A., Magnusdottir, E., and Calame, K., Transcriptional repressor Blimp-1 regulates T cell homeostasis and function, *Nat Immunol* 7(5), 457–465, 2006.

104. Shin, H., Blackburn, S. D., Intlekofer, A. M., Kao, C., Angelosanto, J. M., Reiner, S. L., and Wherry, E. J., A role for the transcriptional repressor Blimp-1 in CD8(+) T cell exhaustion during chronic viral infection, *Immunity* 31(2), 309–320, 2009.

105. Rutishauser, R. L., Martins, G. A., Kalachikov, S., Chandele, A., Parish, I. A., Meffre, E., Jacob, J., Calame, K., and Kaech, S. M., Transcriptional repressor Blimp-1 promotes CD8(+) T cell terminal differentiation and represses the acquisition of central memory T cell properties, *Immunity* 31(2), 296–308, 2009.

106. Kallies, A., Xin, A., Belz, G. T., and Nutt, S. L., Blimp-1 transcription factor is required for the differentiation of effector CD8(+) T cells and memory responses, *Immunity* 31(2), 283–295, 2009.

107. Zhang, S., Rozell, M., Verma, R. K., Albu, D. I., Califano, D., VanValkenburgh, J., Merchant, A., Rangel-Moreno, J., Randall, T. D., Jenkins, N. A., Copeland, N. G., Liu, P., and Avram, D., Antigen-specific clonal expansion and cytolytic effector function of CD8+ T lymphocytes depend on the transcription factor Bcl11b, *J Exp Med* 207(8), 1687–1699, 2010.

108. Crompton, J. G., Narayanan, M., Cuddapah, S., Roychoudhuri, R., Ji, Y., Yang, W., Patel, S. J., Sukumar, M., Palmer, D. C., Peng, W., Wang, E., Marincola, F. M., Klebanoff, C. A., Zhao, K., Tsang, J. S., Gattinoni, L., and Restifo, N. P., Lineage relationship of CD8 T cell subsets is revealed by progressive changes in the epigenetic landscape, *Cell Mol Immunol*, 2015.

109. Cruz-Guilloty, F., Pipkin, M. E., Djuretic, I. M., Levanon, D., Lotem, J., Lichtenheld, M. G., Groner, Y., and Rao, A., Runx3 and T-box proteins cooperate to establish the transcriptional program of effector CTLs, *J Exp Med* 206(1), 51–59, 2009.

110. Zediak, V. P., Johnnidis, J. B., Wherry, E. J., and Berger, S. L., Cutting edge: Persistently open chromatin at effector gene loci in resting memory CD8+ T cells independent of transcriptional status, *J Immunol* 186(5), 2705–2709, 2011.

111. Brehm, M. A., Daniels, K. A., and Welsh, R. M., Rapid production of TNF-alpha following TCR engagement of naive CD8 T cells, *J Immunol* 175(8), 5043–5049, 2005.

112. Chang, J. T., Wherry, E. J., and Goldrath, A. W., Molecular regulation of effector and memory T cell differentiation, *Nat Immunol* 15(12), 1104–1115, 2014.

113. Onodera, A., Yamashita, M., Endo, Y., Kuwahara, M., Tofukuji, S., Hosokawa, H., Kanai, A., Suzuki, Y., and Nakayama, T., STAT6-mediated displacement of polycomb by trithorax complex establishes long-term maintenance of GATA3 expression in T helper type 2 cells, *J Exp Med* 207(11), 2493–2506, 2010.

114. Sarkar, S., Kalia, V., Haining, W. N., Konieczny, B. T., Subramaniam, S., and Ahmed, R., Functional and genomic profiling of effector CD8 T cell subsets with distinct memory fates, *J Exp Med* 205(3), 625–640, 2008.

115. Obar, J. J., Jellison, E. R., Sheridan, B. S., Blair, D. A., Pham, Q. M., Zickovich, J. M., and Lefrancois, L., Pathogen-induced inflammatory environment controls effector and memory CD8+ T cell differentiation, *J Immunol* 187(10), 4967–4978, 2011.

116. Slutter, B., Pewe, L. L., Kaech, S. M., and Harty, J. T., Lung airway-surveilling CXCR3(hi) memory CD8(+) T cells are critical for protection against influenza A virus, *Immunity* 39(5), 939–948, 2013.

117. Cannarile, M. A., Lind, N. A., Rivera, R., Sheridan, A. D., Camfield, K. A., Wu, B. B., Cheung, K. P., Ding, Z., and Goldrath, A. W., Transcriptional regulator Id2 mediates CD8+ T cell immunity, *Nat Immunol* 7(12), 1317–1325, 2006.

118. Kim, M. V., Ouyang, W., Liao, W., Zhang, M. Q., and Li, M. O., The transcription factor Foxo1 controls central-memory CD8+ T cell responses to infection, *Immunity* 39(2), 286–297, 2013.

119. Rao, R. R., Li, Q., Gubbels Bupp, M. R., and Shrikant, P. A., Transcription factor Foxo1 represses T-bet-mediated effector functions and promotes memory CD8(+) T cell differentiation, *Immunity* 36(3), 374–387, 2012.

120. Tejera, M. M., Kim, E. H., Sullivan, J. A., Plisch, E. H., and Suresh, M., FoxO1 controls effector-to-memory transition and maintenance of functional CD8 T cell memory, *J Immunol* 191(1), 187–199, 2013.

121. Hess Michelini, R., Doedens, A. L., Goldrath, A. W., and Hedrick, S. M., Differentiation of CD8 memory T cells depends on Foxo1, *J Exp Med* 210(6), 1189–1200, 2013.

122. Zhou, X., and Xue, H. H., Cutting edge: Generation of memory precursors and functional memory CD8+ T cells depends on T cell factor-1 and lymphoid enhancer-binding factor-1, *J Immunol* 189(6), 2722–2726, 2012.

123. Jeannet, G., Boudousquie, C., Gardiol, N., Kang, J., Huelsken, J., and Held, W., Essential role of the Wnt pathway effector Tcf-1 for the establishment of functional CD8 T cell memory, *Proc Natl Acad Sci USA* 107(21), 9777–9782, 2010.

124. Jenkins, M. R., Kedzierska, K., Doherty, P. C., and Turner, S. J., Heterogeneity of effector phenotype for acute phase and memory influenza A virus-specific CTL, *J Immunol* 179(1), 64–70, 2007.

125. Moffat, J. M., Gebhardt, T., Doherty, P. C., Turner, S. J., and Mintern, J. D., Granzyme A expression reveals distinct cytolytic CTL subsets following influenza A virus infection, *Eur J Immunol* 39(5), 1203–1210, 2009.

126. Peixoto, A., Evaristo, C., Munitic, I., Monteiro, M., Charbit, A., Rocha, B., and Veiga-Fernandes, H., CD8 single-cell gene coexpression reveals three different effector types present at distinct phases of the immune response, *J Exp Med* 204(5), 1193–1205, 2007.

127. Greer, E. L., and Shi, Y., Histone methylation: A dynamic mark in health, disease and inheritance, *Nat Rev Genet* 13(5), 343–357, 2012.

128. Feng, Q., Wang, H., Ng, H. H., Erdjument-Bromage, H., Tempst, P., Struhl, K., and Zhang, Y., Methylation of H3-lysine 79 is mediated by a new family of HMTases without a SET domain, *Curr Biol* 12(12), 1052–1058, 2002.

129. Rea, S., Eisenhaber, F., O'Carroll, D., Strahl, B. D., Sun, Z. W., Schmid, M., Opravil, S., Mechtler, K., Ponting, C. P., Allis, C. D., and Jenuwein, T., Regulation of chromatin structure by site-specific histone H3 methyltransferases, *Nature* 406(6796), 593–599, 2000.

130. Murray, K., The occurrence of epsilon-n-methyl lysine in histones, *Biochemistry* 3, 10–15, 1964.
131. Muntean, A. G., and Hess, J. L., The pathogenesis of mixed-lineage leukemia, *Annu Rev Pathol* 7, 283–301, 2012.
132. Rao, R. C., and Dou, Y., Hijacked in cancer: The KMT2 (MLL) family of methyltransferases, *Nat Rev Cancer* 15(6), 334–346, 2015.
133. Boyer, L. A., Plath, K., Zeitlinger, J., Brambrink, T., Medeiros, L. A., Lee, T. I., Levine, S. S., Wernig, M., Tajonar, A., Ray, M. K., Bell, G. W., Otte, A. P., Vidal, M., Gifford, D. K., Young, R. A., and Jaenisch, R., Polycomb complexes repress developmental regulators in murine embryonic stem cells, *Nature* 441(7091), 349–353, 2006.
134. Bracken, A. P., Kleine-Kohlbrecher, D., Dietrich, N., Pasini, D., Gargiulo, G., Beekman, C., Theilgaard-Monch, K., Minucci, S., Porse, B. T., Marine, J. C., Hansen, K. H., and Helin, K., The Polycomb group proteins bind throughout the INK4A-ARF locus and are disassociated in senescent cells, *Genes Dev* 21(5), 525–530, 2007.
135. Margueron, R., and Reinberg, D., The polycomb complex PRC2 and its mark in life, *Nature* 469(7330), 343–349, 2011.
136. Bantignies, F., and Cavalli, G., Cellular memory and dynamic regulation of polycomb group proteins, *Curr Opin Cell Biol* 18(3), 275–283, 2006.
137. Schuettengruber, B., Chourrout, D., Vervoort, M., Leblanc, B., and Cavalli, G., Genome regulation by polycomb and trithorax proteins, *Cell* 128(4), 735–745, 2007.
138. Lachner, M., O'Carroll, D., Rea, S., Mechtler, K., and Jenuwein, T., Methylation of histone H3 lysine 9 creates a binding site for HP1 proteins, *Nature* 410(6824), 116–120, 2001.
139. Tachibana, M., Sugimoto, K., Fukushima, T., and Shinkai, Y., Set domain-containing protein, G9a, is a novel lysine-preferring mammalian histone methyltransferase with hyperactivity and specific selectivity to lysines 9 and 27 of histone H3, *J Biol Chem* 276(27), 25309–25317, 2001.
140. Tachibana, M., Sugimoto, K., Nozaki, M., Ueda, J., Ohta, T., Ohki, M., Fukuda, M., Takeda, N., Niida, H., Kato, H., and Shinkai, Y., G9a histone methyltransferase plays a dominant role in euchromatic histone H3 lysine 9 methylation and is essential for early embryogenesis, *Genes Dev* 16(14), 1779–1791, 2002.
141. Hamamoto, R., Furukawa, Y., Morita, M., Iimura, Y., Silva, F. P., Li, M., Yagyu, R., and Nakamura, Y., SMYD3 encodes a histone methyltransferase involved in the proliferation of cancer cells, *Nat Cell Biol* 6(8), 731–740, 2004.
142. Schaller, M., Ito, T., Allen, R. M., Kroetz, D., Kittan, N., Ptaschinski, C., Cavassani, K., Carson, W. F. t., Godessart, N., Grembecka, J., Cierpicki, T., Dou, Y., and Kunkel, S. L., Epigenetic regulation of IL-12-dependent T cell proliferation, *J Leukoc Biol* 98(4), 601–613, 2015.
143. Nakata, Y., Brignier, A. C., Jin, S., Shen, Y., Rudnick, S. I., Sugita, M., and Gewirtz, A. M., c-Myb, Menin, GATA-3, and MLL form a dynamic transcription complex that plays a pivotal role in human T helper type 2 cell development, *Blood* 116(8), 1280–1290, 2010.
144. He, S., Xie, F., Liu, Y., Tong, Q., Mochizuki, K., Lapinski, P. E., Mani, R. S., Reddy, P., Mochizuki, I., Chinnaiyan, A. M., Mineishi, S., King, P. D., and Zhang, Y., The histone methyltransferase Ezh2 is a crucial epigenetic regulator of allogeneic T-cell responses mediating graft-versus-host disease, *Blood* 122(25), 4119–4128, 2013.
145. Tumes, D. J., Onodera, A., Suzuki, A., Shinoda, K., Endo, Y., Iwamura, C., Hosokawa, H., Koseki, H., Tokoyoda, K., Suzuki, Y., Motohashi, S., and Nakayama, T., The polycomb protein Ezh2 regulates differentiation and plasticity of CD4(+) T helper type 1 and type 2 cells, *Immunity* 39(5), 819–832, 2013.

146. Tong, Q., He, S., Xie, F., Mochizuki, K., Liu, Y., Mochizuki, I., Meng, L., Sun, H., Zhang, Y., Guo, Y., Hexner, E., and Zhang, Y., Ezh2 regulates transcriptional and post-translational expression of T-bet and promotes Th1 cell responses mediating aplastic anemia in mice, *J Immunol* 192(11), 5012–5022, 2014.

147. He, S., Xie, F., Liu, Y., Tong, Q., Mochizuki, K., Lapinski, P. E., Mani, R. S., Reddy, P., Mochizuki, I., Chinnaiyan, A. M., Mineishi, S., King, P. D., and Zhang, Y., The histone methyltransferase Ezh2 is a crucial epigenetic regulator of allogeneic T cell responses mediating graft-versus-host disease, *Blood* 122(25), 4119–4128, 2013.

148. Xiong, Y., Khanna, S., Grzenda, A. L., Sarmento, O. F., Svingen, P. A., Lomberk, G. A., Urrutia, R. A., and Faubion, W. A., Jr., Polycomb antagonizes p300/CREB-binding protein-associated factor to silence FOXP3 in a Kruppel-like factor-dependent manner, *J Biol Chem* 287(41), 34372–34385, 2012.

149. Arvey, A., van der Veeken, J., Samstein, R. M., Feng, Y., Stamatoyannopoulos, J. A., and Rudensky, A. Y., Inflammation-induced repression of chromatin bound by the transcription factor Foxp3 in regulatory T cells, *Nat Immunol* 15(6), 580–587, 2014.

150. Allan, R. S., Zueva, E., Cammas, F., Schreiber, H. A., Masson, V., Belz, G. T., Roche, D., Maison, C., Quivy, J. P., Almouzni, G., and Amigorena, S., An epigenetic silencing pathway controlling T helper 2 cell lineage commitment, *Nature* 487(7406), 249–253, 2012.

151. Wakabayashi, Y., Tamiya, T., Takada, I., Fukaya, T., Sugiyama, Y., Inoue, N., Kimura, A., Morita, R., Kashiwagi, I., Takimoto, T., Nomura, M., and Yoshimura, A., Histone 3 lysine 9 (H3K9) methyltransferase recruitment to the interleukin-2 (IL-2) promoter is a mechanism of suppression of IL-2 transcription by the transforming growth factor-beta-Smad pathway, *J Biol Chem* 286(41), 35456–35465, 2011.

152. Antignano, F., Burrows, K., Hughes, M. R., Han, J. M., Kron, K. J., Penrod, N. M., Oudhoff, M. J., Wang, S. K., Min, P. H., Gold, M. J., Chenery, A. L., Braam, M. J., Fung, T. C., Rossi, F. M., McNagny, K. M., Arrowsmith, C. H., Lupien, M., Levings, M. K., and Zaph, C., Methyltransferase G9a regulates T cell differentiation during murine intestinal inflammation, *J Clin Invest* 124(5), 1945–1955, 2014.

153. Lehnertz, B., Northrop, J. P., Antignano, F., Burrows, K., Hadidi, S., Mullaly, S. C., Rossi, F. M., and Zaph, C., Activating and inhibitory functions for the histone lysine methyltransferase G9a in T helper cell differentiation and function, *J Exp Med* 207(5), 915–922, 2010.

154. Nagata, D. E., Ting, H. A., Cavassani, K. A., Schaller, M. A., Mukherjee, S., Ptaschinski, C., Kunkel, S. L., and Lukacs, N. W., Epigenetic control of Foxp3 by SMYD3 H3K4 histone methyltransferase controls iTreg development and regulates pathogenic T-cell responses during pulmonary viral infection, *Mucosal Immunol* 8(5), 1131–1143, 2015.

155. Ferrara, J. L., Levine, J. E., Reddy, P., and Holler, E., Graft-versus-host disease, *Lancet* 373(9674), 1550–1561, 2009.

156. He, S., Wang, J., Kato, K., Xie, F., Varambally, S., Mineishi, S., Kuick, R., Mochizuki, K., Liu, Y., Nieves, E., Mani, R. S., Chinnaiyan, A. M., Marquez, V. E., and Zhang, Y., Inhibition of histone methylation arrests ongoing graft-versus-host disease in mice by selectively inducing apoptosis of alloreactive effector T cells, *Blood* 119(5), 1274–1282, 2012.

# 12 Models of Regulatory T Cell Alterations and Systemic Autoimmunity

*Jonathan J. Cho, Kyle J. Lorentsen,
and Dorina Avram*

## CONTENTS

### ABSTRACT

Mice with alterations in specific genes important for regulatory T (Treg) cell development, maintenance, and function develop spontaneous autoimmunity. The supreme manifestation is spontaneous systemic autoimmunity and occurs in the case of genes with critical roles in Treg cells, such as Foxp3. Several mouse models with genetic alterations causing spontaneous systemic auto-immunity are presented, some with germline deletions, others with conditional or inducible deletion in T cells or only in Treg cells. Additionally, the pathology, the immune populations, and the molecular mechanisms underlying the observed pathology in these mice are described, together with the various immunological techniques used to investigate the autoimmune pathogenesis within these mice.

## 12.1 INTRODUCTION

The function of the immune system is to protect the body from foreign invading pathogens. The ability of the immune system to distinguish self from nonself is crucial to this proper function. Paul Ehrlich coined the term *horror autotoxicus* to describe an organism's natural tendency to avoid attacking itself: "We pointed out that the organism possesses certain contrivances by means of which the immunity reaction, so easily produced (induced) by all kinds of cells, is prevented from acting against the organism's own elements and so giving rise to autotoxins ... so that one might be justified in speaking of a 'horror autotoxicus' of the organism."[1,2] Recent studies have characterized the horror autotoxicus that Ehrlich described as self-tolerance, the nonreactivity of the immune system to an organism's own normal cells. Central and peripheral tolerance encompass the main processes that form the basis of immunological self-tolerance.[3] Central tolerance is marked by the deletion of self-reactive lymphocytes and survival of lymphocytes that do not react to self-antigens, a process that focuses the adaptive immune system to recognize foreign pathogens. Central tolerance occurs in the central lymphoid organs, the thymus for developing T lymphocytes, and the bone marrow for developing B lymphocytes, and is based on the reactivity of T cell receptors (TCRs) or B cell receptors (BCRs), respectively, to self-antigens, in the case of T lymphocytes presented in the context of MHC molecules.[4,5] In the thymus, clonal deletion plays a critical role for eliminating thymocytes whose TCRs have high affinity for self-peptides presented by the major histocompatibility complex (MHC). This happens in the thymic cortex for ubiquitous self-antigens and in the medulla for tissue-specific antigens. Medullary thymic epithelial cells (mTECs) express the transcription factor AIRE, which allows expression in these cells of tissue-specific self-antigens. An alternative to clonal deletion is clonal diversion, which allows survival of self-reactive T cells and plays a critical role in development of regulatory T (Treg) cells.[6] The fact that some self-reactive lymphocytes escape central tolerance is partly due to the fact that not all self-antigens are expressed in the central lymphoid organs, thus necessitating peripheral tolerance. Autoreactive T cells that escape thymic selection can be rendered anergic through cell-intrinsic mechanisms that maintain self-immunological tolerance.[7–9] Additionally, Treg cells and tolerogenic dendritic cells play a critical role in peripheral tolerance and immune homeostasis.[10–13] Treg cells are CD4+ T cells derived from the thymus or induced in the periphery that expresses the transcription factor Foxp3, are dependent on IL-2 for survival, and are functionally suppressive of T effector cells and other immune populations (Figure 12.1). They are the main mediators of cell-extrinsic mechanisms underlying immune homeostasis and of peripheral tolerance.[13–15] The broad TCR repertoire of Treg cells is another crucial component in the maintenance of immune homeostasis and immunological self-tolerance.[16] Dysfunctions in central and peripheral tolerance lead to autoimmunity, which is marked by attack of organs by autoreactive immune cells. Defects in cytokines, costimulatory molecules, and transcription factors crucial for central and peripheral tolerance also leads to autoimmunity.[17–20] In developing models to assess autoimmune dysregulations, one may follow methodologies established by literature as exemplified by the following examples of breaches in central and peripheral tolerance. Knockout mouse models that spontaneously develop autoimmune diseases offer invaluable insight into the specific role that a particular gene plays in autoimmune pathogenesis. Several such models are discussed later, with emphasis on models with Treg cell alterations that develop multiorgan inflammation and succumb to

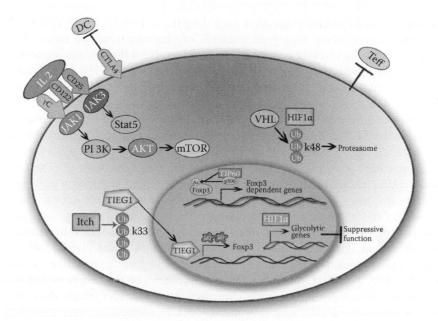

**FIGURE 12.1**    Altered molecular mechanisms in Treg cells that cause systemic autoimmunity. The main players in Treg cell identity and function whose alterations result in systemic autoimmunity. Details are provided in the text.

death early in life. If clinical genetic clues from human patients are available, then the construction of a mouse knockout model may be done to delineate the underlying mechanism that the particular gene controls tolerance and autoimmunity. To investigate the specific role of such a gene in Treg cells, ideally, the gene should be removed conditionally or in an inducible manner in Treg cells. Clinical symptoms and the corresponding mouse phenotype may be agreeable and easily detected, especially in monogenetic diseases with severe multi-organ autoimmune damage.

## 12.2   Foxp3

The key role of Treg cells and the transcription factor Foxp3 in maintaining systemic immune homeostasis is evident in the human loss-of-function genetic disease IPEX (immune dysregulation, polyendocrinopathy, enteropathy, X-linked syndrome), and the mouse mutant strain *scurfy* with X-linked recessive mutation in the *Foxp3* gene (Figure 12.1).[21–24] The human genetic disease IPEX is characterized by a triad of clinical manifestations—diarrhea, eczema, and type 1 diabetes mellitus—within the first months of life in males.[25] IPEX patients may also present with thyroiditis (most commonly with hypothyroidism), cytopenias (hemolytic anemia, thrombocytopenia, and neutropenia), renal disease (most commonly tubulonephropathy and nephrotic syndrome), and splenomegaly and lymphadenopathy.[26–28] The scurfy mouse exhibits similar pathophysiology as human IPEX patients, along with extensive multiorgan infiltration of overproliferative CD4+ T cells, and elevated cytokine expression, including IL-4, IL-6, IL-7, and TNFα.[29–31] The scurfy mouse serves as a good example of characterizing the basis of an autoimmune disease involving Treg cell dysfunction, from the disease phenotype and molecular genetics to cellular and molecular immunology. The first features described included genital papilla reddening and swelling, tail skin dermatitis ("scaliness"), and blepharitis, more precisely "late opening of the eyelids."[29–31] Most scurfy mice succumb to multiorgan inflammation and die at approximately 3 weeks of age.[29] Clinical signs of anemia in scurfy mice were reported as paleness of the feet and tail.[29] Evidence of autoimmune bone marrow failure were seen as reduced numbers of megakaryocytes and apoptotic megakaryocytes.[29] The liver of scurfy mice showed extramedullary hematopoiesis and infiltration of hematopoietic cells, but lack of megakaryocytes, and the blood vessels were noted to be devoid of red blood cells.[29] Hyperplasia of the white and red pulp was present in the spleen of scurfy mice.[29] More recently, the multiorgan inflammation was evaluated in skin, lung, liver, bile duct, pancreas, lacrimal glands, salivary glands, colon, stomach, and accessory reproductive organs (coagulation glands/ seminal vesicle, preputial glands, epididymis, prostate) of the scurfy mouse.[32,33] Additionally, tissue-specific autoantibodies, including liver and biliary duct anti-pyruvate dehydrogenase-E2 and skin anti-keratin-14 antibodies, were identified in scurfy mice.[34,35] Characterizing the organ-infiltrating immune populations and determining which of these populations are responsible for inflammation was the next step in elucidating the mechanisms of multiorgan inflammation in scurfy mice. Flow cytometric analysis of immune markers on organ-infiltrating immune cells allowed identification of the specific immune populations infiltrating the organs. Further in vivo depletion of specific populations was used to establish which populations are responsible for pathogenesis. These studies determined that the effector cells in scurfy mice causative of disease pathogenesis were CD4$^+$ T lymphocytes.[30] Next, using mice with specific gene knockout on scurfy background helped clarify the role of specific molecules, including costimulatory molecules, cytokines, and receptors, in the scurfy multiorgan inflammation pathogenesis. Such examples include Sf.$Cd4^{-/-}$, Sf.$\beta2m^{-/-}$, Sf.$Cd28^{-/-}$, Sf.$Aire^{-/-}$, Sf.$Il2^{-/-}$, Sf.$Itgae^{-/-}$, Sf.$Fas^{lpr/lpr}$, Sf.$Ifng^{-/-}$, Sf.$Il4^{-/-}$, Sf.$Stat6^{-/-}$, Sf.$Ltb4r1^{-/-}$, Sf.$Alox5^{-/-}$, Sf.$Cx3cr1^{gfp/gfp}$, and Sf.$Il10^{-/-}$.[33] Linking the $Foxp3$ loss-of-function mutation in scurfy mice to Treg cell dysfunction is another major step in defining the importance of Treg cells in immunological self-tolerance and immune homeostasis. The $Foxp3$ gene was found to contain loss-of-function mutations in both the $scurfy$ mouse and human IPEX patients with at least 63 different mutations reported thus far.[21–23,25] This suggested that $Foxp3$ may play a role in Treg cell maintenance and function. As expected, CD25$^+$ Treg cells were found to highly express $Foxp3$ at the mRNA and protein level.[36–38] Bone marrow chimeras and adoptive transfers of specific immune populations are the standard experiments used to determine the contribution of a distinct immune population, in addition to germ-line, conditional, and inducible knockout mice later in time. Through such studies, Foxp3 was confirmed as a crucial transcription factor in the differentiation of Treg cells (Figure 12.1).[36] But can the observed pathology in scurfy mice and human IPEX patients be fully attributed to Treg cell dysfunction? Foxp3 protein expression was found to be restricted in a subset of T cells with suppressor function.[39,40] This was accomplished using genetic studies with knock-in mouse expressing fluorescent reporter proteins under the control of the endogenous *Foxp3* regulatory elements.[39,40] The presence or absence of *Foxp3* did not affect negative selection of self-antigen reactive thymocyte, activation and proliferation, or cytokine production by peripheral T cells in a bone marrow transplant chimera study where *Rag2ko* recipients, receiving *Foxp3ko* and *Foxp3wt* bone marrow at a 1:1 ratio, were challenged with bacteria or virus or superantigen,[36,39,41,42] which shows that there are T cell-intrinsic mechanisms of tolerance that do not require *Foxp3*. Last, lympho- and myeloproliferative syndrome of scurfy mice was rescued by adoptive transfer of Treg cells in newborn scurfy mice,[36,43] demonstrating the critical role of Treg cells in the disease. Thus, *Foxp3* is established as the central transcription factor in Treg cell differentiation and a crucial cell-extrinsic element in peripheral tolerance and immune homeostasis (Figure 12.1).

## 12.3 CTLA-4

The protein cytotoxic T lymphocyte antigen-4 (CTLA-4) is an essential negative regulator of immune responses (Figure 12.1). The systemic lack of CTLA-4 triggers a fatal lymphoproliferative disease involving systemic autoimmunity.[20] The fulminant lymphoproliferative disorder and systemic autoimmunity of CTLA-4-deficient mice include the following: myocarditis with interstitial infiltrates of lymphocytes, macrophages, and neutrophils; pancreatitis with mononuclear infiltrates and destruction of islets and glandular portions; lungs and salivary glands with interstitial mononuclear infiltrates; and liver with mononuclear cells aggregation. Additionally, synovitis and vasculitis were observed in some mice, but the kidney and thyroid were not affected.[44,45] These germline CTLA-4-deficient mice die by 3 to 4 weeks of age, similar to scurfy mice.[22,44] The massive multiorgan infiltration of mononuclear cells led to investigation of CTLA-4's role in T cells. To show that CTLA-4 is an essential negative regulator of immune responses and to pinpoint the target effector population regulated by CTLA-4, monoclonal antibody depletion of either CD8+ or CD4+ T cell populations and antibodies blocking costimulation were used.[46] The activated CD4+ T lymphocytes were found to be the mediators of the fatal lymphoproliferative disorder observed in CTLA-4-deficient mice. It was also found that activation of CD4+ T cells was dependent on the costimulatory molecule CD28.[46] In addition, CTLA-4 blockade with specific antibodies was found to enhance antitumor immunity, which further supports the role of CTLA-4 in inhibition of T cell activation.[47] The full elucidation of the exact mechanism underlying the function of CTLA-4 in immune homeostasis became complicated. CTLA-4 is a transmembrane protein expressed on the plasma membrane and it has been shown that CTLA-4 binds to B7 molecules (CD80 and CD86) present on antigen-presenting cells (APCs),[48] competing with CD28.[49] Given this information, it was hypothesized that in the absence of CTLA-4, T cells would become resistant to this major regulatory "checkpoint," leading to uncontrolled activation and proliferation.[50] However, this model failed to account for two observations: (1) the high expression of CTLA-4 on Treg cells and (2) therapeutic administration of a CTLA-4/immunoglobin (CTLA-4Ig) fusion protein still suppresses T cell effector function post-activation, a stage no longer dependent upon CD28.[51] Taken together, these results suggest a secondary suppression mechanism for CTLA-4 which is CD28-independent. Indeed, experiments using CTLA-4Ig in vivo showed a Treg cell-dependent mechanism of suppression.[52] Given the constitutive expression of CTLA-4 in Treg cells,[53–57] focused shifted on the function of CTLA-4 in Treg cells. The finding that CTLA-4 is a target of Foxp3 transcriptome further bolstered the suggestion that CTLA-4 plays a role in proper Treg cell functioning in immune homeostasis (Figure 12.1).[58,59] Depletion of CD25+ T cells further demonstrated that the Treg cell CTLA-4 was responsible for blocking expansion of human T cells in tumors.[60] The indispensable role of CTLA-4 in effective Treg cell response was shown when conditional deletion of CTLA-4 in Foxp3-expressing cells triggered in mice a similar fatal lymphoproliferative disease as in germline CTLA-4-deficient mice, although with delayed onset.[61] Further, using adoptive transfer models, it was shown that loss of CTLA-4 in Treg cells is sufficient to cause Treg cell dysfunction and the inability of Treg cells to regulate pancreatic islet destruction.[62–64] The same studies demonstrated that CTLA-4 acts as a negative regulator in activated conventional T cells to control immune response.[63,64] A recent study challenged the concept that CTLA-4 is crucial for all Treg cell function. Using a flox-inducible Cre recombinase system, whereby the Cre recombinase is under the control of a mutant tamoxifen-responsive estrogen receptor, investigators have removed *Ctla4* in Foxp3-expressing cells in adult mice, and found that no spontaneous autoimmunity developed and that experimental autoimmune encephalomyelitis (EAE) was prevented.[65] The authors found that the mature CTLA-4-deficient Foxp3+ Treg cells possess increased proliferation, increased IL-10, PD-1, and Lag3 expression, all markers of augmented suppressive phenotype.[65] In this model, *Ctla4* is only deleted upon tamoxifen administration either ubiquitously ($UBC^{cre-ERT2}$) or restricted to Foxp3-expressing cells in adult mice.[65] The authors attributed this apparent discrepancy to the fact that there might be a period during neonatal development in which CTLA-4 is essential for proper Tregs function, but that CTLA-4 possesses a cell-intrinsic inhibitory role in Treg cells post development.[65] Interestingly, using confocal microscopy, adoptive transfer, and transgenic mouse models, it was found that CTLA-4 expressed on T cells or Treg cells physically remove CD80 and CD86, two major costimulatory molecules, from APCs through a process termed trans-endocytosis, which established a cell-extrinsic function of CTLA-4.[66] Recently, a human autosomal dominant immune dysregulation syndrome involving *CTLA4* mutations has been characterized.[67,68] Using whole-exome sequencing, genetic linkage studies, and flow cytometric analysis, investigators have found that these patients, although with incomplete penetrance, carry a heterozygous nonsense mutation in the first exon of *CTLA4* and various other haploinsufficient-loss-of-function mutations that decreased the expression of CTLA-4.[67,68] Immune infiltration in lung and gut, and cytopenias, along with diarrhea/enteropathy, hypogammaglobulinemia, and granulomatous lymphocytic interstitial lung disease, were the most common clinical manifestations.[67,68] Treg cell dysfunction was present in these patients: impaired suppressive activity and defective trans-endocytosis of CD80.[67,68] Overall, the molecular mechanisms and temporal control underlying CTLA-4 function are still areas undergoing active investigation.

## 12.4 IL-2/IL-2 RECEPTOR COMPLEX

It was found that low levels of IL-2 are produced by CD4+ T cells during steady-state, and that activated CD4+ and CD8+ T cells highly upregulate IL-2 production, with activated CD4+ T cells producing the most significant IL-2.[69–75] IL2R complex is constituted of IL2Rα (CD25), IL2Rβ (CD122), and IL2rγ (CD132, common gamma chain). The IL-2/IL2R complex system is intimately linked to Treg cell development and optimal Treg function (Figure 12.1). Mice deficient in IL-2, IL2Rα (CD25), or IL2Rβ (CD122) developed lethal systemic autoimmunity.[76–79] IL-2 germline deficient mice were found to develop normally in the first 3 to 4 weeks but soon deteriorated and 50% die by 9 weeks of age.[76] Characteristics of lymphoproliferative disease can be seen in IL-2-deficient mice, namely, splenomegaly, lymphoadenopathy, and severe anemia.[76] The remaining surviving mice developed inflammatory bowel disease with 100% penetrance, with chronic diarrhea, intermittent intestinal bleeding, and rectal prolapse.[76] Infiltration of colonic mucosa by lymphocytes and plasma cells, and amyloidosis in liver, spleen, and kidneys were found in IL-2-deficient mice.[76,77] CD25, the subunit that confers high affinity binding to IL-2, was found to be constitutively expressed on Treg cells and transiently expressed on CD4+ and CD8+ T cells subsequent to TCR activation.[80–85] Thus, IL-2 and CD25 became another focus into the maintenance and function of Treg cells. IL-2 is required for the maintenance, survival, and homeostasis of Treg cells (Figure 12.1), and neutralization of IL-2 inhibits the proliferation of Tregs with concomitant induction of autoimmunity, specifically autoimmune gastritis in BALB/c mice, and early onset of diabetes and gastritis, thyroiditis, sialadenitis, and neuropathy in NOD mice.[85,86] In line with these findings, adoptive transfer of CD4+CD25+ Treg cells into neonatal CD122 (IL2Rβ)-deficient mice prevented the lethal autoimmune phenotype, and it was found that IL-2 and IL2R are required for Treg development in the thymus and critical for Foxp3 induction.[87,88] Furthermore, CD25 expression and the maintenance of high Foxp3 expression, and thus effective suppressive function of Treg cells, was enhanced upon IL-2 signaling.[88,89] Although IL-2 signaling is essential for Treg cell maintenance and homeostasis, Treg cells cannot produce IL-2 autonomously.[90–93] Thus, due to the high level of CD25 and the requirement for IL-2, IL-2 sequestration was proposed as one mechanism through which Treg cells exert their suppressive function (Figure 12.1).[91,94,95] The autoimmune susceptibility locus *Idd3* in NOD mice contains the IL-2 gene, and IL-2 haplo-insufficiency mimics the autoimmune phenotype of NOD mice and of IL-2-deficient mice through the instigation of Treg cell dysfunction.[96,97] Linking the IL-2 or IL2R-deficient mice phenotype to human disease, patients carrying mutations in *CD25* gene were found to exhibit autoimmunity, namely, primary biliary cirrhosis, IPEX-like syndrome, lymphadenopathy, and persistent viral infections.[98–101] These patients also have extensive lymphocytic infiltrations in the lung, liver, gut, and bone, with tissue inflammation and atrophy.[98,99] Moreover, SNPs in *CD25* and *CD122* genes were identified in genome-wide association studies that examined approximately 2000 patients, each with autoimmune diseases, including type 1 diabetes mellitus and rheumatoid arthritis.[102] Thus, these studies collectively highlighted the importance of IL-2 and IL2R in Treg cell and immune homeostasis (Figure 12.1).

## 12.5   TGF-β SYSTEM

Central to the maintenance of immunological self-tolerance and immune homeostasis is the TGF-β system.[103–107] The pleiotropic roles of TGF-β in cell proliferation, differentiation, survival, and migration, affecting various biological processes, such as development, wound healing, immune responses, fibrosis, and carcinogenesis, need to be kept in mind when investigating the TGF-β system in immune homeostasis.[108,109] The significance of TGF-β in immune regulation began with the finding that TGF-β inhibits IL-2-dependent T cell proliferation in a negative feedback manner. Upon IL-2 stimulation, activated T cells upregulate TGF-β receptor and TGF-β mRNA, which represses T cell proliferation.[110] This initial study suggested that TGF-β may act as a negative regulator of T cell immune response. Subsequent removal of TGF-β1 in the germline demonstrated a pivotal role of TGF-β1 in suppressing autoimmune diseases and inflammation.[111–113] TGF-β1-deficient mice showed no gross developmental abnormalities, but exhibited severe wasting disease, quickly followed by death around 3 weeks of age.[111,112] Multiorgan inflammation, involving heart, stomach, liver, lung, pancreas, salivary gland, and striated muscle, was observed in TGF-β1-deficient mice.[111,112] Blood of TGF-β1-deficient mice also contained increased titers of autoantibodies to nuclear antigens such as ssDNA, dsDNA, Sm, and RNPs, and immune complex deposits were found in renal glomeruli.[113] When hematopoietic cells from bone marrow or spleen of TGF-β1-deficient mice were transplanted into a normal irradiated recipient, similar autoimmune phenotype, characterized by elevated autoantibodies titers and inflammatory cell infiltration in tissue, was recapitulated, which indicates that autoimmune disease in the absence of TGF-β1 is caused by the immune system.[113] The most significant source of TGF-β1 in the control of immune homeostasis is from T cells and from nonhematopoietic cells. Restoration of normal blood TGF-β1 level in TGF-β1-deficient mice through liver-specific expression of TGF-β1 was not able to ameliorate the lethal autoimmune phenotype in TGF-β1-deficient mice,[114] suggesting that the T cell, autocrine source of TGF-β1 is essential (Figure 12.1). Using genetic studies involving the expression of a dominant negative TGF-βRII under the control of the *CD4* promoter (CD4-dn TGF-βRII), which abrogated the biological activity of TGF-β in T lymphocytes, the autoimmune phenotype seen in TGF-β1-deficient mice was reproduced.[115] The CD4-dn TGF-βRII mice showed enhanced CD4+ and CD8+ T cell proliferation, prominent lymphocytic infiltration in lung and colon, presence of antinuclear antibodies, increased T helper cell-dependent immunoglobulin (IgG1, IgG2a, and IgA), immune complex deposits in renal glomeruli, spontaneous activation of T lymphocytes (as exhibited by CD44highCD62Llow CD4+ and CD8+ populations), and increased IL-4 and IFNγ production by T cells.[115] Further, *Tgfb1flox/null* CD4-Cre mice, in which TGF-β1 is specifically removed in T cells, also developed lymphocytic infiltration in the colon, liver, and lung; wasting disease; and immune complex deposits in renal glomeruli, similar to TGF-β1-deficient germline mice, but their lifespan was extended until 6 months of age.[116] T cells of these mice were activated, hyperproliferative, differentiated into effector T cells, with a skewing toward increased Th1 phenotype.[116] Additionally,

*Tgfbr2flox/flox CD4*-Cre mice, in which the biological effect of TGF-β1 signaling was investigated specifically in T cells, developed lethal autoimmune manifestations with lymphocytic infiltration in the stomach, lung, liver, pancreas, and thyroid, and death by 5 weeks of age.[117] The cause of the fulminant autoimmunity was traced to enhanced aberrant differentiation of CD4+ and CD8+ pathogenic T cells which express NK1.1.[118] Further efforts to elucidate the role of TGF-β in T cells reached the conclusion that absence of signaling in peripheral effector CD4+ T cells results in rapid type 1 diabetes development on NOD background with elevated Th1 response, and it is the CD4+ T cell-produced TGF-β that exerts the suppressive effect. However, TGF-β signaling was found to be dispensable for Treg cell function, development, and maintenance.[119] Thus the TGF-β system is highly intricate and investigations into its pleiotropic effects are still ongoing, and it is certain that TGF-β plays a role in immune homeostasis.

## 12.6  P300 AND TIP60

It has been shown that Foxp3 is posttranslationally modified and such modifications play a critical role in maintaining its stability or in mediating its function (Figure 12.1).[120] Acetylation was one of the posttranslational modifications which appeared to have an impact on Treg cell function. TAT-interacting protein 60 (TIP60) is a histone acetyl transferase (HAT) known to promote acetylation of Foxp3 together with another HAT, p300.[121,122] *TIP60 flox/flox Foxp3YFP*-Cre mice developed wasting disease, dermatitis, and splenomegaly, and died early in life. This phenotype was associated with a major reduction of Treg cells in the periphery but not in the thymus.[122] *P300 flox/flox Foxp3YFP*-Cre mice also developed dermatitis, lymphadenopathy, and splenomegaly, but later, at 10 weeks of age, associated with increased T cell activation. p300-Deficient Treg cells have reduced survival and reduced suppression activity in the CD45RBhi CD4+ T cell transfer model of colitis in Rag1−/− mice.[121] The critical roles of these two HATs in Treg cells have thus been demonstrated, however, whether the observed phenotype is a direct consequence of Foxp3's altered acetylation in vivo remains to be established, given that the observed phenotype can be due to altered acetylation of other proteins.

## 12.7  Id2/Id3

Id proteins are known to interact with E proteins and block their DNA binding activity, with Id2 and Id3 proteins playing critical roles in T cells.[123] Absence of both Id2 and Id3 conditionally in Treg cells, but neither individually, caused major alterations in maintenance and homing of Treg cells, as well as in formation of follicular T regulatory (Tfr) cells, known to control germinal centers (Figure 12.1).[124] *Id2 flox/flox Id3flox/floxFoxp3*-Cre mice developed dermatitis, splenomegaly, lymphadenopathy, and leukocytic infiltrates in the lung, esophagus, and eyelids, with an overall Th2-mediated inflammation and succumbed within 10 weeks of age,[124] demonstrating that unleashing the E proteins activity can negatively impact Treg cells. Absence of the two Id proteins also affected the suppression activity of Treg cells evaluated by the CD45RBhi CD4+ T cell transfer model of colitis.[124]

## 12.8 TREG CELL UBIQUITIN LIGASES

Several studies indicated that Foxp3 is ubiquitinated, and this modification has an impact on Fopx3's stability and activity (Figure 12.1).[120] Similar to acetylation, in most cases the impact of the specific ubiquitin ligase was demonstrated on Foxp3 by in vitro studies. Additional ubiquitin ligases have been demonstrated to have an impact on Treg cells and in some cases their removal caused systemic autoimmunity; however, in most cases the specific target(s) in Treg cells has not been identified.

## 12.9 VHL

The VHL gene encodes an E3 ubiquitin ligase, and in humans its mutations predispose to several types of cancers.[125] One of the VHL's targets is HIF1$\alpha$. Under normoxic conditions HIF1$\alpha$ is hydroxylated by prolyl hydroxylase domain (PHD) enzymes and then ubiquitinated by VHL for degradation. Hypoxia reduces the activity of the PHD enzymes, and thus HIF1$\alpha$'s recognition by VHL and degradation, which causes accumulation of HIF1$\alpha$ and expression of target genes, including those parts of the glycolytic pathway.[125] It has been recently demonstrated that the HIF1$\alpha$ pathway controls Foxp3 levels[126] and the balance between Th17 and Treg cells (Figure 12.1).[127] Using Treg cell conditional VHL knockout mice, Lee et al.[128] demonstrated that VHL is critical for maintenance of Treg cell stability and function through regulation of the HIF1$\alpha$ pathway. $VHL^{flox/flox}Foxp3$-Cre mice succumbed to systemic inflammation between 6 and 11 weeks of age.[128] VHL-deficient Treg cells had reduced suppression function on CD4$^+$ T cells in vitro and also failed to suppress colitis development in the CD45RB$^{hi}$ CD4$^+$ T cell transfer model in Rag1$^{-/-}$ mice. Additionally, VHL-deficient Treg cells converted into IFN$\gamma$-producing-, glycolytic-reprogrammed effector T cells, in a manner dependent on HIF1$\alpha$ (Figure 12.1).[128] Thus these results demonstrate that VHL is required to restrict HIF1$\alpha$ in Treg cells and maintain integrity of Treg cell metabolism and function (Figure 12.1).

## 12.10 ITCH

Another E3 ubiquitin ligase with a critical role in Treg cells is Itch.[129] As the name indicates, Itch-deficient mice develop a skin-scratching phenotype, along with lymphadenopathy, splenomegaly, and inflammation in the lungs and digestive tract.[130] Itch was found to ubiquitinate the transcription factor TIEG1 thus leading to TIEG1-dependent upregulation of Foxp3 (Figure 12.1).[131] Treg specific depletion of Itch resulted also in a lymphoproliferative disorder, pulmonary inflammation, and skin lesions starting with 6 weeks of age, together with high mortality.[129] Itch-deficient Treg cells had normal suppression function in vitro and in vivo, but Itch-deficient Treg cells acquired Th2-like properties believed to be responsible for the observed phenotype.[129]

## 12.11 CONCLUSION

The genetic models presented in this chapter have alterations in specific genes critical for Treg cell identity and function and display a systemic autoimmune phenotype

(Figure 12.1). The underlying defective mechanisms causative of systemic autoimmunity can be studied in these models, together with other genes implicated in the same pathway. However, the fulminant display and timing for some of these disease models may preclude in-depth investigation into the exact molecular and cellular pathogenesis due to the bystander activation and a deregulated systemic cytokine environment that impact Treg cells. In some instances the use of inducible models bypasses the early systemic autoimmunity. Additionally, the use of the conditional reporter deleter *Foxp3*-Cre-*YFP* model in the mosaic females *Foxp3*-Cre-*YFP*[+/WT] allows one to study the Treg cell population with the specific deletion in the presence of wild-type Treg cells.[132] Such mosaic females are likely to not develop systemic inflammation due to the random inactivation of X chromosome-linked genes (i.e., *Foxp3YFP*-Cre), and the presence of both wild type and KO cells in the same mouse. Thus such alternative avenues can be employed to decipher the molecular mechanisms responsible for Treg cell alterations in such mice with autoimmune phenotypes.

## ACKNOWLEDGMENTS

This work was supported by National Institutes of Health (NIH) grants RO1AI067846 and RO1AI078273, and by University of Florida Gatorade Trust to Dorina Avram. In addition, we thank Mr. Adrian Avram for support with graphical presentation.

The United States Navy, United States Armed Forces Health Professions Scholarship Program funds the medical education of Jonathan J. Cho. The views expressed in this article are those of the authors and do not necessarily reflect the official policy or position of the Department of the Navy, Department of Defense, nor the U.S. government.

## REFERENCES

1. Ehrlich, P., and Himmelweit, F. *Collected Papers; Including a Complete Bibliography.* (Pergamon, 1956).
2. Steinman, R. M., and Nussenzweig, M. C. Avoiding horror autotoxicus: The importance of dendritic cells in peripheral T cell tolerance. *Proc Natl Acad Sci USA* **99**, 351–358, doi:10.1073/pnas.231606698 (2002).
3. Bluestone, J. A. Mechanisms of tolerance. *Immunol Rev* **241**, 5–19, doi:10.1111/j.1600-065X.2011.01019.x (2011).
4. Hogquist, K. A., Baldwin, T. A., and Jameson, S. C. Central tolerance: Learning self-control in the thymus. *Nat Rev Immunol* **5**, 772–782, doi:10.1038/nri1707 (2005).
5. Goodnow, C. C., Sprent, J., Fazekas de St Groth, B., and Vinuesa, C. G. Cellular and genetic mechanisms of self tolerance and autoimmunity. *Nature* **435**, 590–597, doi:10.1038/nature03724 (2005).
6. Xing, Y., and Hogquist, K. A. T-cell tolerance: Central and peripheral. *Cold Spring Harb Perspect Biol* **4**, doi:10.1101/cshperspect.a00695710.1101/cshperspect.a006957 (2012).
7. Fathman, C. G., and Lineberry, N. B. Molecular mechanisms of CD4+ T-cell anergy. *Nat Rev Immunol* **7**, 599–609, doi:10.1038/nri2131 (2007).
8. Krammer, P. H., Arnold, R., and Lavrik, I. N. Life and death in peripheral T cells. *Nat Rev Immunol* **7**, 532–542, doi:10.1038/nri2115 (2007).

9. Wells, A. D. New insights into the molecular basis of T cell anergy: Anergy factors, avoidance sensors, and epigenetic imprinting. *J Immunol* **182**, 7331–7341, doi:10.4049 /jimmunol.0803917 (2009).

10. Wing, K., and Sakaguchi, S. Regulatory T cells exert checks and balances on self tolerance and autoimmunity. *Nat Immunol* **11**, 7–13, doi:10.1038/ni.1818 (2010).

11. Hsieh, C. S., Lee, H. M., and Lio, C. W. Selection of regulatory T cells in the thymus. *Nat Rev Immunol* **12**, 157–167, doi:10.1038/nri3155 (2012).

12. Mueller, D. L. Mechanisms maintaining peripheral tolerance. *Nat Immunol* **11**, 21–27, doi:10.1038/ni.1817 (2010).

13. Sakaguchi, S., Yamaguchi, T., Nomura, T., and Ono, M. Regulatory T cells and immune tolerance. *Cell* **133**, 775–787, doi:10.1016/j.cell.2008.05.009 (2008).

14. Rudensky, A. Y. Regulatory T cells and Foxp3. *Immunol Rev* **241**, 260–268, doi:10.1111 /j.1600-065X.2011.01018.x (2011).

15. Sakaguchi, S., Vignali, D. A., Rudensky, A. Y., Niec, R. E., and Waldmann, H. The plasticity and stability of regulatory T cells. *Nat Rev Immunol* **13**, 461–467, doi:10.1038 /nri3464 (2013).

16. Feng, Y. et al. A mechanism for expansion of regulatory T-cell repertoire and its role in self-tolerance. *Nature* **528**, 132–136, doi:10.1038/nature16141 (2015).

17. Mosser, D. M., and Zhang, X. Interleukin-10: New perspectives on an old cytokine. *Immunol Rev* **226**, 205–218, doi:10.1111/j.1600-065X.2008.00706.x (2008).

18. Avram, D., and Califano, D. The multifaceted roles of Bcl11b in thymic and peripheral T cells: Impact on immune diseases. *J Immunol* **193**, 2059–2065, doi:10.4049 /jimmunol.1400930 (2014).

19. Ouyang, W., Rutz, S., Crellin, N. K., Valdez, P. A., and Hymowitz, S. G. Regulation and functions of the IL-10 family of cytokines in inflammation and disease. *Annu Rev Immunol* **29**, 71–109, doi:10.1146/annurev-immunol-031210-101312 (2011).

20. Walker, L. S., and Sansom, D. M. The emerging role of CTLA4 as a cell-extrinsic regulator of T cell responses. *Nat Rev Immunol* **11**, 852–863, doi:10.1038/nri3108 (2011).

21. Chatila, T. A. et al. JM2, encoding a fork head-related protein, is mutated in X-linked autoimmunity-allergic disregulation syndrome. *J Clin Invest* **106**, R75–R81, doi:10.1172 /JCI11679 (2000).

22. Brunkow, M. E. et al. Disruption of a new forkhead/winged-helix protein, scurfin, results in the fatal lymphoproliferative disorder of the scurfy mouse. *Nat Genet* **27**, 68–73, doi:10.1038/83784 (2001).

23. Bennett, C. L. et al. The immune dysregulation, polyendocrinopathy, enteropathy, X-linked syndrome (IPEX) is caused by mutations of FOXP3. *Nat Genet* **27**, 20–21, doi:10.1038/83713 (2001).

24. Wildin, R. S. et al. X-linked neonatal diabetes mellitus, enteropathy and endocrinopathy syndrome is the human equivalent of mouse scurfy. *Nat Genet* **27**, 18–20, doi:10.1038/83707 (2001).

25. Barzaghi, F., Passerini, L., and Bacchetta, R. Immune dysregulation, polyendocrinopathy, enteropathy, x-linked syndrome: A paradigm of immunodeficiency with autoimmunity. *Front Immunol* **3**, 211, doi:10.3389/fimmu.2012.00211 (2012).

26. Gambineri, E. et al. Clinical and molecular profile of a new series of patients with immune dysregulation, polyendocrinopathy, enteropathy, X-linked syndrome: Inconsistent correlation between forkhead box protein 3 expression and disease severity. *J Allergy Clin Immunol* **122**, 1105–1112.e1101, doi:10.1016/j.jaci.2008.09.027 (2008).

27. Wildin, R. S., Smyk-Pearson, S., and Filipovich, A. H. Clinical and molecular features of the immunodysregulation, polyendocrinopathy, enteropathy, X linked (IPEX) syndrome. *J Med Genet* **39**, 537–545 (2002).

28. Burroughs, L. M. et al. Stable hematopoietic cell engraftment after low-intensity nonmyeloablative conditioning in patients with immune dysregulation, polyendocrinopathy, enteropathy, X-linked syndrome. *J Allergy Clin Immunol* **126**, 1000–1005, doi:10.1016/j.jaci.2010.05.021 (2010).

29. Lyon, M. F., Peters, J., Glenister, P. H., Ball, S., and Wright, E. The scurfy mouse mutant has previously unrecognized hematological abnormalities and resembles Wiskott-Aldrich syndrome. *Proc Natl Acad Sci USA* **87**, 2433–2437 (1990).

30. Blair, P. J. et al. CD4+CD8- T cells are the effector cells in disease pathogenesis in the scurfy (sf) mouse. *J Immunol* **153**, 3764–3774 (1994).

31. Kanangat, S. et al. Disease in the scurfy (sf) mouse is associated with overexpression of cytokine genes. *Eur J Immunol* **26**, 161–165, doi:10.1002/eji.1830260125 (1996).

32. Sharma, R. et al. IL-2-controlled expression of multiple T cell trafficking genes and Th2 cytokines in the regulatory T cell-deficient scurfy mice: Implication to multiorgan inflammation and control of skin and lung inflammation. *J Immunol* **186**, 1268–1278, doi:10.4049/jimmunol.1002677 (2011).

33. Ju, S. T., Sharma, R., Gaskin, F., Kung, J. T., and Fu, S. M. The biology of autoimmune response in the scurfy mice that lack the CD4+Foxp3+ regulatory T-cells. *Biology (Basel)* **1**, 18–42, doi:10.3390/biology1010018 (2012).

34. Zhang, W. et al. Deficiency in regulatory T cells results in development of antimitochondrial antibodies and autoimmune cholangitis. *Hepatology* **49**, 545–552, doi:10.1002/hep.22651 (2009).

35. Huter, E. N., Natarajan, K., Torgerson, T. R., Glass, D. D., and Shevach, E. M. Autoantibodies in scurfy mice and IPEX patients recognize keratin 14. *J Invest Dermatol* **130**, 1391–1399, doi:10.1038/jid.2010.16 (2010).

36. Fontenot, J. D., Gavin, M. A., and Rudensky, A. Y. Foxp3 programs the development and function of CD4+CD25+ regulatory T cells. *Nat Immunol* **4**, 330–336, doi:10.1038/ni904 (2003).

37. Hori, S., Nomura, T., and Sakaguchi, S. Control of regulatory T cell development by the transcription factor Foxp3. *Science* **299**, 1057–1061, doi:10.1126/science.1079490 (2003).

38. Khattri, R., Cox, T., Yasayko, S. A., and Ramsdell, F. An essential role for Scurfin in CD4+CD25+ T regulatory cells. *Nat Immunol* **4**, 337–342, doi:10.1038/ni909 (2003).

39. Fontenot, J. D. et al. Regulatory T cell lineage specification by the forkhead transcription factor foxp3. *Immunity* **22**, 329–341, doi:10.1016/j.immuni.2005.01.016 (2005).

40. Wan, Y. Y., and Flavell, R. A. Identifying Foxp3-expressing suppressor T cells with a bicistronic reporter. *Proc Natl Acad Sci USA* **102**, 5126–5131, doi:10.1073/pnas.0501701102 (2005).

41. Hsieh, C. S., Zheng, Y., Liang, Y., Fontenot, J. D., and Rudensky, A. Y. An intersection between the self-reactive regulatory and nonregulatory T cell receptor repertoires. *Nat Immunol* **7**, 401–410, doi:10.1038/ni1318 (2006).

42. Chen, Z., Benoist, C., and Mathis, D. How defects in central tolerance impinge on a deficiency in regulatory T cells. *Proc Natl Acad Sci USA* **102**, 14735–14740, doi:10.1073/pnas.0507014102 (2005).

43. Huter, E. N. et al. TGF-beta-induced Foxp3+ regulatory T cells rescue scurfy mice. *Eur J Immunol* **38**, 1814–1821, doi:10.1002/eji.200838346 (2008).

44. Tivol, E. A. et al. Loss of CTLA-4 leads to massive lymphoproliferation and fatal multiorgan tissue destruction, revealing a critical negative regulatory role of CTLA-4. *Immunity* **3**, 541–547 (1995).

45. Waterhouse, P. et al. Lymphoproliferative disorders with early lethality in mice deficient in Ctla-4. *Science* **270**, 985–988 (1995).

46. Chambers, C. A., Sullivan, T. J., and Allison, J. P. Lymphoproliferation in CTLA-4-deficient mice is mediated by costimulation-dependent activation of CD4+ T cells. *Immunity* **7**, 885–895 (1997).
47. Leach, D. R., Krummel, M. F., and Allison, J. P. Enhancement of antitumor immunity by CTLA-4 blockade. *Science* **271**, 1734–1736 (1996).
48. Linsley, P. S. et al. CTLA-4 is a second receptor for the B cell activation antigen B7. *J Exp Med* **174**, 561–569 (1991).
49. Linsley, P. S. et al. Human B7-1 (CD80) and B7-2 (CD86) bind with similar avidities but distinct kinetics to CD28 and CTLA-4 receptors. *Immunity* **1**, 793–801 (1994).
50. Tivol, E. A. et al. Loss of CTLA-4 leads to massive lymphoproliferation and fatal multiorgan tissue destruction, revealing a critical negative regulatory role of CTLA-4. *Immunity* **3**, 541–547, doi:10.1016/1074-7613(95)90125-6 (1995).
51. Deppong, C. M., Parulekar, A., Boomer, J. S., Bricker, T. L., and Green, J. M. CTLA4-Ig inhibits allergic airway inflammation by a novel CD28-independent, nitric oxide synthase-dependent mechanism. *Eur J Immunol* **40**, 1985–1994, doi:10.1002/eji.200940282 (2010).
52. Deppong, C. M. et al. CTLA4Ig inhibits effector T cells through regulatory T cells and TGFβ1. *J Immunol* **191**, 3082–3089, doi:10.4049/jimmunol.1300830 (2013).
53. Walunas, T. L. et al. CTLA-4 can function as a negative regulator of T cell activation. *Immunity* **1**, 405–413 (1994).
54. Linsley, P. S. et al. Coexpression and functional cooperation of CTLA-4 and CD28 on activated T lymphocytes. *J Exp Med* **176**, 1595–1604 (1992).
55. Takahashi, T. et al. Immunologic self-tolerance maintained by CD25(+)CD4(+) regulatory T cells constitutively expressing cytotoxic T lymphocyte-associated antigen 4. *J Exp Med* **192**, 303–310 (2000).
56. Read, S., Malmström, V., and Powrie, F. Cytotoxic T lymphocyte-associated antigen 4 plays an essential role in the function of CD25(+)CD4(+) regulatory cells that control intestinal inflammation. *J Exp Med* **192**, 295–302 (2000).
57. Mead, K. I. et al. Exocytosis of CTLA-4 is dependent on phospholipase D and ADP ribosylation factor-1 and stimulated during activation of regulatory T cells. *J Immunol* **174**, 4803–4811 (2005).
58. Wu, Y. et al. FOXP3 controls regulatory T cell function through cooperation with NFAT. *Cell* **126**, 375–387, doi:10.1016/j.cell.2006.05.042 (2006).
59. Gavin, M. A. et al. Foxp3-dependent programme of regulatory T-cell differentiation. *Nature* **445**, 771–775, doi:10.1038/nature05543 (2007).
60. Manzotti, C. N. et al. Inhibition of human T cell proliferation by CTLA-4 utilizes CD80 and requires CD25+ regulatory T cells. *Eur J Immunol* **32**, 2888–2896, doi:10.1002/1521-4141(2002010)32:10<2888::AID-IMMU2888>3.0.CO;2-F (2002).
61. Wing, K. et al. CTLA-4 control over Foxp3+ regulatory T cell function. *Science* **322**, 271–275, doi:10.1126/science.1160062 (2008).
62. Schmidt, E. M. et al. Ctla-4 controls regulatory T cell peripheral homeostasis and is required for suppression of pancreatic islet autoimmunity. *J Immunol* **182**, 274–282 (2009).
63. Ise, W. et al. CTLA-4 suppresses the pathogenicity of self antigen-specific T cells by cell-intrinsic and cell-extrinsic mechanisms. *Nat Immunol* **11**, 129–135, doi:10.1038/ni.1835 (2010).
64. Jain, N., Nguyen, H., Chambers, C., and Kang, J. Dual function of CTLA-4 in regulatory T cells and conventional T cells to prevent multiorgan autoimmunity. *Proc Natl Acad Sci USA* **107**, 1524–1528, doi:10.1073/pnas.0910341107 (2010).
65. Paterson, A. M. et al. Deletion of CTLA-4 on regulatory T cells during adulthood leads to resistance to autoimmunity. *J Exp Med* **212**, 1603–1621, doi:10.1084/jem.20141030 (2015).

66. Qureshi, O. S. et al. Trans-endocytosis of CD80 and CD86: A molecular basis for the cell-extrinsic function of CTLA-4. *Science* **332**, 600–603, doi:10.1126/science.1202947 (2011).

67. Kuehn, H. S. et al. Immune dysregulation in human subjects with heterozygous germline mutations in CTLA4. *Science* **345**, 1623–1627, doi:10.1126/science.1255904 (2014).

68. Schubert, D. et al. Autosomal dominant immune dysregulation syndrome in humans with CTLA4 mutations. *Nat Med* **20**, 1410–1416, doi:10.1038/nm.3746 (2014).

69. Minasi, L. E., Kamogawa, Y., Carding, S., Bottomly, K., and Flavell, R. A. The selective ablation of interleukin 2-producing cells isolated from transgenic mice. *J Exp Med* **177**, 1451–1459 (1993).

70. Yui, M. A., Sharp, L. L., Havran, W. L., and Rothenberg, E. V. Preferential activation of an IL-2 regulatory sequence transgene in TCR gamma delta and NKT cells: Subset-specific differences in IL-2 regulation. *J Immunol* **172**, 4691–4699 (2004).

71. Mier, J. W., and Gallo, R. C. Purification and some characteristics of human T-cell growth factor from phytohemagglutinin-stimulated lymphocyte-conditioned media. *Proc Natl Acad Sci USA* **77**, 6134–6138 (1980).

72. Villarino, A. V. et al. Helper T cell IL-2 production is limited by negative feedback and STAT-dependent cytokine signals. *J Exp Med* **204**, 65–71, doi:10.1084/jem.20061198 (2007).

73. Yui, M. A., Hernández-Hoyos, G., and Rothenberg, E. V. A new regulatory region of the IL-2 locus that confers position-independent transgene expression. *J Immunol* **166**, 1730–1739 (2001).

74. Sojka, D. K., Bruniquel, D., Schwartz, R. H., and Singh, N. J. IL-2 secretion by CD4+ T cells in vivo is rapid, transient, and influenced by TCR-specific competition. *J Immunol* **172**, 6136–6143 (2004).

75. Huse, M., Lillemeier, B. F., Kuhns, M. S., Chen, D. S., and Davis, M. M. T cells use two directionally distinct pathways for cytokine secretion. *Nat Immunol* **7**, 247–255, doi:10.1038/ni1304 (2006).

76. Sadlack, B. et al. Ulcerative colitis-like disease in mice with a disrupted interleukin-2 gene. *Cell* **75**, 253–261 (1993).

77. Sadlack, B. et al. Generalized autoimmune disease in interleukin-2-deficient mice is triggered by an uncontrolled activation and proliferation of CD4+ T cells. *Eur J Immunol* **25**, 3053–3059, doi:10.1002/eji.1830251111 (1995).

78. Willerford, D. M. et al. Interleukin-2 receptor alpha chain regulates the size and content of the peripheral lymphoid compartment. *Immunity* **3**, 521–530 (1995).

79. Suzuki, H. et al. Deregulated T cell activation and autoimmunity in mice lacking interleukin-2 receptor beta. *Science* **268**, 1472–1476 (1995).

80. Wang, X., Rickert, M., and Garcia, K. C. Structure of the quaternary complex of interleukin-2 with its $\alpha$, $\beta$, and $\gamma_c$ receptors. *Science* **310**, 1159–1163, doi:10.1126/science.1117893 (2005).

81. Stauber, D. J., Debler, E. W., Horton, P. A., Smith, K. A., and Wilson, I. A. Crystal structure of the IL-2 signaling complex: Paradigm for a heterotrimeric cytokine receptor. *Proc Natl Acad Sci USA* **103**, 2788–2793, doi:10.1073/pnas.0511161103 (2006).

82. Rickert, M., Wang, X., Boulanger, M. J., Goriatcheva, N., and Garcia, K. C. The structure of interleukin-2 complexed with its alpha receptor. *Science* **308**, 1477–1480, doi:10.1126/science.1109745 (2005).

83. Kim, H. P., Imbert, J., and Leonard, W. J. Both integrated and differential regulation of components of the IL-2/IL-2 receptor system. *Cytokine Growth Factor Rev* **17**, 349–366, doi:10.1016/j.cytogfr.2006.07.003 (2006).

84. Kim, H. P., Kim, B. G., Letterio, J., and Leonard, W. J. Smad-dependent cooperative regulation of interleukin 2 receptor alpha chain gene expression by T cell receptor and transforming growth factor-beta. *J Biol Chem* **280**, 34042–34047, doi:10.1074/jbc.M505833200 (2005).

85. Setoguchi, R., Hori, S., Takahashi, T., and Sakaguchi, S. Homeostatic maintenance of natural Foxp3(+) CD25(+) CD4(+) regulatory T cells by interleukin (IL)-2 and induction of autoimmune disease by IL-2 neutralization. *J Exp Med* **201**, 723–735, doi:10.1084/jem.20041982 (2005).

86. Yu, A., Zhu, L., Altman, N. H., and Malek, T. R. A low interleukin-2 receptor signaling threshold supports the development and homeostasis of T regulatory cells. *Immunity* **30**, 204–217, doi:10.1016/j.immuni.2008.11.014 (2009).

87. Malek, T. R., Yu, A., Vincek, V., Scibelli, P., and Kong, L. CD4 regulatory T cells prevent lethal autoimmunity in IL-2Rbeta-deficient mice. Implications for the nonredundant function of IL-2. *Immunity* **17**, 167–178 (2002).

88. Fontenot, J. D., Rasmussen, J. P., Gavin, M. A., and Rudensky, A. Y. A function for interleukin 2 in Foxp3-expressing regulatory T cells. *Nat Immunol* **6**, 1142–1151, doi:10.1038/ni1263 (2005).

89. Barron, L. et al. Cutting edge: Mechanisms of IL-2-dependent maintenance of functional regulatory T cells. *J Immunol* **185**, 6426-6430, doi:10.4049/jimmunol.0903940 (2010).

90. Rubtsov, Y. P. et al. Stability of the regulatory T cell lineage in vivo. *Science* **329**, 1667–1671, doi:10.1126/science.1191996 (2010).

91. Thornton, A. M., and Shevach, E. M. CD4+CD25+ immunoregulatory T cells suppress polyclonal T cell activation in vitro by inhibiting interleukin 2 production. *J Exp Med* **188**, 287–296 (1998).

92. Schallenberg, S., Tsai, P. Y., Riewaldt, J., and Kretschmer, K. Identification of an immediate Foxp3(–) precursor to Foxp3(+) regulatory T cells in peripheral lymphoid organs of nonmanipulated mice. *J Exp Med* **207**, 1393–1407, doi:10.1084/jem.20100045 (2010).

93. Chen, Q., Kim, Y. C., Laurence, A., Punkosdy, G. A., and Shevach, E. M. IL-2 controls the stability of Foxp3 expression in TGF-beta-induced Foxp3+ T cells in vivo. *J Immunol* **186**, 6329–6337, doi:10.4049/jimmunol.1100061 (2011).

94. Pandiyan, P., Zheng, L., Ishihara, S., Reed, J., and Lenardo, M. J. CD4+CD25+Foxp3+ regulatory T cells induce cytokine deprivation-mediated apoptosis of effector CD4+ T cells. *Nat Immunol* **8**, 1353–1362, doi:10.1038/ni1536 (2007).

95. Létourneau, S., Krieg, C., Pantaleo, G., and Boyman, O. IL-2- and CD25-dependent immunoregulatory mechanisms in the homeostasis of T-cell subsets. *J Allergy Clin Immunol* **123**, 758–762, doi:10.1016/j.jaci.2009.02.011 (2009).

96. Yamanouchi, J. et al. Interleukin-2 gene variation impairs regulatory T cell function and causes autoimmunity. *Nat Genet* **39**, 329–337, doi:10.1038/ng1958 (2007).

97. Lyons, P. A. et al. Congenic mapping of the type 1 diabetes locus, Idd3, to a 780-kb region of mouse chromosome 3: Identification of a candidate segment of ancestral DNA by haplotype mapping. *Genome Res* **10**, 446–453 (2000).

98. Sharfe, N., Dadi, H. K., Shahar, M., and Roifman, C. M. Human immune disorder arising from mutation of the alpha chain of the interleukin-2 receptor. *Proc Natl Acad Sci USA* **94**, 3168–3171 (1997).

99. Roifman, C. M. Human IL-2 receptor alpha chain deficiency. *Pediatr Res* **48**, 6–11, doi:10.1203/00006450-200007000-00004 (2000).

100. Aoki, C. A. et al. IL-2 receptor alpha deficiency and features of primary biliary cirrhosis. *J Autoimmun* **27**, 50–53, doi:10.1016/j.jaut.2006.04.005 (2006).

101. Caudy, A. A., Reddy, S. T., Chatila, T., Atkinson, J. P., and Verbsky, J. W. CD25 deficiency causes an immune dysregulation, polyendocrinopathy, enteropathy, X-linked-like syndrome, and defective IL-10 expression from CD4 lymphocytes. *J Allergy Clin Immunol* **119**, 482–487, doi:10.1016/j.jaci.2006.10.007 (2007).

102. Consortium, W. T. C. C. Genome-wide association study of 14,000 cases of seven common diseases and 3,000 shared controls. *Nature* **447**, 661–678, doi:10.1038/nature05911 (2007).

103. Li, M. O., Wan, Y. Y., Sanjabi, S., Robertson, A. K., and Flavell, R. A. Transforming growth factor-beta regulation of immune responses. *Annu Rev Immunol* **24**, 99–146, doi:10.1146/annurev.immunol.24.021605.090737 (2006).

104. Wahl, S. M., Orenstein, J. M., and Chen, W. TGF-beta influences the life and death decisions of T lymphocytes. *Cytokine Growth Factor Rev* **11**, 71–79 (2000).

105. Li, M. O., and Flavell, R. A. TGF-beta: A master of all T cell trades. *Cell* **134**, 392–404, doi:10.1016/j.cell.2008.07.025 (2008).

106. Massague, J. TGFbeta signalling in context. *Nat Rev Mol Cell Biol* **13**, 616–630, doi:10.1038/nrm343410.1038/nrm3434. Epub Sep 20 (2012).

107. Travis, M. A., and Sheppard, D. TGF-beta activation and function in immunity. *Annu Rev Immunol* **32**, 51–82, doi:10.1146/annurev-immunol-032713-12025710.1146 /annurev-immunol-032713-120257. Epub 2013 Dec 2. (2013).

108. Blobe, G. C., Schiemann, W. P., and Lodish, H. F. Role of transforming growth factor beta in human disease. *N Engl J Med* **342**, 1350–1358, doi:10.1056/NEJM200005043421807 (2000).

109. Gordon, K. J., and Blobe, G. C. Role of transforming growth factor-beta superfamily signaling pathways in human disease. *Biochim Biophys Acta* **1782**, 197–228, doi:10.1016/j.bbadis.2008.01.006 (2008).

110. Kehrl, J. H. et al. Production of transforming growth factor beta by human T lympho cytes and its potential role in the regulation of T cell growth. *J Exp Med* **163**, 1037–1050 (1986).

111. Shull, M. M. et al. Targeted disruption of the mouse transforming growth factor-beta 1 gene results in multifocal inflammatory disease. *Nature* **359**, 693–699, doi:10.1038/359693a0 (1992).

112. Kulkarni, A. B. et al. Transforming growth factor beta 1 null mutation in mice causes excessive inflammatory response and early death. *Proc Natl Acad Sci USA* **90**, 770–774 (1993).

113. Yaswen, L. et al. Autoimmune manifestations in the transforming growth factor-beta 1 knockout mouse. *Blood* **87**, 1439–1445 (1996).

114. Longenecker, G. et al. Endocrine expression of the active form of TGF-beta1 in the TGF-beta1 null mice fails to ameliorate lethal phenotype. *Cytokine* **18**, 43–50 (2002).

115. Gorelik, L., and Flavell, R. A. Abrogation of TGFbeta signaling in T cells leads to spontaneous T cell differentiation and autoimmune disease. *Immunity* **12**, 171–181 (2000).

116. Li, M. O., Wan, Y. Y., and Flavell, R. A. T cell-produced transforming growth factor-beta1 controls T cell tolerance and regulates Th1- and Th17-cell differentiation. *Immunity* **26**, 579–591, doi:10.1016/j.immuni.2007.03.014 (2007).

117. Li, M. O., Sanjabi, S., and Flavell, R. A. Transforming growth factor-beta controls development, homeostasis, and tolerance of T cells by regulatory T cell-dependent and -independent mechanisms. *Immunity* **25**, 455–471, doi:10.1016/j.immuni.2006.07.011 (2006).

118. Marie, J. C., Liggitt, D., and Rudensky, A. Y. Cellular mechanisms of fatal early-onset autoimmunity in mice with the T cell-specific targeting of transforming growth factor-beta receptor. *Immunity* **25**, 441–454, doi:10.1016/j.immuni.2006.07.012 (2006).

119. Ishigame, H. et al. Excessive Th1 responses due to the absence of TGF-beta signaling cause autoimmune diabetes and dysregulated Treg cell homeostasis. *Proc Natl Acad Sci USA* **110**, 6961–6966, doi:10.1073/pnas.130449811010.1073/pnas.1304498110. Epub 2013 Apr 8. (2013).

120. van Loosdregt, J., and Coffer, P. J. Post-translational modification networks regulating FOXP3 function. *Trends Immunol* **35**, 368–378, doi:10.1016/j.it.2014.06.00510.1016/j .it.2014.06.005. Epub Jul 18 (2014).

121. Liu, Y. et al. Inhibition of p300 impairs Foxp3(+) T regulatory cell function and promotes antitumor immunity. *Nat Med* **19**, 1173–1177, doi:10.1038/nm.3286 (2013).
122. Xiao, Y. et al. Dynamic interactions between TIP60 and p300 regulate FOXP3 function through a structural switch defined by a single lysine on TIP60. *Cell Rep* **7**, 1471–1480, doi:10.1016/j.celrep.2014.04.02110.1016/j.celrep.2014.04.021. Epub May 15. (2014).
123. Belle, I., and Zhuang, Y. E proteins in lymphocyte development and lymphoid diseases. *Curr Top Dev Biol* **110**, 153–187, doi:10.1016/b978-0-12-405943-6.00004-x10.1016/B978-0-12-405943-6.00004-X. (2014).
124. Miyazaki, M. et al. Id2 and Id3 maintain the regulatory T cell pool to suppress inflammatory disease. *Nat Immunol* **15**, 767–776, doi:10.1038/ni.2928 (2014).
125. Kaelin, W. G., Jr. The von Hippel-Lindau tumour suppressor protein: O2 sensing and cancer. *Nat Rev Cancer* **8**, 865–873, doi:10.1038/nrc250210.1038/nrc2502. Epub Oct 16 (2008).
126. Clambey, E. T. et al. Hypoxia-inducible factor-1 alpha-dependent induction of FoxP3 drives regulatory T-cell abundance and function during inflammatory hypoxia of the mucosa. *Proc Natl Acad Sci USA* **109**, E2784–2793, doi:10.1073/pnas.120236610910.1073/pnas.1202366109. Epub 2012 Sep 17. (2012).
127. Dang, E. V. et al. Control of T(H)17/T(reg) balance by hypoxia-inducible factor 1. *Cell* **146**, 772–784, doi:10.1016/j.cell.2011.07.03310.1016/j.cell.2011.07.033. Epub Aug 25 (2011).
128. Lee, J. H., Elly, C., Park, Y., and Liu, Y. C. E3 ubiquitin ligase VHL regulates hypoxia-inducible factor-1alpha to maintain regulatory T cell stability and suppressive capacity. *Immunity* **42**, 1062–1074, doi:10.1016/j.immuni.2015.05.01610.1016/j.immuni.2015.05.016 (2015).
129. Jin, H. S., Park, Y., Elly, C., and Liu, Y. C. Itch expression by Treg cells controls Th2 inflammatory responses. *J Clin Invest* **123**, 4923–4934, doi:10.1172/jci6935510.1172/JCI69355. Epub Oct 25 (2013).
130. Liu, Y. C. The E3 ubiquitin ligase Itch in T cell activation, differentiation, and tolerance. *Semin Immunol* **19**, 197–205, doi:10.1016/j.smim.2007.02.003 (2007).
131. Venuprasad, K. et al. The E3 ubiquitin ligase Itch regulates expression of transcription factor Foxp3 and airway inflammation by enhancing the function of transcription factor TIEG1. *Nat Immunol* **9**, 245–253, doi:10.1038/ni156410.1038/ni1564. Epub 2008 Feb 17. (2008).
132. Rubtsov, Y. P. et al. Regulatory T cell-derived interleukin-10 limits inflammation at environmental interfaces. *Immunity* **28**, 546–558, doi:S1074-7613(08)00113-1 [pii] 10.1016/j.immuni.2008.02.017 (2008).

# Index

Page numbers followed by f and t indicate figures and tables, respectively.

Printed and bound by CPI Group (UK) Ltd, Croydon, CR0 4YY

24/10/2024

01778308-0007